LINEAR ALGEBRA

HOLDEN-DAY SERIES IN MATHEMATICS

Earl A. Coddington and Andrew Gleason, Editors

LINEAR ALGEBRA

Burton W. Jones

The Mathematics Department
The University of Colorado

HOLDEN-DAY, INC.
San Francisco
Düsseldorf *Johannesburg* *London*
Panama *Singapore* *Sydney* *Toronto*

LINEAR ALGEBRA

500 Sansome Street, San Francisco, California 94111

Library of Congress Catalog Card Number: 72-83244
ISBN: 0-8162-4544-4

Printed in the United States of America

234567890 MP 8079876543

PREFACE

This is an introductory text in linear algebra for students who have had experience with beginning college mathematics and have some acquaintance with proof, including a little mathematical induction. No previous experience with linear algebra or matrices is assumed, but it would be helpful if the student had some familiarity with the ideas of a field and a group. However, for those without such information, appendixes are provided which review, with some proofs, the salient ideas needed. The teacher who wishes to confine attention to the fields of real and complex numbers can do so by exercising a little care. However, I feel that the student should gradually progress toward thinking in terms of a general field, if only from the point of view that the field postulates really are a formulation of the rules of manipulation.

In the process of developing the important definitions and results of linear algebra, I have tried to demonstrate to the student what more abstract mathematics is like. Thus motivation is of fundamental importance. A set of postulates should arise from experience. The student should be told where we are going and why. Interrelationships should be stressed. Terminology should be introduced only when it is about to be used. It is important that the student know the pedigree of a concept so that he may become a partner in the development.

The exercises are, of course, a very important part of the book. A student can test his knowledge first by working through the routine numerical ones. He is then given the opportunity of proving some of the theorems and extending the theory so that he may experience the joy of being a mathematician and have an important part in the development of the subject. On occasion, exercises lead into material which is to follow. However, important results whose proofs are left as exercises are stated in the body of the text for continuity and easy reference.

The first three chapters constitute the basic material on vector spaces, matrices, and linear transformations. We begin with a short preview of where we are going and why. Then we trace the evolution of the idea of a vector from that of a physical "arrow" to an n-tuple of numbers. We next codify the properties of an n-tuple and, on the basis of this experience, set up an abstract definition of a vector space; at the same time showing the advantages of this point of view. The abstract definition is basis free. In fact, polynomials in a single variable are shown to constitute an infinite-dimensional vector space. However, though the abstract definition is extensively used, almost all the spaces dealt with in this book are finite dimensional.

The second chapter introduces matrices through their use in connection with sets of linear equations. The development of the echelon form is shown to be a natural abbreviation of the usual method of solution by elimination. Cues for the definitions of the operations on matrices are provided by our experience with vectors and by applications we should like to make.

The first nine sections of Chapter 3 deal with linear transformations without the use of matrices or determinants; this is to stress their inherent properties without being matrix bound. However, we reach a point, in Section 3.10, where matrices are needed to give us information about transformations. (Certainly it would be difficult to define the trace of a transformation without a matrix.) Then, too, from this point on, the interplay between matrices and linear transformations adds enrichment to both developments. Very fundamental, too, is the idea of certain invariants of a matrix or transformation under change of basis. In this connection, I have found that one gains conceptually by postponing the use of determinants. So without determinants it is shown that the minimum polynomial of a transformation of a vector space into itself has degree not greater than the dimension of the space. The proof is a constructive one and paves the way for the general theorems on the Jordan normal form in Chapter 6.

In the fourth chapter, determinants are developed to introduce the characteristic equation, prove the Cayley-Hamilton theorem, and give a more efficient means of calculating characteristic roots. No previous knowledge of determinants is assumed. Indeed the approach is essentially that of Weierstrass; that is, we decide what we want a determinant to do and how we wish it to behave, and then we define it so that it does and behaves as we want. This not only has the advantage of motivation, but also results in some economy of development.

The fifth chapter opens with the idea of a dot product of vectors and, after some development of its properties, leads into the more abstract concept of a vector product function. Here progress is from the general bilinear form, through the idea of a nonsingular form, to the inner product space and quadratic forms. Connections with and applications to Euclidean transformations are stressed to the point where it is shown that any Euclidean transformation can be represented as a product of symmetries. Dual spaces are not introduced as such, though they are inherent in the development of the adjoint transformation. It seemed that without the prospect of dealing more systematically with duals, there was not much point in introducing them explicitly.

In the sixth chapter, the Jordan form is developed. It begins with a review of some results already obtained and, for motivation purposes, a problem in stochastic matrices of order 2. Then, after the general form is developed, applications are given to a method of computing the characteristic polynomial, to differential equations for those with some experience in that subject, and to positive stochastic matrices.

The first three appendixes constitute material on groups, fields, and polynomials which is used in the book but which some students may not already know. Also, in the case of permutation groups, relegation to an appendix avoids breaking the continuity of the development of determinants. The fourth appendix is a curious and interesting result from Artin's *Geometric Algebra*.

From time to time throughout the book some ideas of projective geometry are introduced not only for historical reasons but because the applications are especially close. However, no prior knowledge of projective geometry is assumed.

For students without previous experience with linear algebra, the first three chapters should provide enough for a semester's course, meeting three times a week. Not much of this material could be omitted except perhaps for affine transformations and projective geometry. On the other hand, for students with some previous knowledge of the subject, one should be able to cover much of the material in Chapter 5 in a semester's course, by going lightly over some of the chapter on determinants.

The level of sophistication rises throughout the book as the student acquires experience. In a number of cases a difficult proof is approached by a preview of the method, and afterward it is illustrated by a numerical example. In particular, special cases of the Jordan normal form are proved in previous chapters by essentially the same method as that used for the general case in Chapter 6.

I wish to give special thanks to Professor Andrew M. Gleason, without whose careful reading and most perceptive comments the book might have been written in half the time but would have been half as good. Thanks are also due to Edward Millman for his editorship, to Holden-Day, Inc., for the production, and to Mrs. Mae Jean Ruehlman for the typing of the manuscript.

Burton W. Jones

CONTENTS

LINEAR ALGEBRA

1

VECTOR SPACES

1.1 INTRODUCTION

What is linear algebra? Often in mathematics we begin with a simple idea, work it and knead it and stretch it until the final product has very little resemblance to what we started with. The only clue which remains is, perhaps, a word in the name. This process, of course, happens outside mathematics as well. Who would think that a beautiful glass *objet d'art* was made mostly of sand.

The word *linear* should have something to do with a line, and indeed it has in the beginning. We know that in analytic geometry the equation of a line has the form $ax + by + c = 0$. So we call $ax + by + c$ a *linear expression.* A property of this expression is that x and y each occur to the first degree. Taking our cue from this, we say that $ax + by$ is *linear in x and y* and that $ax + by + c = 0$ is a *linear equation.* On the other hand, we do not call the expression $2x + xy$ linear in x *and* y since the degree of xy is $1 + 1 = 2$ in x and y, though it is linear in x alone.

The next step in our generalization is to remove the restriction to two variables. We also call $ax + by + cz$ a linear expression, where a, b, c are thought of as numbers and x, y, z as variables. We are no longer concerned with a line, for the equation

$$ax + by + cz + d = 0$$

is satisfied by points (x,y,z) on a plane. And with an expression of this kind in four or more variables, we leave the realm of visual geometry completely but still, as in an afterimage, think of

$$ax + by + cz + dt + e = 0$$

for instance, as representing a "surface" in four dimensions which is in some vague sense "flat." Again $ax + by + cz + dt$ would be called a linear function or linear expression in the variables x, y, z, and t.

In mathematics, as well as outside of it, it is often easier and certainly more productive to define a concept not just by how it looks but by how it behaves. In the description above we had to use words like *degree* and *term* which are essentially dependent on what an expression looks like. If we consider instead a function, with which you are also familiar, we can concentrate on the behavior. To be somewhat specific, let us think in terms of a function of three variables: $f(x,y,z)$. Now

$$f(x,y,z) = ax + by + cz$$

looks linear. It has the property that

$$(ax + by + cz) + (ax' + by' + cz') = a(x + x') + b(y + y') + c(z + z')$$

In functional notation this shows up a little better. We have

Property 1 If $f(x,y,z)$ is a linear function, then

$$f(x,y,z) + f(x',y',z') = f(x + x', y + y', z + z')$$

Notice one consequence. If we let $x = x'$, $y = y'$, and $z = z'$, we have

$$f(x,y,z) + f(x,y,z) = f(2x,2y,2z)$$

that is,
$$2f(x,y,z) = f(2x,2y,2z)$$

This is a special instance of a second property defined as follows:

Property 2 If $f(x,y,z)$ is a linear function, then

$$f(cx,cy,cz) = cf(x,y,z)$$

for all numbers c. The operation involved is called *multiplication by a scalar*, the *scalar* being the number c.

We have shown above that if $c = 2$ this property follows from Property 1. Indeed it can be seen that Property 1 implies Property 2 for any positive integer c. But we could not use this method if c were $\sqrt{2}$ or -3, for instance. This shows, intuitively, that Property 1 is not sufficient for linearity. We can show, more rigorously, that Property 2 is not sufficient by the following example, where Property 2 holds but Property 1 does not. Let

$$f(x,y,z) = \sqrt{x^2 + y^2 + z^2}$$

This satisfies Property 2. But the following shows that it does not satisfy Property 1:

$$\sqrt{x^2 + y^2 + z^2} + \sqrt{x'^2 + y'^2 + z'^2} \neq \sqrt{(x + x')^2 + (y + y')^2 + (z + z')^2}$$

On the other hand, $f(x,y,z) = ax + by + dz$ does have Property 2 by the distributive property. Thus we formulate our definition of linearity of a function.

Definition A function $f(x,y,z)$ is said to be *linear* if it satisfies Properties 1 and 2.

This definition can be extended to any number of variables as follows:

Definition Let $f(x_1, x_2, \ldots, x_n)$ be a function of n variables. We call it *linear* if it satisfies the following two properties:

1. $f(x_1, x_2, \ldots, x_n) + f(y_1, y_2, \ldots, y_n) = f(x_1 + y_1, \ldots, x_n + y_n)$
2. $f(cx_1, cx_2, \ldots, cx_n) = cf(x_1, x_2, \ldots, x_n)$

It would be an oversimplification to say that linear algebra is the study of linear functions. But it would be not far from the truth to say that linear algebra is a study of mathematical systems which have the two properties above.

For those who prefer to think geometrically, we can also associate linearity with flatness as mentioned earlier. A plane can be characterized by the fact that if P and Q are any two points of the plane, then the line determined by P and Q lies entirely in the plane. It is not hard to make the connection, by means of analytic geometry, between this geometrical concept and linear functions, but we do this more efficiently later (see Exercise 8 of Sec. 1.7).

Why does one study linear algebra? One reason is that linear functions are simpler than functions which are quadratic or of higher degree. This means that linearity gives us a powerful tool for getting results, both within and outside of mathematics. As we develop the theory we shall point out some of its uses. (In fact, the uses which one makes of a theory often determine the direction in which it is to progress.) Also, the development has an intrinsic interest of its own.

1.2 VECTORS

Of all the examples of mathematical systems which have the property of linearity, vectors seem the most convenient to begin with since they are useful in considering other linear systems. Let us begin by discussing them very intuitively and recalling their properties. We think of a vector as a line segment with a point on one end or as an arrow without a tail. The length of a vector might be a measure of the magnitude of a force, say, and its direction shows the direction of the force. We could represent a vector in a plane by two points, say (1,3) and (6,9), together with the line segment between them and the designation of one as the initial point (the blunt end) and the other as the terminal point (the sharp end).

Now the vector from (1,3) to (6,9) has the same magnitude and direction as that from (0,0) to $(6 - 1, 9 - 3)$, that is, from (0,0) to (5,6) (see Fig. 1.1). We call these two vectors *equivalent*. In general two vectors are defined to be equivalent if they have the same length and direction. The idea of direction is hard to pin down rigorously; and since we are operating on an intuitive level at this point it is probably best to leave it undefined.

There are a number of operations that can be performed on a vector. One is to change its length or magnitude; another is to change its direction; a third is to "move it around," that is, replace it by an equivalent vector as defined in the above paragraph. Consider now the operation of addition of two vectors. Here we can take our cue from the physical situation and then

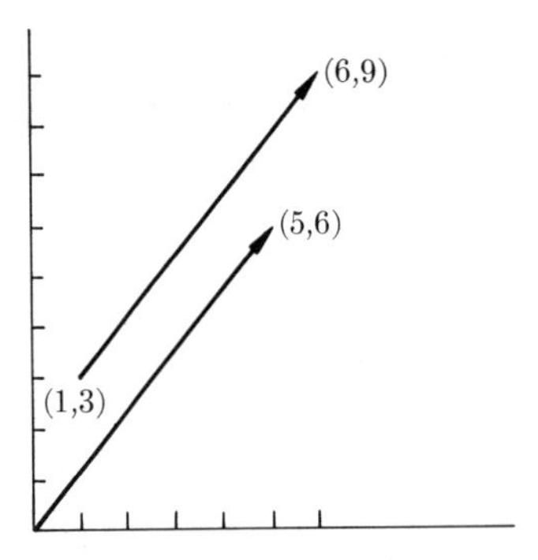

Figure 1.1

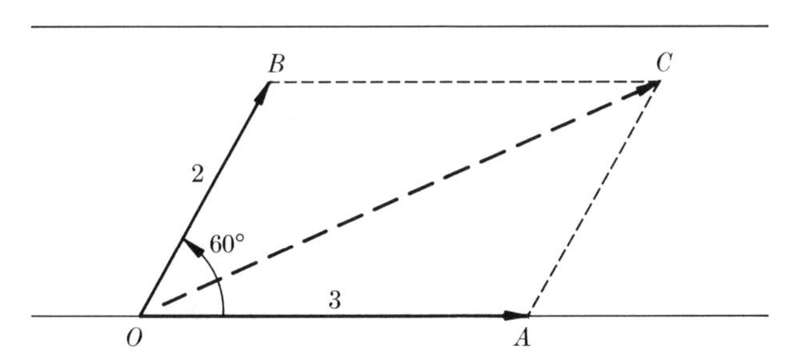

Figure 1.2

see what this amounts to algebraically. To be more specific, suppose a stream is flowing eastward at a rate of 3 miles per hour, and a person is rowing across it at an angle of 60° with the bank and at a rate of 2 miles per hour. We can ask the question: At what rate and in what direction is he approaching the opposite shore? The theory of vectors in physics informs us that we can find the answer by drawing a horizontal vector 3 units long to represent the direction and speed of the water and a vector of length 2 at an angle of 60° with it to represent the direction and speed (velocity) of the rower (see Fig. 1.2). Then if we "complete the parallelogram" as shown, the length of the vector OC gives the rower's rate across the stream and its direction determines his direction. The vector OC is called the *resultant* or *sum* of the vectors OA and OB. One can also find the resultant by completing a triangle as indicated in Fig. 1.3.

In this particular case, let us see what this vector addition amounts to algebraically. Consider the first figure placed on coordinate axes so that point A has the coordinates (3,0), and B is denoted by (1,2). Then it is not hard to see that C has the coordinates $(3 + 1, 0 + 2) = (4,2)$. We merely add the corresponding coordinates.

1.3 VECTORS AS n-TUPLES

We can preserve most of the intuitive properties of vectors and become much more precise if we largely confine ourselves to vectors whose initial points are at the origin of coordinates; any vector is equivalent to one of these. One advantage of doing this is that we can then identify any such vector by the

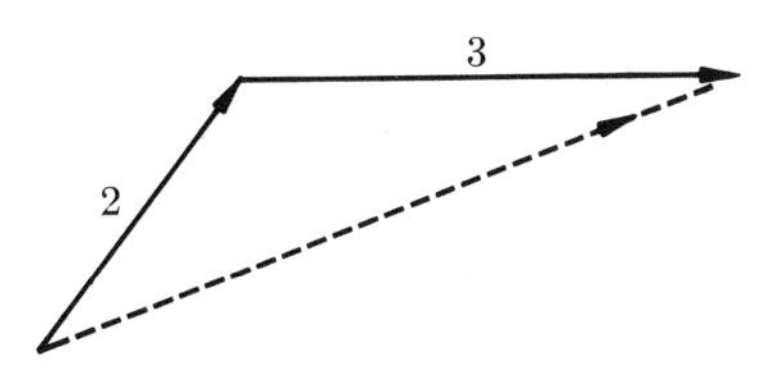

Figure 1.3

coordinates of its terminal point. In this sense we can then think of the ordered pair (a,b) as being the vector whose initial point is the origin and whose terminal point is that defined by the pair (a,b). In three dimensions, we similarly call the ordered triple (a,b,c) a vector, thinking of the physical vector that starts at the origin and has the point (a,b,c) as its terminal point. What we lose by this point of view is the opportunity to consider vectors whose initial points are not at the origin. We shall see that the gain much more than counterbalances the loss. So we have the following definition:

Definition A *vector* is a sequence of n numbers $(a_1, a_2, \ldots, a_n)$, for n some positive integer, that is, an ordered n-tuple of numbers. The a's are called the *components* of the vector. (Note that, for instance, (1,2,3) and (3,2,1) are different vectors.)

We may think of these numbers as being rational, real, complex, or, in fact, numbers of any field (see Appendix A). In this book we generally use lowercase Greek letters to represent vectors, and lowercase Roman letters to represent numbers. Thus we might write $\alpha = (a_1, a_2, \ldots, a_n)$.

1.4 OPERATIONS ON VECTORS

When adding vectors, we use the algebraic equivalent of the physical vector sum illustrated in the previous section. Thus the *resultant* or *sum* of the two vectors from the origin to the points (a_1,a_2) and (b_1,b_2) is the vector from the origin to $(a_1 + b_1, a_2 + b_2)$, as may be seen geometrically. So we define the sum of two vectors in the following way:

Vector Addition If $\alpha = (a_1, a_2, \ldots, a_n)$ and $\beta = (b_1, b_2, \ldots, b_n)$, then

$$\alpha + \beta = (a_1 + b_1, a_2 + b_2, \ldots, a_n + b_n)$$

Note that we can add two vectors only if the number of components of one is the same as the number of components of the other; to add two vectors we add corresponding components.

We can stretch or contract a vector, that is, change its length and/or change its direction. Algebraically this amounts to multiplying each component of the n-tuple by a number. So our next operation is

Multiplication by a Scalar If $\alpha = (a_1, a_2, \ldots, a_n)$ and c is a number, then

$$c\alpha = \alpha c = (ca_1, ca_2, \ldots, ca_n)$$

The number c is, in such a connection, often called a *scalar*. We usually write it to the left of the vector, but sometimes (see Sec. 1.7) it is more convenient to put it on the right. The two vectors are the same since the number c is commutative with the a's.

Two vectors are equal if and only if they are the same. For instance, $(2,3) \neq (3,2)$, but $(\frac{4}{2},\frac{6}{3}) = (2,2)$. The vectors all of whose components are zero, that is, $(0, 0, \ldots, 0)$, are called *zero* or *null* vectors. We use the symbol θ to denote the null vector regardless of the number of components, with the understanding that if we write, for example, $\alpha + \theta$, then α and θ have the same number of components. From the definition of vector addition, we have

$$\alpha + \theta = \theta + \alpha = \alpha$$

The null vector is the additive identity.

The additive inverse of α is

$$-\alpha = (-1)(a_1, a_2, \ldots, a_n) = (-a_1, -a_2, \ldots, -a_n)$$

since $\alpha + (-\alpha) = (-\alpha) + \alpha = \theta$.

It should be noted that there is again no conflict between multiplication by a scalar and vector addition. For example, as we saw above, $\alpha + \alpha = 2\alpha$ whether by vector addition or multiplication by a scalar. It is important to notice that our definitions of addition of vectors and multiplication by a scalar have been precisely in accord with the requirements of linearity imposed in the first section of this chapter. That is, if in the two properties of linearity which we stated in Sec. 1.1 we omit the symbol f, we have vector addition and multiplication by a scalar.

We have thus set up a kind of model for vectors. Just as the vectors themselves are models for physical concepts like force and velocity, so our n-tuples are models for vectors. We have preserved all the properties of "physical" vectors in our mathematical model except equivalence. (This could be recovered if needed.) What have we gained, besides preciseness and a simple method for adding vectors? The answer to this question should become clear in later sections.

EXERCISES

1. In each case below when the indicated sum exists, express it as a single vector; when it does not exist, explain why.

(*a*) $(1,2,3) + (3,4,5)$ (*b*) $3(1,0,5) + 2(1,0)$
(*c*) $3(1,0,5) + 2(\frac{1}{2},\sqrt{3},4)$ (*d*) $5(1,0,7) - \sqrt{3}(4,\frac{1}{2},7)$
(*e*) $(1,0,-6,7) + 0(1,2,3)$ (*f*) $2(1,0,5,7) + (2,3,-1,7) + 0(5,6,7)$

2. Find the vector α satisfying each of the following equations:

(*a*) $\alpha + (3,4,5) = (7,1,-7)$ (*b*) $3\alpha = (1,2,3)$
(*c*) $\alpha + 2(3,1,0) = (2,0,3)$ (*d*) $5\alpha + (7,-1,3) = 7\alpha$

3. Find the pairs of vectors α, β which satisfy the following pairs of equations when such vectors exist. When they do not exist, explain why not.

(*a*) $3\alpha + 4\beta = (1,0,5)$ and $\alpha + 2\beta = (2,3,0)$
(*b*) $2\alpha + \beta = (1,2)$ and $4\alpha + 2\beta = (2,3)$
(*c*) $3\alpha + \beta = (4,4)$ and $6\alpha + 2\beta = (8,8)$

4. Prove that the line segment from the origin to the point $(a + b, c + d)$ is the diagonal of the parallelogram, two of whose sides are the line segments from the origin to the points (a,c) and (b,d).

5. What properties of the components of the vectors below are used in verifying the following equalities?

(*a*) $(a,b) + (c,d) = (c,d) + (a,b)$
(*b*) $[(a,b) + (c,d)] + (e,f) = (a,b) + [(c,d) + (e,f)]$
(*c*) $r[(a,b) + (c,d)] = r(a,b) + r(c,d)$

1.5 EXAMPLES OF VECTORS

To see what is gained by considering vectors to be n-tuples, let us describe various examples. An inventory is really a vector from our point of view. Consider

$$\begin{matrix} A & B & C & D \\ (14, & 15, & 79, & 80) \end{matrix}$$

This might indicate that in a given store there were 14 items of product A, 15 of product B, and so forth. To double the inventory one would double each item. If there were two stores, to find the total inventory we would add the corresponding numbers for each product. So such an inventory not only can be made to look like a vector but behaves like one as well.

We could let the vector (a,b,c) correspond to the linear expression $ax + by + cz$; multiplication by a scalar and vector sum would then have their usual meaning. Or we can associate with the quadratic expression $ax^2 + bx + c$ the triple of its coefficients (a,b,c). Then the sum of two such expressions will be associated with the sum of the triples, and the product $d(ax^2 + bx + c)$ with the product $d(a,b,c)$. We shall develop this further in Sec. 1.9.

In representing the complex number $x + iy$, where x and y are real, we use the point (x,y). Part of the usefulness of complex numbers stems from the fact that adding two complex numbers is equivalent to adding the two vectors which they may be used to represent.

We shall have occasion to carry these and other examples further as we progress.

1.6 VECTOR SPACES OF DIMENSION 1 AND 2

Let $\alpha = (a_1,a_2)$, where a_1 and a_2 are real numbers not both zero; that is, α is not the zero vector. Then the scalar multiples of α consist of all those vectors whose components are coordinates of points on the line connecting the point (a_1,a_2) with the origin. This is an example of what we call a *one-dimensional vector space.* In fact, we define all the scalar multiples of a single nonzero vector to be a one-dimensional vector space, regardless of the number of components.

Suppose $\beta = (b_1,b_2)$ is another vector different from zero. Consider the set of vectors $r\alpha + s\beta$, where r and s are numbers. Here there are two possibilities:

1. The vector β is a scalar multiple of α, say $\beta = t\alpha$. Then $r\alpha + s\beta = (r + st)\alpha$ and β contributes nothing new, because the set of vectors $r\alpha + s\beta$ is the same as the set of all scalar multiples of α. In this case we say that α and β are *linearly dependent.* That is, if $\beta = t\alpha$ then α and β are called linearly dependent, regardless of the number of components.

2. The vector β is not a scalar multiple of α. In that case not all vectors of the form $r\alpha + s\beta$ are scalar multiples of α (e.g., for $r = 0$ and $s = 1$). Here we call α and β *linearly independent* and say that the set of vectors $r\alpha + s\beta$, where r and s range over all numbers, constitute a vector space of dimension 2.

In fact, in case 2 every vector (c_1,c_2) can be expressed in the form $r\alpha + s\beta$ as may be seen either geometrically or algebraically. Let us first show it geometrically. Let A be the point (a_1,a_2), B the point (b_1,b_2), and C the point (c_1,c_2), and suppose Fig. 1.4 is as shown. Then let A' be the point

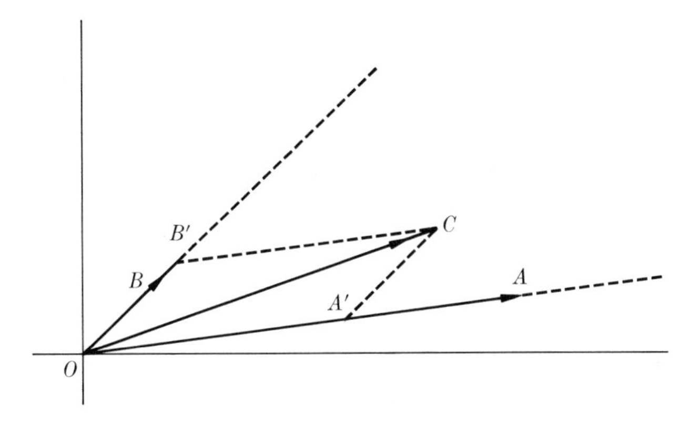

Figure 1.4

where the line through C parallel to the line OB intersects the line OA, and let B' be the point where the line through C parallel to the line OA intersects the line OB. Then the vector OB' is a scalar multiple of β and the vector OA' is a scalar multiple of α, while the vector OC is the sum (resultant) of vectors OA' and OB', that is, is of the form $r\alpha + s\beta$.

In fact, suppose OA and OB are two vectors α and β in three dimensions. They will determine a plane if they are linearly independent and, by the same method as above, it can be shown that any vector OC *in this plane* can be represented in the form $r\alpha + s\beta$.

We can accomplish the same result algebraically for the two-dimensional case as follows. The problem is to show that there are numbers x and y such that $x\alpha + y\beta = \gamma$, where $\gamma = (c_1,c_2)$. This means

$$x(a_1,a_2) + y(b_1,b_2) = (c_1,c_2)$$

that is,

$$(xa_1 + yb_1, xa_2 + yb_2) = (c_1,c_2)$$

Equating corresponding components, we have

$$\begin{aligned} a_1x + b_1y &= c_1 \\ a_2x + b_2y &= c_2 \end{aligned} \tag{1.1}$$

Recall that $\alpha \neq \theta$ and hence not both a_1 and a_2 can be zero. Suppose $a_1 \neq 0$ and let $t = a_2/a_1$. Then, by the usual process, which we shall justify more formally in a general context later, we multiply the first equation by t and subtract from the second to get

$$(b_2 - b_1t)y = c_2 - c_1t \tag{1.2}$$

Now α and β are linearly dependent if and only if $\beta = k\alpha$ for some number k, that is, if and only if $(b_1,b_2) = k(a_1,a_2) = (ka_1,ka_2)$. This equality holds if and only if $k = b_1/a_1$ and $b_2 = ka_2 = (b_1/a_1)a_2 = b_1t$. Thus α and β are linearly dependent if and only if $b_2 - b_1t = 0$. Hence if α and β are linearly independent, $b_2 - b_1t \neq 0$, Equation (1.2) can be solved for y, and we may use the first equation of (1.1) to determine x. The case $a_2 \neq 0$ can be handled similarly. Hence we have shown geometrically and algebraically the following

Theorem 1.1 If $\alpha = (a_1,a_2)$ and $\beta = (b_1,b_2)$ are linearly independent vectors, then every vector (c_1,c_2) can be expressed in the form $r\alpha + s\beta$.

In Fig. 1.4 notice that OB' is the projection of OC on the line OB along a line parallel to OA, and OA' is the projection of OC on the line OA along a line parallel to OB. Then the vector OC is the vector sum of these two projections.

In this connection let us consider an application of linear dependence not so closely bound in with the fabric of mathematics. Suppose a store has two cans of an oil and gasoline mixture for power lawn mowers. If can A contains a mixture of 1 quart of oil to 3 of gasoline and can B contains 4 quarts of oil to 12 of gasoline, we could represent the mixtures by the vectors (1,3) and (4,12), respectively. A disadvantage in these proportions is that no matter how the contents of the two cans are combined, the resulting mixture always has one part of oil to three of gasoline. The two vectors are linearly dependent. But if can B had 2 quarts of oil to 14 of gasoline, the vectors are linearly independent, and a wider variety of mixtures would be possible. This is developed further in some of the exercises of the next set.

EXERCISES

1. Which of the following pairs of vectors are linearly dependent and which are independent?

(*a*) (4,6) and (6,9) (*b*) (7,3) and (1,6)
(*c*) $(\frac{1}{2},-1)$ and $(-4,8)$ (*d*) (2,4) and (6,3)

2. Show that the pair of vectors $\alpha = (1,3)$, $\beta = (5,7)$ is linearly independent and find numbers r and s such that $r\alpha + s\beta = (-1,5)$.

3. Show graphically the meaning of Exercise 2.

4. For what numbers x will the pair of vectors $(1,4)$, $(x,3)$ be linearly dependent?

5. In the following sets let a and b range over all real numbers. Which of the sets are vector spaces? Give reasons for your answers.

(*a*) $\{(a, a + 1)\}$†
(*b*) $\{(a,b)\}$ where $a + b = 0$
(*c*) $\{(a,b)\}$ where $a^3 + b^3 = 0$

6. (This exercise has as its object the broadening of the concept of a vector.) Consider the set of numbers $a + b\sqrt{3}$, where a and b are rational numbers. We can make the following one-to-one correspondence between such numbers and vectors:

Number	$a + b\sqrt{3}$	$(a + b\sqrt{3}) + (c + d\sqrt{3})$	$r(a + b\sqrt{3})$
Vector	(a,b)	$(a,b) + (c,d)$	$r(a,b)$

where the letters refer to rational numbers. We can define two numbers of the form $a + b\sqrt{3}$ to be linearly independent if their corresponding vectors are linearly independent. Prove that if $a + b\sqrt{3}$ and $c + d\sqrt{3}$ are linearly independent then any number $e + f\sqrt{3}$, where e and f are rational numbers, can be expressed in the form

$$x(a + b\sqrt{3}) + y(c + d\sqrt{3})$$

where x and y are rational numbers.

† We use braces to indicate sets.

7. Prove that if the pair of vectors (a,b) and (c,d) are linearly dependent, if (a,b) and (e,f) are linearly dependent, and if a and b are not both zero, then (c,d) and (e,f) are linearly dependent.

8. If throughout Exercise 7 the word *dependent* is replaced by *independent*, is the conclusion still true?

9. Let L be the set of vectors (x,y) such that $ax + by + c = 0$. For what values of a, b, and c will L be a vector space?

10. If in Exercise 9, a, b, and c are such that L is a vector space, will there be a linearly independent pair of vectors of L? Give reasons for your answer.

11. Let a 1-gallon can be filled with a 40% solution of alcohol and another with a 70% solution. Prove that one may get a gallon of any strength between 40% and 70% by a proper mixture of the contents of the two cans. Why is it not possible to get a strength of 30%?

12. Generalize the result of Exercise 11 to an a% solution and a b% solution.

13. Let α and β be two linearly independent vectors having two components each, and consider the set of vectors $\alpha r + \beta s$, where $r + s = 1$. Show that all the points corresponding to this set of vectors lie on the line determined by the points corresponding to the vectors α and β.

14. In Exercise 13 under the conditions on r and s, what points correspond to r and s both positive? To r positive and s negative? To r negative and s positive?

15. (This exercise relates to Exercises 13 and 14 but is a little far afield from the main track of this course.) Let $ax + by = c$ be some line not parallel to the line referred to in Exercise 13; let A be the point corresponding to α, and B that corresponding to β. Prove that the maximum and minimum values of $ax + by$ for (x,y) on the segment $\overline{AB}$† are at A and B. This has applications to linear programming.

1.7 VECTOR SPACES

In this section we extend to n-tuples some of the ideas of the previous section. But first we should look a little more carefully at the n-tuples themselves. We have not said much about what the components of an n-tuple are. They are numbers, but what kinds of numbers? If we are to add n-tuples, we must be able to add the numbers, for a product by a scalar we must multiply, and, as we shall see, it is convenient on occasion to divide. Hence we assume that the components of our vectors are elements of some field F (see Appendix A).

† The overbar indicates that it is the *segment*.

There is no great harm, for the most part, in considering F to be the field of complex numbers or that of real numbers or the rational numbers. The chief purpose in discussing a field is to record the kind of manipulations which we can use without restraint. Once we choose some field F, called the *ground field,* we consider the vectors or n-tuples as being "over the field F," that is, having components in F. We can then manipulate the vector components subject only to the properties of the field. Though in many situations we shall not mention the field, it is always there and assumed.

Let $\mathcal{S}$ be the following set of vector n-tuples:

$$\alpha_1, \alpha_2, \ldots, \alpha_r \tag{1.3}$$

We call any vector of the form

$$c_1\alpha_1 + c_2\alpha_2 + \cdots + c_r\alpha_r \tag{1.4}$$

where the c_i are in a field F, a *linear combination* of the vectors of $\mathcal{S}$ over F.

The set of all vectors of the form (1.4) with coefficients in a field F we call a *vector space* over F and say that the set (1.3) *spans the space.* We designate the space spanned by the α's as

$$\text{sp}(\alpha_1, \alpha_2, \ldots, \alpha_r)$$

Notice that the order in which the α's appear is immaterial. Also r can be any positive integer.

In particular, suppose $r = n$ and α_i is the n-tuple in which all the components are 0 except for a 1 in the ith place. Then the vector space spanned by the α's is the set of *all* n-tuples over the field F. This is the "largest" space of n-tuples in the sense that any space spanned by a set of n-tuples is a subspace of this "largest" space.

We dealt briefly in previous pages with the fundamental idea of linear dependence and independence, which occurs many times in various fields of mathematics. Now is the time to deal more completely with this concept. We give two equivalent definitions, since sometimes one is easier to apply and sometimes the other. We state them first, then give examples, and finally prove their equivalence. We have

Definition 1 A sequence of vectors (1.3) is said to be *linearly dependent* if some vector of the sequence is a linear combination of the others. If no vector of the sequence is a linear combination of the others, we call the sequence *linearly independent.*

Definition 2 A sequence of vectors (1.3) is said to be *linearly dependent* if there is a sequence of numbers $c_1, c_2, \ldots, c_r$ in F not all zero such that

$$c_1\alpha_1 + c_2\alpha_2 + \cdots + c_r\alpha_r = \theta$$

the zero vector. If the equality holds only when all the c's are zero, we call the sequence (1.3) *linearly independent.*

Note that we have changed from the word *set* to *sequence* in our two definitions. This is done for two reasons. One is to provide for the possibility that two vectors of (1.3) may be the same. Another is to help the transition toward the idea of a basis, below, where the use of *sequence* instead of *set* is more important. However, it is important to notice that if no two vectors of (1.3) are the same, the property of being linearly dependent or independent is a property of the *set.* That is, if two sequences (1.3) differ only in the order in which the vectors are written, they are both linearly dependent or both linearly independent. So when we are discussing linear dependence or independence we shall use the words *set* and *sequence* interchangeably. When we are dealing with a basis, defined below, we shall confine ourselves to the word *sequence.* Now let us consider some examples.

First, let $\alpha = (1,2)$ and $\beta = (3,6)$. Then by Definition 1 they are linearly dependent since $\beta = 3\alpha$. By Definition 2 they are also linearly dependent since $3\alpha - \beta = \theta$.

Second, if $\alpha = (1,2)$ and $\beta = (3,5)$, β is not a scalar multiple of α nor is α a multiple of β. Hence they are linearly independent by Definition 1. For Definition 2, write $c_1(1,2) + c_2(3,5) = (0,0)$. This is equivalent to the pair of equations

$$\begin{aligned} c_1 + 3c_2 &= 0 \\ 2c_1 + 5c_2 &= 0 \end{aligned}$$

Since the only solution is $c_1 = c_2 = 0$, the pair are linearly independent by Definition 2.

Third, let $\alpha = (1,2,3)$, $\beta = (2,3,4)$, and $\gamma = (3,5,7)$. These are linearly dependent by Definition 1 since $\gamma = \alpha + \beta$. They are linearly dependent by Definition 2 since $\alpha + \beta - \gamma = \theta$.

Now we prove that the two definitions are equivalent.

Theorem 1.2 Definitions 1 and 2 of linear dependence and independence are equivalent.

Proof Suppose set (1.3) is linearly dependent by Definition 1; then one of the vectors is a linear combination of the others. For instance, suppose $\alpha_1 = d_2\alpha_2 + \cdots + d_n\alpha_n$. Then

$$1 \cdot \alpha_1 - d_2\alpha_2 - d_3\alpha_3 - d_4\alpha_4 - \cdots - d_n\alpha_n = \theta$$

which shows that they are linearly dependent by Definition 2.

Suppose set (1.3) is linearly dependent by Definition 2. Then, for a sequence of numbers c_i not all zero,

$$c_1\alpha_1 + c_2\alpha_2 + \cdots + c_n\alpha_n = \theta$$

If some c_i are not zero we can divide by c_i and solve this equation for α_i, thereby expressing α_i as a linear combination of the rest of the set. Since some c_i are not zero, some α_i are a linear combination of the others. This shows that the second definition implies the first. This completes the proof.

It should be noted that when a set of vectors is linearly dependent it is not necessarily true that *every* vector is a linear combination of the others. For instance, let $\alpha = (1,2,3)$, $\beta = (2,4,6)$, and $\gamma = (3,4,7)$. Here γ is not a linear combination of the first two, but each of the first two is a linear combination of the others.

Suppose the sequence $(\alpha_1, \alpha_2, \ldots, \alpha_n)$ spans a vector space V and the sequence is linearly dependent. Then at least one of the vectors must be a linear combination of the others. Suppose α_1 is a linear combination of the rest. Then V is spanned by $(\alpha_2, \alpha_3, \ldots, \alpha_n)$ since any linear combination of $(\alpha_1, \alpha_2, \ldots, \alpha_n)$ is also a linear combination of $(\alpha_2, \alpha_3, \ldots, \alpha_n)$. This may be seen notationally as follows. Suppose

$$\alpha_1 = a_2\alpha_2 + a_3\alpha_3 + \cdots + a_n\alpha_n$$

Then

$$c_1\alpha_1 + c_2\alpha_2 + c_2\alpha_3 + \cdots + c_n\alpha_n = (c_1a_2 + c_2)\alpha_2 + \cdots + (c_1a_n + c_n)\alpha_n$$

Now if the sequence $(\alpha_2, \alpha_3, \ldots, \alpha_n)$ is linearly dependent, we can reduce by one more the number of α's needed to span the space. This process may be continued until we have a sequence which is "minimal" in the sense that no shorter (proper) subsequence will span the space V. As long as we have a subsequence which is linearly dependent, we can reduce the number of α's required to span V. When we arrive at a linearly independent sequence of these α's, we cannot further decrease the number of these α's spanning V. We then call the sequence a *basis* of V. More formally we have the following:

Definition If a sequence of one or more vectors spanning a vector space V is linearly independent, it is said to form a *basis* of V. The number of vectors in this linearly independent sequence is called the *dimension* of the space V. We denote the dimension by $\dim(V)$.

For instance, let $\alpha_1 = (1,0,1)$, $\alpha_2 = (0,1,1)$, and $\alpha_3 = (1,1,2)$ span a space V. Then, since α_3 is a linear combination of α_1 and α_2, V is also spanned by α_1 and α_2. Since α_1 and α_2 are linearly independent and span V, they form a basis of this vector space and its dimension is 2.

We should note two things about this definition. Here the word *sequence* is important, for we will find it convenient to think of the sequences $(\alpha_1,\alpha_2,\alpha_3)$ and $(\alpha_2,\alpha_3,\alpha_1)$, for example, as two different bases.

Second, it is important to notice that the words *the dimension* require justification. The word *the* would seem to indicate that the dimension depends only on the space, and not on the particular basis which we choose. Thus we must prove soon that if two sets of linearly independent vectors span the same space, then the two sets contain the same number of vectors. The proof is given in the next section.

We should briefly mention another important concept, that of subspace. Suppose V and W are two vector spaces. It may be that every vector of W is in V. In that case we call W a *subspace* of V and, using the usual notation for sets, write $W \subseteq V$. If W is a subspace of V and V contains some vector not in W, we call W a *proper subspace* of V and write $W \subset V$. It is also true that

$$W \subseteq V \quad \text{and} \quad V \subseteq W$$

implies

$$V = W$$

There is an immediate application of what we have been doing to the solution of simultaneous linear equations. Consider the set

$$\begin{aligned} 3x + 4y + 6z &= 7 \\ 2x + + 3z &= 8 \\ -x + 3y - 2z &= 5 \end{aligned}$$

If we write our vectors in vertical instead of horizontal form, the above set can be written as follows:

$$\begin{bmatrix} 3 \\ 2 \\ -1 \end{bmatrix} x + \begin{bmatrix} 4 \\ 0 \\ 3 \end{bmatrix} y + \begin{bmatrix} 6 \\ 3 \\ -2 \end{bmatrix} z = \begin{bmatrix} 7 \\ 8 \\ 5 \end{bmatrix}$$

where now the vectors, written in vertical form, are the columns of coefficients in the set of equations. Notice that it is natural here to write the scalars x, y, z to the right of the vectors. (The student should perform the indicated multiplications and additions to convince himself that this vector equation is equivalent to the set of equations originally given.) To say that the set of equations has a solution is thus the same as saying that the vector of the constant terms is a linear combination of the column vectors of coefficients; that is, the vector of constants is in the vector space spanned by the vectors of columns of coefficients. It turns out to be true that this is not any great help in actually finding solutions, but, as we shall see, it can help us to acquire information about whether or not solutions exist and, if they do exist, what their nature is.

Any set of three linear equations can be expressed in the form

$$\alpha^T x + \beta^T y + \gamma^T z = \delta^T$$

where the superscript T is used to indicate that the vectors are written in vertical form. In fact this same notation could be used for any number of linear equations in any number of unknowns.

EXERCISES

1. Which of the following sets of vectors are linearly dependent and which are linearly independent?

(*a*) (1,4,5), (−1,3,6), (2,0,7)

(*b*) (2,4,6), (1,0,7), (0,1,−2)

(*c*) (1,0,0,0), (0,1,0,0), (0,0,1,0), $(a,b,c,0)$

2. Let α_1, α_2, α_3, and α_4 be four vectors with four components each. Suppose the first component of α_1 is different from zero, the first two components of α_2 are 0 and 2, the first three of α_3 are 0, 0, 3, and the first three components of α_4 are all zero. Under what conditions will the set be linearly dependent? Under what conditions will it be linearly independent?

3. Prove that every vector with three components is a linear combination of the following three vectors: (1,0,0), (0,0,1), and (0,1,0).

4. Does the set of vectors $r\alpha + s\beta$, with $r + s = 1$, constitute a vector space?

5. Prove that if α and β are two vectors in a vector space, then the following are also in the space:

$$2\alpha - \alpha \qquad \alpha + \beta \qquad \alpha - \beta$$

6. Prove that if α, β, and γ are linearly independent, so are $\alpha + k\beta$, β, and γ for any number k.

7. Does the set of vectors $r\alpha + s\beta + t\gamma$, where $r + s + t = 0$, form a vector space?

8. Let $\xi_1 = (x_1,y_1,z_1)$ and $\xi_2 = (x_2,y_2,z_2)$ be two vectors such that

$$ax_i + by_i + cz_i = d \qquad \text{for } i = 1, 2$$

Prove that for all numbers r the components of the vector $r\xi_1 + (1 - r)\xi_2$ also satisfy the equation. (Geometrically this can be interpreted to mean that if two points lie on the plane $ax + by + cz = 0$, then the line determined by the two points lies on the plane.)

9. Let α and β be two linearly independent vectors. State and prove necessary and sufficient conditions on r, s, t, u so that the vectors $r\alpha + s\beta$ and $t\alpha + u\beta$ are linearly independent.

10. Here are two other possible definitions of a basis of a vector space. Prove that each is equivalent to one given in this section.

(*a*) A sequence of vectors forms a basis of a vector space if (and only if) every vector of the space can be expressed *uniquely* as a linear combination of the vectors of the sequence.

(*b*) A sequence of vectors spanning a vector space V is a basis of V if (and only if) it is minimal in the sense that it contains no proper subspace that spans V.

11. Prove that if a sequence of vectors is linearly independent, so is any subsequence.

12. Let S and T be two sequences of vectors such that S is a subsequence of T. If T is linearly dependent, must S be? If S is linearly dependent, must T be? Give reasons for your conclusions.

13. Which of the following sets of vectors form a vector space? Give reasons for your conclusions.

(*a*) $\{(x,y,z)\}$ where $x + y + z = 1$
(*b*) $\{(x,y,z)\}$ where $x + y + z = 0$
(*c*) $\{(x + y, z)\}$ where $x + 2y + 3z = 0$
(*d*) $\{(x,y,z)\}$ where $x^2 + y^2 + z^2 = 1$

14. Let $f(x)$ and $g(x)$ be two polynomials of degree less than 4. To what vectors can they be made to correspond? (See the example of a quadratic expression in Sec. 1.5.) Is this correspondence preserved under scalar multiplication and addition; that is, if α corresponds to $f(x)$ and β to $g(x)$, what vectors correspond to $cf(x)$ and $f(x) + g(x)$? Do f and g have to be of the same degree?

15. Let α_1, α_2 and α_1, α_3 be two pairs of linearly dependent vectors, and suppose α_1 is not the null vector. Prove that α_2 and α_3 are linearly dependent. (Compare Exercise 7 of Sec. 1.6.)

16. Suppose α_1, α_2, α_3 and α_1, α_2, α_4 are two sets of linearly dependent vectors and that α_1 and α_2 are linearly independent. Prove that any set of three vectors chosen from α_1, α_2, α_3, α_4 is linearly dependent.

17. Generalize Exercise 16 to sets of r vectors.

18. Let (x_1,y_1,z_1) and (x_2,y_2,z_2) be two points lying on the plane $ax + by + cz + d = 0$, that is, satisfying the equation. Note that as in Exercise 13 of Sec. 1.6 all the points of the line connecting the given points have coordinates

$$r(x_1,y_1,z_1) + s(x_2,y_2,z_2) = (rx_1 + sx_2,\ ry_1 + sy_2,\ rz_1 + sz_2)$$

with $r + s = 1$. Show algebraically that every such point lies on the plane.

19. Let V be the vector space spanned by the linearly independent vectors α and β, over the field of real numbers. Suppose γ is a vector of V such that $\alpha + \beta + \gamma = \theta$. Prove that every vector in V can be expressed uniquely as a linear combination of two of α, β, γ with nonnegative coefficients.

20. Given a set of vectors α, β, γ. The set of all linear combinations of these three with *integer* coefficients is called a *lattice*. Linear dependence and independence can be defined as on page 13 but now restrict the coefficients to be integers. Are the two definitions equivalent for lattices?

1.8 THE COMPLETION AND BASIS THEOREMS

Before proving that any two bases of a vector space have the same number of vectors, it is convenient because of the method of proof to prove another result which turns out to be useful in many other situations as well. It is important enough to have a name:

***Theorem 1.3* (*Completion Theorem*)** If $\alpha_1, \alpha_2, \ldots, \alpha_n$ is a basis of a vector space V and if $\beta_1, \beta_2, \ldots, \beta_s$ is a sequence of linearly independent vectors of V, then there is a basis of V consisting of all the β's and some of the α's, except that if the β's form a basis, none of the α's are needed.

Proof The proof consists essentially in adjoining α's until we get a basis. Let W denote the space spanned by the β's. If all the α's are linear combinations of the β's, then every vector of V is in W; that is, V is a subspace of W. But, since all the β's are in V, we know that W is a subspace of V. This shows that if all the α's are linear combinations of the β's, then $V = W$ and the theorem holds trivially. If this is not the case, there must be some α which is not a linear combination of the β's. To simplify notation, call it α_1 and denote by S_1 the space spanned by the β's and α_1.

Next we show that the sequence

$$(\alpha_1, \beta_1, \beta_2, \ldots, \beta_s) \tag{1.5}$$

is linearly independent. Using the second definition of linear independence, suppose

$$d\alpha_1 + c_1\beta_1 + c_2\beta_2 + \cdots + c_s\beta_s = \theta \tag{1.6}$$

Now if $d \neq 0$, we can solve (1.6) for α_1 as a linear combination of the β's contrary to our supposition, while if $d = 0$, the linear independence of the sequence of β's shows that all the c's are zero. This establishes the linear independence of the sequence (1.5).

If $S_1 = V$ we are through since V has (1.5) as a base. If not there is a vector α_i which is not a linear combination of α_1 and β's, that is, is not in S_1. Call it α_2 and S_2 the space spanned by the sequence

$$(\alpha_1, \alpha_2, \beta_1, \beta_2, \ldots, \beta_s) \tag{1.7}$$

Just as in the case above we can show that this sequence is linearly independent.

We continue adjoining α's until we have a sequence

$$(\alpha_1, \alpha_2, \ldots, \alpha_k, \beta_1, \beta_2, \ldots, \beta_s)$$

which spans V. This process must stop short of n, that is, $k < n$, because

$$(\alpha_1, \alpha_2, \ldots, \alpha_n, \beta_1, \beta_2, \ldots, \beta_s)$$

is a linearly dependent set. This completes the proof.

Now we can prove the dimensionality theorem:

Theorem 1.4 If $(\alpha_1, \alpha_2, \ldots, \alpha_r)$ and $(\beta_1, \beta_2, \ldots, \beta_s)$ are two bases of a vector space V, then $r = s$.

Proof Suppose $r < s$. Here we do not actually use the result of Theorem 1.3 but we use the same method. That is, we gradually replace the α's by the β's. Since β_1 is in V which is spanned by the α's, we know that β_1 is a linear combination of the α's; that is, for some numbers c_i

$$\beta_1 + c_1\alpha_1 + c_2\alpha_2 + \cdots + c_r\alpha_r = \theta \tag{1.8}$$

Since $\beta_1 \neq \theta$, not all the c_i in (1.8) can be zero. By change of notation, if necessary, take $c_1 \neq 0$ and see that, if (1.8) is to hold, then α_1 is a linear combination of the sequence

$$(\beta_1, \alpha_2, \ldots, \alpha_r) \tag{1.9}$$

Furthermore, since each of the α's is a linear combination of the sequence (1.9), that sequence spans V. Since β_2 is in V, it is a linear combination of the vectors of (1.9) and hence

$$\beta_2 + d_1\beta_1 + d_2\alpha_2 + d_3\alpha_3 + \cdots + d_r\alpha_r = \theta \tag{1.10}$$

for some numbers d_i in F. Now $d_2, d_3, \ldots, d_r$ are not all zero since β_1 and β_2 are linearly independent. Hence we may take $d_2 \neq 0$ and see that α_2 is a linear combination of the sequence

$$(\beta_1, \beta_2, \alpha_3, \alpha_4, \ldots, \alpha_r) \tag{1.11}$$

As above, the sequence (1.11) spans V.

So we may continue introducing β's and taking out α's until we have removed all the α's. Since, by our assumption, $r < s$, we will have a sequence

$$(\beta_1, \beta_2, \ldots, \beta_r) \tag{1.12}$$

which spans V. This is impossible since β_s in V would then imply that β_s is a linear combination of the vectors of (1.12). This shows that our assumption that r is less than s is false.

If r were greater than s, we could interchange the roles of the α's and β's and arrive at a similar contradiction. Thus $r = s$ and our proof is complete.

We now have the right to speak of *the* dimension of a vector space. A vector space V may have many bases, but they all include the same number of vectors. Let ϵ_i denote the vector with n components all of which are zero except for a 1 in the ith place. The sequence of n vectors $(\epsilon_1, \epsilon_2, \ldots, \epsilon_n)$ is sometimes called the *canonical basis.* The vectors of this set are linearly independent since

$$c_1\epsilon_1 + c_2\epsilon_2 + \cdots + c_n\epsilon_n = (c_1, c_2, \ldots, c_n)$$

and this sum is the zero vector if and only if all the c's are zero. Furthermore, every n-tuple of numbers is a linear combination of the ϵ_i. This shows, as we would expect, that

The vector space consisting of all n-tuples of numbers has dimension n. We denote this space by V^n.

Since any n-tuple is a linear combination of the canonical basis, we know that it must be in V^n, and hence any vector space W spanned by a set of n-tuples must be a subspace of V^n. The relationship between being a subspace and dimensionality is given in

Theorem 1.5 If W is a subspace of a vector space V, then $\dim(W) \leq \dim(V)$ and if the equality holds, $W = V$.

Proof Let $\alpha_1, \alpha_2, \ldots, \alpha_r$ be a basis of W, where $r = \dim(W)$. If W is a proper subspace of V, then V contains some vector not in W. Now the vectors of the basis of W are all in V, and hence by the completion theorem V has a basis which properly contains the basis of W. This shows that the dimension of V is greater than that of W. If, on the other hand, $r = \dim(V)$, then the basis of W is also a basis of V. This completes the proof.

There is an application of the above to the set of equations considered at the end of the previous section. If the column vectors are linearly independent, we know that they span V^3, and hence, no matter what the vector of the constants is, there is a solution. On the other hand, if the set of column vectors is linearly dependent, then there will not be a solution for all vectors of constants.

Though we shall confine ourselves in this book, for the most part, to finite-dimensional vector spaces, we should exhibit by contrast at least one vector space which is not. Let S be the set of all polynomials in x with coefficients in a field F. We can associate any polynomial with an n-tuple as follows:

$$\begin{array}{cccccc} a_0 + & a_1x + & a_2x^2 + \cdots & + a_{n-2}x^{n-2} & + a_{n-1}x^{n-1} & \\ (a_0, & a_1, & a_2, \ldots, & a_{n-2}, & a_{n-1} &) \end{array}$$

The correspondence is "preserved under addition and multiplication by a scalar"; that is, the vector corresponding to the sum of the two polynomials is the sum of the vectors which correspond to the polynomials, and multiplication of a polynomial by a number yields a vector which is a product by the same scalar. There are two complications. If the polynomials are of different degrees, we merely adjoin zero coefficients to the one of lower degree. A more important contingency is that, while every polynomial corresponds in the above manner to an n-tuple for some n, there is no n which will do for all polynomials. Thus there is no finite basis for the set of all polynomials. We shall have more to say about this at the end of the next section and later in the book.

EXERCISES

1. Let $\alpha = (1,3,5)$ and $\beta = (2,12,20)$. Find for which values of i the three vectors α, β, ϵ_i form a basis of V^3, where the ϵ_i form a canonical set as defined in the previous section.

2. Find a basis for the space spanned by each set of vectors below:

(*a*) (1,2,3), (1,0,5), (1,1,4), (3,2,1)
(*b*) (1,2,3), (2,4,6), (1,0,1), (2,2,4)
(*c*) (3,0,5), (1,2,7), (2,1,6), $(1,-1,-1)$

3. Show that any subset of three of the following four vectors is a linearly independent subset.

$$(1,0,0) \qquad (0,1,0) \qquad (0,0,1) \qquad (1,1,1)$$

4. Consider the set of vectors $\{(x,y,z)\}$ such that $x + y + z = 0$. Show that these vectors form a vector space and find a basis of the space.

5. Consider the set of vectors $\{(x,y,z)\}$ such that

$$x + y + z = 0 \qquad \text{and} \qquad x + 2y - z = 0$$

Show that these form a vector space and find a basis of this space.

6. Let $\alpha_1 = (2,1,1)$, $\alpha_2 = (0,-1,1)$, $\alpha_3 = (0,0,1)$, and $\beta_1 = (1,0,1)$, $\beta_2 = (1,1,0)$. Show that the set of α's is a basis of V^3. For what i will β_1, β_2, α_i be a basis of V^3?

7. In Exercise 6 will β_1 and any two of the α's be a basis of V^3? Give reasons for your answer.

8. Find the dimension of the space spanned by the vectors of coefficients in Exercise 5, that is, by $(1,1,1)$ and $(1,2,-1)$.

9. Consider the set of vectors $\{(x,y,z)\}$ such that

$$x + y + z = 0 \qquad x + 2y - z = 0 \qquad \text{and} \qquad y - 2z = 0$$

Show that this set is a vector space and find its dimension. What is the dimension of the vector space spanned by the three triples of coefficients?

10. Show that the three polynomials $x - 2$, $x^2 + x + 1$, and $x^2 + 2$ form a basis for the set of all quadratic polynomials.

11. If α, β, γ form a basis of V^3, show that $\alpha + k\beta$, β, γ also form a basis of V^3 for every number k.

12. If α and β are linearly independent, prove that $a\alpha + b\beta$ and $c\alpha + d\beta$ are also linearly independent if and only if $ad - bc \neq 0$ (see Exercise 9 in Sec. 1.7).

1.9 AN ABSTRACT DEFINITION OF A VECTOR SPACE AND ISOMORPHIC SPACES

Now that we are acquainted with vector spaces, it is time for an abstract definition. Such a formulation has a number of advantages. It codifies the properties of vectors so that we can manipulate the symbols without having to think each time of an n-tuple. It allows us to consider as vectors entities which are not n-tuples but which just behave like them. Furthermore, it permits us to include vector spaces of infinite dimension and, indeed, some which might not have any basis at all, even though in this course we shall not make much use of this freedom.

In setting up the abstract definition, we think of the properties of n-tuples as guides, but after the definition is formulated we may throw away the crutch by which we got it.

Definition A vector space V over any field F is a nonempty set of entities $\mathcal{S}$, designated by α, β, γ, . . . on which two operations are defined: addition, $\alpha + \beta$, and scalar multiplication, $c\alpha$, for c in the field, with the following properties:

1. The set $\mathcal{S}$ is an *Abelian group under addition* (see Appendix A), that is,
   (*a*) $\alpha + \beta$ is in $\mathcal{S}$ for all α and β in $\mathcal{S}$ (*closure*).
   (*b*) $(\alpha + \beta) + \gamma = \alpha + (\beta + \gamma)$ (*associativity*).

(*c*) There is an *additive identity* θ, the null vector, such that

$$\theta + \alpha = \alpha + \theta = \alpha \qquad \text{for all } \alpha \text{ in } \mathcal{S}$$

(*d*) Each member α of $\mathcal{S}$ has an *additive inverse* $(-\alpha)$ such that

$$(-\alpha) + \alpha = \alpha + (-\alpha) = \theta$$

(*e*) Addition is commutative: $\alpha + \beta = \beta + \alpha$.

2. If c is in F and α in $\mathcal{S}$, then $c\alpha = \alpha c$ is in $\mathcal{S}$ (*closure and commutativity of multiplication by a scalar*).

3. For c and c' in F, $c(\alpha + \beta) = c\alpha + c\beta$ and $(c + c')\alpha = c\alpha + c'\alpha$ (*distributivity for multiplication by a scalar*).

4. For c and c' in F, $(cc')\alpha = c(c'\alpha)$ (*associativity for multiplication by a scalar*).

5. $1 \cdot \alpha = \alpha$.

You will find that in some treatments of vector spaces it is not assumed that $c\alpha = \alpha c$, but we do make this assumption here. Notice that the first part of property 3 is part of the property of linearity (Property 2 described in Sec. 1.1), where the function referred to there is now scalar multiplication. The other properties may seem so reasonable that you may wonder why they are even mentioned, but a little thought may convince you that they are necessary. Note that there is no mention of a product of two vectors. Later (in Chap. 5) we shall be considering such products, but at this point we have enough to do without dealing with that complication.

There are two properties of a vector space which follow immediately from the five listed above. We list and then prove them:

6. $0 \cdot \alpha = \theta$ for all α in V.
7. $(-1)\alpha = (-\alpha)$.

To show property 6, write the following sequence of equations:

$$\alpha + 0 \cdot \alpha = 1 \cdot \alpha + 0 \cdot \alpha = (1 + 0)\alpha = 1 \cdot \alpha = \alpha$$

This shows that if $0 \cdot \alpha$ is added to α, the result is α. It is left as an exercise to show that the additive identity must be unique. This is all that is needed to complete the proof of property 6. To prove property 7, write the equations

$$\alpha + (-1)\alpha = (1 - 1)\alpha = 0 \cdot \alpha = \theta$$

By showing the uniqueness of the additive inverse (left as an exercise), we complete the proof of property 7.

In this more abstract context we can carry over with almost no change the concepts developed for n-tuples in Sec. 1.7, namely, linear combination, spanning, vector space, linear dependence and independence, basis, dimension. The only difference to this point which needs to be reckoned with is that we have not required that a vector space have a basis. Almost all the vector spaces dealt with in this book do have a finite basis. At the end of this section we shall give an example of one which has a basis but not a finite one.

One example of a vector space which is not really a set of n-tuples, though it behaves like it, is the set of translations in the Euclidean plane: call it $\mathfrak{T}$. Let us see why. As you know, a translation replaces each point (x,y) by $(x + h, y + k)$ for some h and k. We can designate this translation by $T(h,k)$ and say that the "image" of (x,y) under $T(h,k)$ is the point (x',y') where

$$x' = x + h \qquad y' = y + k$$

Then we can call $cT(h,k)$ the translation $T(ch,ck)$, where each of h and k is multiplied by c. It would be natural to consider $T(h_1,k_1) + T(h_2,k_2)$ as the result of translating first according to h_1, k_1 and then according to h_2, k_2. If we let (x',y') be the image of (x,y) under the first translation and (x'',y'') be the image of (x',y') under the second translation, we have the two sets of equations

$$\begin{aligned} x' &= x + h_1 \\ y' &= y + k_1 \end{aligned} \qquad \text{and} \qquad \begin{aligned} x'' &= x' + h_2 \\ y'' &= y' + k_2 \end{aligned}$$

From this it follows that

$$\begin{aligned} x'' &= x + (h_1 + h_2) \\ y'' &= y + (k_1 + k_2) \end{aligned}$$

This shows that $T(h_1,k_1) + T(h_2,k_2)$ should be defined to be $T(h_1 + h_2, k_1 + k_2)$. So the sum of two translations is a translation (see Fig. 1.5). It is easy to check the five properties above to show that the set of translations in the plane is a vector space. Of course, it is not hard to see that without the T in front of the number pair we would just have an ordered pair. However, a translation is not an ordered pair even though in many respects it behaves like one.

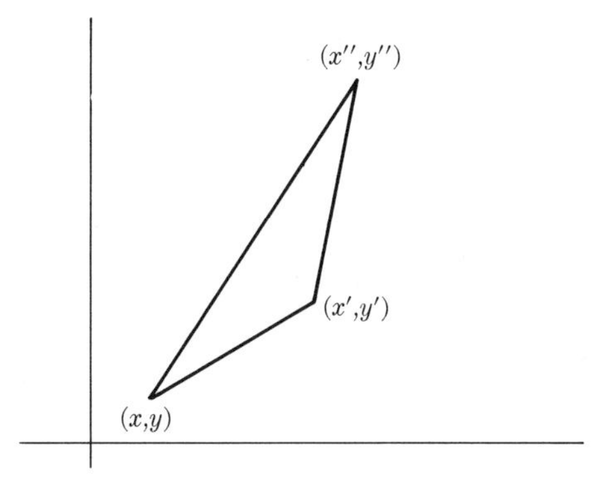

Figure 1.5

On the other hand, not all number pairs behave like vectors. Suppose we denote by $S(a,b)$ the stretching

$$x' = ax \qquad y' = by$$

Then we can define $cS(a,b)$ to be $S(ca,cb)$ and ask how we should define $S(a,b) + S(c,d)$. Suppose we use the same method as above and define the sum to be what we get by applying first one stretching and then another as follows:

$$\begin{matrix} x' = ax \\ y' = by \end{matrix} \quad \text{followed by} \quad \begin{matrix} x'' = cx' \\ y'' = dy' \end{matrix}$$

Combining these, we get

$$x'' = cax \qquad \text{and} \qquad y'' = dby$$

Hence we would be justified in defining the sum as follows:

$$S(a,b) + S(c,d) = S(ac,bd)$$

However, for such definitions, the set of all $S(a,b)$ does not constitute a vector space since a little computation, left as an exercise, shows that

$$2[S(a,b) + S(c,d)] \neq 2S(a,b) + 2S(c,d)$$

which denies the first part of property 3 of a vector space.

Suppose α_1, α_2, α_3 is a basis of V. Then any vector of this space can be expressed in the form $x_1\alpha_1 + x_2\alpha_2 + x_3\alpha_3$, and any vector of this form is in the space. Thus, by this means, any triple (x_1,x_2,x_3) defines a vector in V, and any vector in V, for the basis given, determines a triple. This correspondence carries through the various operations of the vector space. To see this, let us use two parallel columns:

$$\begin{aligned} \xi &= x_1\alpha_1 + x_2\alpha_2 + x_3\alpha_3 && \leftrightarrow (x_1,x_2,x_3) \\ \eta &= y_1\alpha_1 + y_2\alpha_2 + y_3\alpha_3 && \leftrightarrow (y_1,y_2,y_3) \\ \xi + \eta &= (x_1 + y_1)\alpha_1 + (x_2 + y_2)\alpha_2 + (x_3 + y_3)\alpha_3 && \leftrightarrow (x_1 + y_1,\ x_2 + y_2, x_3 + y_3) \\ c\xi &= cx_1\alpha_1 + cx_2\alpha_2 + cx_3\alpha_3 && \leftrightarrow c(x_1,x_2,x_3) = (cx_1,cx_2,cx_3) \end{aligned}$$

That is, for the third equation, the following computations give the same result: (1) Add ξ and η and then find the triple which corresponds to the sum; (2) add the triples which correspond to ξ and η. The same kind of property is seen to hold in the fourth equation as well.

It is important to note, however, that the triple depends on which basis is used. For instance, suppose ξ is the vector $(1,2,3)$. If $\alpha_1 = (1,0,0)$, $\alpha_2 = (0,1,0)$, and $\alpha_3 = (0,0,1)$ then

$$\xi = 1 \cdot \alpha_1 + 2 \cdot \alpha_2 + 3 \cdot \alpha_3$$

corresponds to the triple (1,2,3). But if the basis is $\beta_1 = (1,1,0)$, $\beta_2 = (1,0,1)$, and $\beta_3 = (0,1,1)$ then

$$\xi = 0 \cdot \beta_1 + 1 \cdot \beta_2 + 2 \cdot \beta_3$$

corresponds to the triple (0,1,2).

As far as results go, it really does not matter whether we write a vector as a linear combination of the α's in F or as a triple of coefficients in F. We call the set of linear combinations of the α's and the set of triples *isomorphic* because they behave the same way with respect to the operations of a vector space. We make this more precise in the following

Definition Two vector spaces V and V' are said to be *isomorphic* if there is a one-to-one correspondence $\xi \leftrightarrow \xi'$ between the vectors of the two spaces, such that, if $\xi \leftrightarrow \xi'$ and $\eta \leftrightarrow \eta'$, then

1. $(\xi + \eta)' = \xi' + \eta'$
2. $(c\xi)' = c(\xi')$

The set of translations $\{T(a,b)\}$ defined above is isomorphic to the two-dimensional vector space $\{(a,b)\}$. In fact, we have shown above for V^3 the following general result.

Theorem 1.6 Two finite-dimensional vector spaces are isomorphic if and only if they are of the same dimension.

Proof First suppose the spaces V and V' have the same dimension. Let $\alpha_1, \alpha_2, \ldots, \alpha_n$ be a basis of V and $\alpha_1', \alpha_2', \ldots, \alpha_n'$ a basis of V'. We set up the correspondence:

$$\alpha_1 \leftrightarrow \alpha_1', \ \alpha_2 \leftrightarrow \alpha_2', \ \ldots, \ \alpha_n \leftrightarrow \alpha_n'$$

Then we define a one-to-one correspondence between the vectors of V and those of V' by specifying that any linear combination of the α_i shall correspond to the same linear combination of the α_i'. Since no two distinct linear combinations of the elements of a basis can be the same vector, it follows that the correspondence between vectors of V and V' is one to one. Furthermore, by this choice, if α and α' correspond, so will $c\alpha$ and $c\alpha'$, while if α and β correspond to α' and β', respectively, then $\alpha + \beta$ will correspond to $\alpha' + \beta'$. This shows that V and V' are isomorphic.

Conversely if V and V' are isomorphic, let $\alpha_1, \alpha_2, \ldots, \alpha_n$ be a basis of V. We know there is a one-to-one correspondence and so let α_1' be that vector of V' which corresponds to α_1; let α_2' be that vector of V' which corresponds to α_2; etc. Notice that the null vectors of the two spaces must correspond, for let θ' denote the vector in V' which corresponds to θ in V. Then, because of the isomorphism, $\alpha + \theta = \theta + \alpha = \alpha$ for all α in V implies that $\alpha' + \theta' = \theta' +$

$\alpha' = \alpha'$ for the corresponding α' in V'. This shows that θ' is the null vector of V'.

It seems reasonable to suppose that the α_i' turn out to be a basis for V'. To prove this we need to show two things: (1) the α_i' are linearly independent, and (2) the α_i' span V'. For the first, the natural thing is to apply the definition. Suppose $c_1\alpha_1' + c_2\alpha_2' + \cdots + c_n\alpha_n' = \theta'$ for some set $(c_1, c_2, \ldots, c_n)$. Then, since we have an isomorphism, the same linear combination of the α's will be the null vector in V. But the α's are linearly independent, which implies that all the c_i are zero. Thus we have proved that the α_i' form a linearly independent set.

To prove that the α_i' span V', let β' be any vector of V'. Because of the one-to-one correspondence, there is a vector β in V to which β' corresponds. Now β, being in V, can be expressed as a linear combination of the basis vectors α_i. Denote by β'' the same linear combination of the α_i'. So β'' corresponds, under the isomorphism, to β. But since the correspondence is one to one, β' and β'' must be the same vector. Thus β' is a linear combination of the α_i'. This shows that V' is spanned by the α_i' and completes the proof.

The next result should not be very startling since we defined a vector space so that it would behave like n-tuples. In fact, since the set of all n-tuples of numbers in a field F forms a vector space of dimension n, we have the following important corollary:

Corollary Every n-dimensional vector space over a field F is isomorphic to the set of all n-tuples of numbers of F.

This shows that to find the properties of any finite-dimensional vector space we need merely find the properties of n-tuples. But the advantages of the abstract formulation remain: ease of manipulation, concentration on the fundamental properties, more latitude in possible examples.

A few comments are in order. One should notice that the correspondence by which an isomorphism is established can be set up in any number of ways. For instance, let $\alpha_1, \alpha_2, \ldots, \alpha_n$ and $\beta_1, \beta_2, \ldots, \beta_n$ be two bases of V. The natural choice of a one-to-one correspondence is to let α_i correspond to β_i for $1 \leq i \leq n$. But another choice is the following:

$$\alpha_1 \leftrightarrow \beta_n,\ \alpha_2 \leftrightarrow \beta_{n-1},\ \ldots,\ \alpha_{n-1} \leftrightarrow \beta_2,\ \alpha_n \leftrightarrow \beta_1$$

To show isomorphism, we need only show that there is *one* one-to-one correspondence which has the necessary properties.

Then, we need to take note of the null space. Every vector space contains a null vector. A set consisting of the single vector θ satisfies all the conditions for a vector space which we gave, though in a trivial manner. We could add another requirement to exclude the set consisting of the null vector and no more. If we did this, it would make the formulation of some later results a little more awkward. So we admit the vector space consisting only of the null vector and call it the *null space*. Notice that it is not a null set since it contains a vector, namely, the null vector.

Having made this choice, we must make two others. What is the dimension of the null space? By our definition of dimension, this is the number of vectors in a maximal set of linearly independent vectors which it contains. This forces us to answer the question: Is the set consisting of the single vector θ a linearly independent set? To apply the second definition of linear independence we look at $c\theta$ and ask if it can be θ without c being zero? The answer is certainly "yes." Hence the set consisting of θ alone is linearly dependent. (In fact this is the only set consisting of a single vector for which this is true.) Thus, since the null space contains no set of linearly independent vectors, we must, to be consistent, define its dimension to be zero. It is not a null set, but its dimension is zero. The first definition of linear dependence is meaningless in this context, since there are no vectors in the null space other than θ. The null space is not really very important, but we cannot afford to ignore it.

Consider again the polynomials in x over a field F, discussed at the close of the previous section. We need not now consider n-tuples but can take as a basis of the set of all polynomials the following:

$$1,\ x,\ x^2,\ x^3,\ .\ .\ .$$

namely, the sequence of all nonnegative integral powers of x.

EXERCISES

1. Which of the following pairs of sets are isomorphic? Where they are, set up a one-to-one correspondence and show why it satisfies the requirements of an isomorphism (a, b, c, r, s are real numbers):

(*a*) $S = \{(a,b)\}$, $T = \{(r + s, s)\}$.

(*b*) $S = \{(a,a,b)\}$, $T = \{(r,s)\}$.

(*c*) $S = \{(a,b,c)\}$, $T = \{(r\alpha + s\beta)\}$, where α and β are two linearly independent vectors of V^2.

(*d*) $S = \{(0,a)\}$; $T = \{(b,0)\}$.

(*e*) $S = \{(a + b, a - b)\}$; $T = \{(a,b)\}$.

2. In parts (*a*), (*d*), and (*e*) of Exercise 1, show what the correspondence means geometrically.

3. Let S be the set of all cubic polynomials in x with rational coefficients. To what vector space is this isomorphic?

4. Let α, β, γ, δ be a set of vectors, and suppose that 3 is the dimension of the space which they span. Let W be the set of 4-tuples (a,b,c,d) for which $a\alpha + b\beta + c\gamma + d\delta = \theta$. Prove that W is a vector space of dimension 1.

5. Do Exercise 4 for the case in which the spanned space is of dimension 2. What is now dim (W)?

6. Complete the proof of properties 6 and 7; that is, show that in a vector space the additive identity and inverse are unique.

7. Show that the set of translations $T(a,b)$ is a vector space.

8. Verify the inequality which shows that the set of stretching transformations S, with addition and scalar multiplication as defined in this section, is not a vector space.

9. Prove that if vector spaces V and V' are isomorphic and if V' and V'' are isomorphic, then V and V'' are also.

10. Let $f(x,y,z)$ be a polynomial in x, y, and z. If the correspondence $f(x,y,z) \leftrightarrow (x,y,z)$ is an isomorphism for the vector space consisting of triples (x,y,z), what must the form of this polynomial be?

11. If the polynomial of Exercise 10 is isomorphic to the space V^2, what possibilities are there for the form of the polynomial?

12. Prove that if α is a vector, then $-(-\alpha) = \alpha$.

13. Let α_1, α_2, α_3 and β_1, β_2, β_3 be two bases of V^3. In how many ways can a one-to-one correspondence be set up between the two bases?

14. Consider the following one-to-one correspondence between two bases of V^3.

$$(1,0,0) \leftrightarrow (0,1,1);\ (0,1,0) \leftrightarrow (1,0,1);\ (0,0,1) \leftrightarrow (1,1,0)$$

The vector (a,b,c) is a linear combination of the first basis. Write it as a linear combination of the second basis.

15. Let α_1, α_2, α_3 be a basis of a vector space V and let $\gamma = a_1\alpha_1 + a_2\alpha_2 + a_3\alpha_3$, for some a_i in F. Prove that $\gamma - \alpha_1$, $\gamma - \alpha_2$, $\gamma - \alpha_3$ is a basis of V if and only if $a_1 + a_2 + a_3 \neq 1$.

16. In Exercise 15 suppose $\gamma - \alpha_i$, for $i = 1, 2, 3$, is a basis and set up the one-to-one correspondence $\alpha_i \leftrightarrow \gamma - \alpha_i$, for $i = 1, 2, 3$. What vectors correspond to their negatives under this correspondence?

17. Let α_1, α_2, α_3 be a basis of a vector space V and β_1, β_2, β_3 a set of three vectors of V, not necessarily linearly independent. Set up the correspondence

$$x_1\alpha_1 + x_2\alpha_2 + x_3\alpha_3 \rightarrow x_1\beta_1 + x_2\beta_2 + x_3\beta_3$$

(Notice that the arrow points in one direction only, indicating only a one-way correspondence for all numbers x_i in F.) Prove that the β_i are a basis of V if and only if, under this correspondence,

$$a_1\alpha_1 + a_2\alpha_2 + a_3\alpha_3 \mapsto \theta$$

implies that

$$a_1 = a_2 = a_3 = 0$$

18. Referring to Exercise 17, prove that the correspondence is one to one if and only if the β_i are a basis of V.

1.10 SUM AND INTERSECTION OF SUBSPACES

It is possible in a number of ways to construct vector spaces from other vector spaces or to decompose a vector space into subspaces. Suppose S and T are two vector spaces, what can be said about their intersection, that is, the set of vectors common to S and T? We represent this set by the usual notation: $S \cap T$. It may be that S and T have no vectors in common. This will happen, for instance, if S is the set of pairs (a,b) and T the set of triples (c,d,e), where a, b, c, d, e are in a field F. However, if S and T are subspaces of a vector space V, they must have the null vector in common since all vector spaces contain *a* null vector which in this case must be *the* null vector of V.

We first wish to prove that the intersection of two subspaces of V is a vector space. But before doing this, we can reduce our labor by looking back at the abstract definition of a vector space in the previous section. We let $\mathcal{S}$ denote a subset of the vector space V and consider in turn the various properties of a vector space. There are some which the elements of $\mathcal{S}$ possess by virtue of being in V (sometimes referred to as *hereditary properties*). These are the associative and commutative properties of addition as well as properties 3 to 5, the distributive and associative properties of multiplication by a scalar, and the fact that $1 \cdot \alpha = \alpha$. So these we do not have to check. On the other hand, closure under vector addition and multiplication by a scalar must certainly be checked. The two remaining properties are the existence of an additive identity and an additive inverse. We prove below that we need not check these two. In fact, we have

Theorem 1.7 If $\mathcal{S}$ is a nonempty subset of a vector space V, then $\mathcal{S}$ is a vector space if and only if $\mathcal{S}$ is closed under vector addition and multiplication by a scalar.

Proof By the definition of a vector space, it should be clear that closure is a necessary condition. Now suppose that $\mathcal{S}$ is closed under the two operations. Then if α is in $\mathcal{S}$, we know that $0 \cdot \alpha$ and $(-1)\ \alpha$ are in $\mathcal{S}$. But properties 6

and 7 of a vector space which we proved in Sec. 1.9 imply that $0 \cdot \alpha = \theta$ and $(-1)\alpha = -\alpha$. This completes the proof.

We apply Theorem 1.7 to prove

Theorem 1.8 If S and T are two subspaces of a vector space V, then their intersection is a vector space.

Proof Suppose α and β are two vectors in both S and T. Then their sum is in S since S is a vector space, and their sum is in T for the corresponding reason; hence their sum is in both, that is, in their intersection. Similarly one can prove closure for multiplication by a scalar. So the proof is completed.

As we noted above any two subspaces of V have a vector in common, the null vector. If this is their only common vector, their intersection is the null space described in the previous section. In fact, one of the principal reasons for the introduction of the concept of a null space is to avoid having to make exceptions in the statement of Theorem 1.8. If the intersection of two subspaces S and T is the null space, we say that S and T are *disjoint*. You should notice that this is a small departure from the use of this term with regard to sets, where disjoint means "having no elements in common." This departure should cause no confusion.

We have dealt now with the intersection of two subspaces. What about their union? Let S denote the vector space of pairs $(a,0)$ and T that of pairs $(0,b)$, where a and b are in a field F. Then S and T are subspaces of V^2 and their union is the set of all pairs in which one of the components is zero. But this set is not closed under addition, since

$$(a,0) + (0,b) = (a,b)$$

which is not in the union unless one of a and b is zero. So, in place of union, we define what we call the *sum* of two subspaces of V to consist of all vectors of V which can be expressed as the sum of a vector of S and one of T. That is, we have the

Definition If S and T are subspaces of a vector space V, then their *sum*, written $S + T$, is the set of all vectors of V which can be expressed in the form $\alpha + \beta$, where α is in S and β in T.

Notice that S and T must be subspaces of the same vector space to permit us to add vectors. We leave as an exercise the proof of

Theorem 1.9 The sum of two subspaces of a vector space V is a vector space.

We also leave as an exercise the proof of the following theorem, which looks much like a formula for sets with finite numbers of elements.

Theorem 1.10 Let S and T be two subspaces of a vector space V, then

$$\dim(S + T) = \dim(S) + \dim(T) - \dim(S \cap T)$$

Another useful concept is the direct sum defined as follows:

Definition A vector space V is said to be a *direct sum* of subspaces S and T if $V = S + T$ and $S \cap T = \theta$. In this case, we write $V = S \oplus T$.

From Theorem 1.10, if $V = S \oplus T$, then $\dim(V) = \dim(S) + \dim(T)$. Notice that V^2, for instance, is the direct sum of the space $\{(a,0)\}$ and $\{(0,b)\}$.

EXERCISES

1. Let S be the vector space spanned by (1,0,0) and (0,1,0), while T is the vector space spanned by (0,0,1) and (1,1,1). Find the intersection, union, and sum of S and T. Exhibit one vector which is in the sum but not in the union.

2. Let S be the vector space spanned by (1,2,3) and (1,−1,0), while T is V^3. Find the intersection, union, and sum of S and T. Is there an element of the sum which is not in the union?

3. Let S and T be two subspaces of a vector space V. Find a necessary and sufficient condition that the sum of S and T is not their union.

4. Let $S = \{(a,a,b)\}$ and $T = \{(c,d,d,)\}$. Find the intersection and sum of these two spaces. Find the dimension of the sum.

5. Define S and T as in Exercise 4 and define $U = \{(2r + 3s,\ 2r + 3t,\ 2u + 3t)\}$. Find the dimension of $S + T + U$.

6. Finish the proof of Theorem 1.8; that is, show closure for multiplication by a scalar.

7. Prove Theorem 1.9.

8. Define S and T as in Exercise 4 and let $W = \{(r,s,0)\}$. Find the dimension of $S + T + W$.

9. Does the set of vectors $\{(a,\ a + 2)\}$ form a subspace of V^2?

10. Let α and β be two distinct vectors of V^2. Does the set $\alpha + \beta + (a,0)$, where a ranges over the set of real numbers, form a subspace?

11. Let W and V be vector spaces of dimensions r and s, respectively, with $r < s$. Show that W is isomorphic to a subspace of V.

12. If V and W are two vector spaces, prove that every subspace contained in both V and W is contained in $V \cap W$.

13. If V and W are two vector spaces, prove that every vector space which contains both as subspaces also contains the space $V + W$.

14. If $V = M + N$, where M and N are subspaces of V, prove that V is the direct sum of M and N if and only if the sum of the dimensions of M and N is equal to the dimension of V.

15. Does $S \oplus T = S \oplus W$ imply $T = W$? Prove it or give an example in which it is not so. State the geometric meaning of this when S, T, and W are all one-dimensional subspaces of a two-dimensional space V.

16. Let α_1, α_2, α_3 be a basis of a vector space V and define

$$S = \text{sp}(\alpha_1,\alpha_2) \qquad T = \text{sp}(\alpha_3)$$

Show that $V = S \oplus T$. Does $V = S \oplus T'$ imply that $T = T'$?

17. Let S, T, and T' be three subspaces of V such that

$$V = S \oplus T = S \oplus T'$$

Prove that if τ is any vector of T, there exists a vector τ' of T' such that $\tau' - \tau$ is in S.

18. Prove Theorem 1.10 and the statement at the close of Sec. 1.10.

19. Let A, B, and C denote three finite sets in which the numbers of elements are denoted by $n(A)$, $n(B)$, $n(C)$, respectively. The following formula is known to hold:

$$n(A \cup B \cup C) = n(A) + n(B) + n(C) - n(A \cap B) - n(A \cap C) - n(B \cap C) + n(A \cap B \cap C)$$

If we replace n by dim and consider A, B, and C to be subspaces of V and if "union" is replaced by "sum," does the corresponding formula hold? Prove it or exhibit a case in which the formula is false.

20. Suppose in Theorem 1.8 we merely assume that S and T are vector spaces without assuming them subspaces of a vector space. Must their intersection be a vector space? Explain.

21. Let W, W', W'' be three subspaces of a vector space V. Does either of the following "distributive" properties necessarily hold? Give reasons.

(a) $W \cap (W' + W'') = (W \cap W') + (W \cap W'')$

(b) $W + (W' \cap W'') = (W + W') \cap (W + W'')$

22. If in either case in Exercise 21 the equality does not universally hold, state conditions on the subspaces sufficient to make the equality hold.

1.11 AN APPLICATION TO PROJECTIVE GEOMETRY

Historically, much of linear algebra and the theory of matrices was developed as a tool for analytic projective geometry. The interplay between the algebra and geometry adds depth to both. To see this one must of course have a little knowledge of projective geometry. So the object of this section is to develop the subject to the point where the connections can be seen and appreciated. Later in the book, as occasion arises, we shall discuss further connections. At this point, we shall confine ourselves largely to two-dimensional analytic projective geometry. We shall limit ourselves to the field of real numbers, since we are dealing with geometry, but much of the material is independent of the field.

Let us start with the equation of a straight line in the familiar Euclidean analytic geometry: $ax + by + c = 0$. We can describe this line by the triple $[a,b,c]$ which we could call *coordinates* of the line, using square brackets instead of parentheses to emphasize the fact that they are coordinates of a line, rather than a point. Now since the equations

$$ax + by + c = 0 \qquad \text{and} \qquad dax + dby + dc = 0 \qquad d \neq 0$$

represent the same line, the corresponding triples

$$[a,b,c] \qquad \text{and} \qquad [da,db,dc] \tag{1.13}$$

represent the same line. Thus *two triples represent the same line if and only if they are proportional.* We call such triples *homogeneous line coordinates.*

To every line in the plane corresponds a triple of homogeneous line coordinates, that is, a triple of real numbers. Turning it around, we can ask, Will every triple represent a line? Yes, with two exceptions. One is the triple $[0,0,0]$. From the point of view of Euclidean analytic geometry, this is no line since all points (x,y) in the plane satisfy the equation $0x + 0y + 0 = 0$. From the point of view of the triples, it is awkward, since taking $d = 0$ in (1.13) would seem to make the triple $[a,b,c]$ represent the same line as $[0,0,0]$. So we *exclude the triple consisting of zeros.*

A second exception is the triple $[0,0,1]$. This cannot correspond to a line in Euclidean geometry either since

$$0x + 0y + 1 = 0$$

is not satisfied by *any* pair (x,y); that is, the set of solutions of the equation contains no pairs at all. We can actually devise a new geometry by being stubborn and postulating that there *shall be* a line whose homogeneous coordinates are $[0,0,1]$. We could give this line a name, say, *ideal line* or *line at infinity.* So as our first step in building up a projective geometry, we have the following:

Definition To every line of analytic projective plane geometry corresponds a triple of real numbers $[a,b,c]$ in which a, b, c are not all zero, and to any such triple corresponds a line. Two triples correspond to the same line if and only if they are proportional.

You might wonder why we do not define the triple to be the line. There are two difficulties with this. In the first place, several triples can correspond to the same line. In the second place, if we are thinking in geometric terms we like to consider a line as being a set of points. So from this point of view the triple should correspond to a set of points.

Now what about the points of this new geometry? Since we have homogeneous coordinates for the lines, it is convenient to have them for the points as well. So we can set up the following correspondence between the usual (Euclidean) coordinates of a point and triples of numbers:

$$(x,y) \leftrightarrow (z_1,z_2,z_3)$$

where $$x = \frac{z_1}{z_3} \qquad y = \frac{z_2}{z_3} \qquad \text{for } z_3 \neq 0$$

Then to each point (x,y) corresponds a triple of z's, and two triples of z's which are proportional correspond to the same point (x,y). We call such triples *homogeneous point coordinates.* Then we have the

Definition A point (z_1,z_2,z_3) lies on the line $[a,b,c]$ if (and only if)

$$az_1 + bz_2 + cz_3 = 0 \tag{1.14}$$

Notice that if $z_3 = 1$ this reduces to the familiar $ax + by + c = 0$. Equation (1.14) is called an *incidence relationship.* We have the

Definition A *line* denoted by $[a,b,c]$ consists of all the points whose homogeneous point coordinates satisfy the incidence relationship (1.14).

Now, as was the case for lines, there are two exceptional triples for points. We exclude the triple (0,0,0) for the same reasons as for lines: The triples in which $z_3 = 0$ do not correspond to points in the Euclidean plane. But if the line has the homogeneous coordinates [0,0,1], we see that the points $(z_1,z_2,0)$ do satisfy the incidence relationship. So to set up plane projective geometry, we adjoin to the points in the Euclidean plane new points corresponding to homogeneous coordinates $(z_1,z_2,0)$ which we can call *ideal points* or *points at infinity.* The ideal line is the set of ideal points.

Let us summarize our development of analytic plane projective geometry to this point:

Definition The points of plane projective geometry correspond to triples of real numbers (z_1,z_2,z_3) where the triple $(0,0,0)$ is excluded, and two triples correspond to the same point if and only if they are proportional. A line, designated by the homogeneous triple $[a,b,c]$, consists of all points whose coordinates satisfy the incidence relationship

$$az_1 + bz_2 + cz_3 = 0 \tag{1.14}$$

Notice that in the definition there is no mention of ideal points of lines. Such concepts are for making the transition from Euclidean to projective geometry. They are not, however, like scaffolding which is torn down after a structure is set up, but more like an outside stairway connecting two floors. We shall find, strangely enough, that while there is essentially only one stairway up from Euclidean to projective geometry, there are many "downspouts" by which we can slide down.

Now we can prove a few basic properties analogous to those in Euclidean geometry.

Theorem 1.11 Any two points in plane projective geometry *determine a line;* that is, given two points, they are contained in exactly one line.

Proof Let the two points correspond to (z_1,z_2,z_3) and (z'_1,z'_2,z'_3). Since the points are distinct, their coordinates are not proportional. This means, in the language of vectors, that the two triples are linearly independent. We seek to solve the following pair of equations for a, b, and c:

$$\begin{aligned} az_1 + bz_2 + cz_3 &= 0 \\ az'_1 + bz'_2 + cz'_3 &= 0 \end{aligned} \tag{1.15}$$

This can be written in matrix form as

$$a(z_1,z'_1) + b(z_2,z'_2) + c(z_3,z'_3) = (0,0) \tag{1.16}$$

The vector space spanned by the three vectors on the left of (1.16) is of dimension not greater than 2. If it were of dimension 1, two of the vectors would have to be scalar multiples of a third; for instance,

$$(z_2,z'_2) = r(z_1,z'_1) \qquad \text{and} \qquad (z_3,z'_3) = s(z_1,z'_1)$$

for some numbers r and s. Then we would have

$$\begin{aligned} (z_1,z_2,z_3) &= (z_1,rz_1,sz_1) = z_1(1,r,s) \\ (z'_1,z'_2,z'_3) &= (z'_1,rz'_1,sz'_1) = z'_1(1,r,s) \end{aligned}$$

This is impossible since the points represented by the triples of z's are distinct. Hence the three vectors on the left of (1.16) span a two-dimensional vector space. That is, one vector is a unique linear combination of the other two. To simplify notation, suppose (z_1,z_1') and (z_2,z_2') are linearly independent, then

$$(z_3,z_3') = t(z_1,z_1') + u(z_2,z_2') \tag{1.17}$$

for a unique pair of numbers t and u. Thus, in virtue of (1.17), one solution of (1.16) is $a = -t$, $b = -u$, $c = 1$, and hence the line with coordinates $[-t,-u,1]$ contains the two given points.

It remains to show that the line with coordinates $[-t,-u,1]$ containing the given points is unique. Suppose the line with coordinates $[a,b,c]$ does contain the points. Substitution of (1.17) in (1.16) gives us

$$(a + ct)(z_1,z_1') + (b + cu)(z_2,z_2') = (0,0)$$

Since (z_1,z_1') and (z_2,z_2') are linearly independent, $a + ct = b + cu = 0$; that is,

$$a = -ct \qquad b = -cu \qquad c = c$$

This shows that a, b, c are proportional to $-t$, $-u$, 1, and hence the line through the given points is unique. This completes the proof.

We leave as an exercise the proof of the following theorem, which may at first seem surprising.

Theorem 1.12 Any two (distinct) lines intersect in a unique point, that is, have exactly one point in common.

A closer look reveals that this result is not as strange as it at first seems, for suppose we consider the Euclidean lines

$$ax + by + c = 0 \qquad \text{and} \qquad ax + by + c' = 0 \qquad \text{for } c \neq c'$$

These lines are parallel in Euclidean geometry. But in projective geometry, the point $(-b,a,0)$ satisfies the incidence relationships

$$a(-b) + b(a) + c(0) = 0$$

and

$$a(-b) + b(a) + c'(0) = 0$$

and hence is on both lines. The point $(-b,a,0)$ is on the ideal line. Thus if m and m' are two lines in projective geometry which correspond to parallel lines in Euclidean geometry, then m and m' intersect in a point which is on the ideal line.

In projective geometry we cannot "add points" as we can add vectors. To see this, consider the points (1,2,3) and (4,5,6). The sum of these vectors, (5,7,9), corresponds to a point. But the triple (2,4,6) represents the same point as does (1,2,3), whereas the sum of the vectors (2,4,6) and (4,5,6) is (6,9,12), which corresponds to quite a different point from the vector (5,7,9). Thus using two different triples for a point can lead to nonproportional sums. But, the set of all linear combinations of (1,2,3) and (4,5,6) is the same as the set of all linear combinations of (2,4,6) and (4,5,6). In fact, the set of all linear combinations of vectors corresponding to two given points in projective geometry consists of vectors corresponding to the points of the line determined by the given points. We prove this in the following

Theorem 1.13 Given two points P and Q whose coordinates are the vectors (p_1,p_2,p_3) and (q_1,q_2,q_3). Every linear combination of these two vectors, except the zero vector, corresponds to a point on the line PQ, and every point on the line PQ has as its corresponding vector a linear combination of the two given vectors.

Proof Let $[a,b,c]$ be the line coordinates of the line determined by the points P and Q. Then

$$ap_1 + bp_2 + cp_3 = 0 \qquad \text{and} \qquad aq_1 + bq_2 + cq_3 = 0$$

imply that for any r and s

$$r(ap_1 + bp_2 + cp_3) + s(aq_1 + bq_2 + cq_3) = a(rp_1 + sq_1) + b(rp_2 + sq_2) + c(rp_3 + sq_3) = 0$$

Hence, for any r and s, $(rp_1 + sq_1, rp_2 + sq_2, rp_3 + sq_3)$ is on the line.

We complete the proof by showing that any set of three vectors representing points on a given line is linearly dependent. To this end suppose

$$\begin{aligned} ap_1 + bp_2 + cp_3 &= 0 \\ aq_1 + bq_2 + cq_3 &= 0 \\ au_1 + bu_2 + cu_3 &= 0 \end{aligned}$$

We suppose that the vectors (p_1,p_2,p_3), (q_1,q_2,q_3), (u_1,u_2,u_3) are linearly independent and arrive at a contradiction. If they are linearly independent, they must span V^3, and hence the vector (0,0,1) must be some linear combination of the three. From the first part of this proof, we know that this vector must correspond to a point on the given line; that is, $a(0) + b(0) + c(1) = 0$. This shows that $c = 0$. Similarly we can prove that b and a are both zero. This gives us the desired contradiction and completes the proof of the theorem.

We now have the equipment to prove a basic theorem of projective geometry called *Desargues' theorem*. It is convenient to use some familiar terms from geometry. Three points on the same line are called *collinear* and three lines through the same point are called *concurrent*. Now we state and prove

***Theorem 1.14* (*Desargues' Theorem*)** If, in projective plane geometry, A, B, C are three (distinct) points and A', B', C' is another set of three points

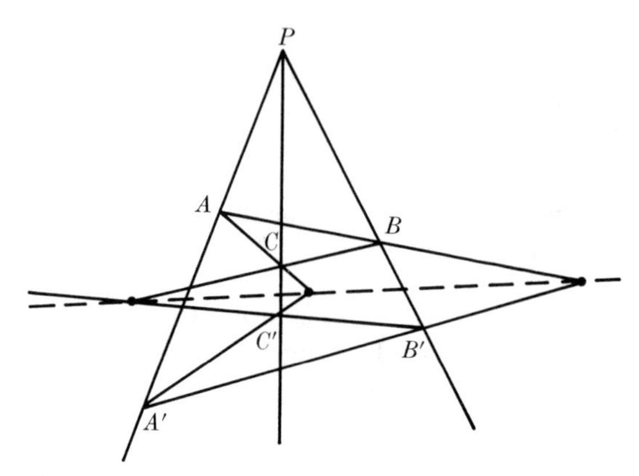

Figure 1.6

such that A and A' are distinct, B and B' are distinct, and C and C' are distinct, and if the lines AA', BB', and CC' are concurrent, then the intersections

$$(1.18) \qquad AB \cap A'B' \qquad AC \cap A'C' \qquad BC \cap B'C'$$

are collinear (see Fig. 1.6).

Proof Notice first that it is possible that the lines AB and $A'B'$, for example, might be the same line. In that case, if the other two intersections are points, the line AB will intersect the line determined by the other two intersections, and the theorem is trivially true. If the lines AB and $A'B'$ are the same as well as AC and $A'C'$, then B and C as well as B' and C' are on the line AA', and the theorem is true even more trivially. So, in effect, we may assume that the intersections (1.18) are three points, though this need not be mentioned in the hypothesis of the theorem.

Let P be the point of intersection of the lines AA', BB', and CC'. (These are lines since the pairs of points are distinct by hypothesis.) Call π a vector corresponding to P and denote vectors (triples) corresponding to the other points by corresponding Greek letters; that is, α is a vector corresponding to A, α' a vector corresponding to A', and so on. Then, since P is on each of the lines, we have

$$(1.19) \qquad \pi = a\alpha + a'\alpha' = b\beta + b'\beta' = c\gamma + c'\gamma'$$

for properly chosen numbers a, a', b, b', c, c'. Then from these three equations we have

$$(1.20) \qquad a\alpha - b\beta = b'\beta' - a'\alpha'$$

$$(1.21) \qquad b\beta - c\gamma = c'\gamma' - b'\beta'$$

$$(1.22) \qquad c\gamma - a\alpha = a'\alpha' - c'\gamma'$$

In Equation (1.20), the vector on the left side represents a point on AB and that on the right side a point on $A'B'$. Since they are equal, the vector $a\alpha - b\beta$ corresponds to the intersection of the lines AB and $A'B'$. Similarly Equation (1.21) gives us a vector that corresponds to the point of intersection of BC and $B'C'$, while that of (1.22) corresponds to the point of intersection of AC and $A'C'$. But these three points are linearly dependent since the sum of the vectors on the left (or right) sides is the zero vector. Thus the points of intersection are collinear.

This is as far as we will carry the application to projective geometry in this chapter, but it is perhaps useful to look back over the construction of projective geometry to see what we have gained and what we have lost by the transition. One of the most important gains is a kind of symmetry between the concept of line and point. Consider the following true statement about points and lines in a plane:

Two points determine a line.

This is true both in Euclidean geometry and in projective geometry. But if we interchange the words *point* and *line*, the statement is not true in Euclidean geometry but is true in projective geometry. To be sure, as we set up our projective geometry, the treatment was not symmetrical since we defined a line to be a set of points. But this was to make the imagery clearer. We could have made the treatment symmetrical at the cost of more terminology by defining lines in terms of triples within square brackets, and points as triples within parentheses, and using the word *incidence*. That is, we could have written "the line $[a,b,c]$ is incident with the point (z_1,z_2,z_3) if $az_1 + bz_2 + cz_3 = 0$" and understood that if a line is incident with a point, then the point is incident with the line. This symmetry of treatment is called *duality* and has the effect that, roughly speaking, when there is a theorem in plane projective geometry about lines and points, there is another theorem about points and lines. In an exercise you will be asked to state and prove the dual of Desargues' theorem (Theorem 1.14). The dual of this theorem turns out to be the same as the converse.

Another advantage of projective geometry is that special cases often do not have to be considered separately. One can get several Euclidean theorems from Desargues' theorem. For instance, if the point P is, in the transition, an ideal point, this means that the lines AA', BB', and CC' are parallel. Still the intersections (1.18) are collinear. Another possibility is that one of the intersections of (1.18) might be an ideal point. You will be asked in an exercise to state what this means in terms of Euclidean geometry.

However, we pay a price for the advantages of projective geometry, namely, the loss of the idea of distance. We shall not attempt to show this rigorously, but it should seem reasonable from the following consideration.

One approach to Euclidean geometry is at some point to postulate a one-to-one correspondence between the set of real numbers and the points of a line. This means that we have available no real number corresponding to the new point, that is, the ideal point; and the distance from any point to the ideal point is not defined.

We have stressed the transition from Euclidean geometry to projective geometry. What about going the other way? Suppose we start with projective plane geometry and pick out *any* line m. Then delete this line, that is, the set of points on the line. Call the resulting geometry G. We call two lines in G "parallel" if in the projective geometry they intersect on the line m. The parallel postulate holds in G since if P is a point in G not on a line k, let Q be the point where k intersects m. The point Q is unique, and PQ is the unique line through P which is, in G, parallel to k. Geometry G still has no distance function, but it has many of the characteristics of Euclidean geometry. This geometry G is called an *affine geometry*.

But we should not travel too far afield from linear algebra. There is an intimate connection between plane analytic projective geometry and a vector space of three dimensions. Let α and β be nonzero vectors, which are linearly independent. Then $\text{sp}(\alpha)$ corresponds to a point A in projective geometry, and $\text{sp}(\alpha,\beta)$ corresponds to a line AB in projective geometry. This means that if we start with V^3, eliminate the null vector, and decrease the dimension by 1, we have analytic projective plane geometry. That is, a one-dimensional vector space corresponds to a point in projective geometry, and a two-dimensional vector space corresponds to a line in projective geometry. This can be extended to any number of dimensions by starting with V^n, deleting the null vector, and letting a vector subspace of dimension r correspond to a projective geometry of dimension $r - 1$.

Why is the geometry we have introduced in this section called *projective geometry?* While a complete answer would lead us too far afield, a little insight can be given by contrasting it with Euclidean plane geometry. In the latter we have points and lines which can be described in terms of pairs of coordinates in the plane; and the properties we are interested in are those which remain unchanged under "rigid motions," that is, transformations which leave distance unchanged. For projective plane geometry we have points and lines which are designated by triples of coordinates. We do not have distance, but we are interested in those properties of projective points and lines which remain unchanged under "projective" transformations in place of rigid motions. A projective transformation has only a distant connection with our usual idea of a projection. But it can be most simply described by affirming that it is one which takes projective lines into projective lines, that is, which takes any three collinear points into three points which are collinear. Such transformations are also called *collineations;* we deal with them briefly in the last section of Chap. 3 as a further application of linear algebra. A student who wishes to explore projective geometry further should read one or more of the many texts on the subject.

EXERCISES

1. Suppose instead of the correspondence used in this section we agreed on the following correspondence between Euclidean coordinates and the projective homogeneous coordinates:

$$(x,y) \leftrightarrow (z_1,z_2,z_3)$$

where $$x = \frac{z_2}{z_1} \qquad y = \frac{z_3}{z_1} \qquad \text{for } z_1 \neq 0$$

Then what will be the homogeneous coordinates of the ideal line?

2. Find the homogeneous coordinates of the line in projective geometry through the two points (1,0,1) and (0,1,0).

3. Find the point of intersection of the two lines [1,0,1] and [0,1,0].

4. In the statement of Desargues' theorem (Theorem 1.14) interchange the words *point* and *line*. Show that this is the same as the converse of Desargues' theorem.

5. Prove Theorem 1.12.

6. Prove the converse of Desargues' theorem (see Exercise 4).

7. Consider the vector space V^4, and in projective geometry of three dimensions make the following definitions:

(*a*) A point is a vector subspace of dimension 1.
(*b*) A line is a vector subspace of dimension 2.
(*c*) A plane is a vector subspace of dimension 3.

Prove the following:

(1) If two points of a line are in a plane, the line is in the plane.
(2) Two distinct planes always have a unique line in common.
(3) Three noncollinear points determine a plane.
(4) If two lines have a point in common, they lie in a unique plane.

8. Suppose in Desargues' theorem that the lines AB and $A'B'$ intersect in an ideal point. What does the theorem tell us in that case?

2

LINEAR EQUATIONS AND MATRICES

2.1 AN EXAMPLE

Let us begin by recalling, in terms of a specific example, the method of solving a set of linear equations. We shall find a number of connections with the material of the first chapter. Suppose, for instance, we have the following set:

$$\begin{aligned} 2x + 7y + 9z &= 16 \\ x + 2y + 3z &= 5 \\ 3x + 8y + bz &= c \end{aligned}$$

where we have inserted two symbols, b and c, so that we may discuss various possibilities for the outcome.

The method of solution is to replace this set by a "simpler" set, replace that by a still "simpler" one, and so forth, until we have a "simple" set of equations having the same set of solutions as that with which we started. More specifically, we try to derive from the given set a pair of equations in only two unknowns; then, working with this pair, get a single equation in one unknown.

We shall first carry through the process without much thought of why it works and then later look at it more critically. The first step is to get two equations in y and z alone. This is done by:

Step 1 Add -2 times the second equation to the first and add -3 times the second equation to the last. This gives us

$$\begin{aligned} 3y + 3z &= 6 \\ x + 2y + 3z &= 5 \\ 2y + (b - 9)z &= c - 15 \end{aligned}$$

Now the first and third equations contain only y and z. Our next step can be

Step 2 Divide the first equation by 3, obtaining

$$\begin{aligned} y + z &= 2 \\ x + 2y + 3z &= 5 \\ 2y + (b - 9)z &= c - 15 \end{aligned}$$

To "get rid of y," we take the next step.

Step 3 Add -2 times the first equation to the third. This gives us

$$\begin{aligned} y + z &= 2 \\ x + 2y + 3z &= 5 \\ (b - 11)z &= c - 19 \end{aligned}$$

Now see what may happen according to the values of b and c. First, if $b - 11 \neq 0$, we can solve the last equation for z, substitute this into the first equation and solve for y; then, substituting the values of y and z in the second equation, solve for x.

Second, if $b = 11$, then the last equation, and hence the whole set, can have no solution unless $c - 19 = 0$. If $b = 11$ *and* $c = 19$, we have only two equations left. We can solve the first for $y = 2 - z$, substitute in the second to get $x = 1 - z$, and, no matter what z is: $x = 1 - z$, $y = 2 - z$, and $z = z$ will be a solution of the given set of equations.

Of course, the process which we used is not the only one. We could, in fact, continue and after Step 3 subtract twice the first equation from the second to get

$$\begin{aligned} y + z &= 2 \\ x + z &= 1 \\ (b - 11)z &= c - 19 \end{aligned}$$

This would make substituting for z easier.

Now, let us look at the operations which we used on the equations.

Operation 1 Add a multiple of one equation to another (Step 1).

Operation 2 Multiply an equation by a nonzero number. (In Step 2 we multiplied the first equation by $\frac{1}{3}$.)

There is a third operation which is not really necessary but which in a certain sense makes "things look better" and also makes a process easier to describe:

Operation 3 Interchange two equations.

These three operations are called the *elementary operations on equations.* To justify our procedure, we need to show that none of these operations changes the set of solutions, sometimes called *the solution set;* that is, that the set of solutions of the final set of equations is the same as the set of solutions of the given equations. We state this formally as

Theorem 2.1 The elementary operations on a set of linear equations do not change the set of solutions.

Proof It is easiest to show this result for Operation 3, for interchanging two equations does not change the set of equations and hence cannot change the set of solutions.

To deal with Operation 2, represent one of the equations by $f = c$, where f is a linear expression in $x_1, x_2, \ldots, x_n$. Any set of numbers in place of the x_i which satisfy the equation $f = c$ will certainly satisfy $kf = kc$ and vice versa if $k \neq 0$. Hence Operation 2 does not change the set of solutions.

To deal with Operation 3, let $f_1, f_2, \ldots, f_r$ be r linear expressions in $x_1, x_2, \ldots, x_n$, and compare the following two sets of equations:

$$f_1 = c_1, f_2 = c_2, \ldots, f_r = c_r \tag{2.1}$$

$$f_1 + kf_2 = c_1 + kc_2, f_2 = c_2, f_3 = c_3, \ldots, f_r = c_r \tag{2.2}$$

The two sets are the same except that to get the first equation of (2.2) we have added k times the second equation of (2.1) to the first equation. We shall now show that the two sets of equations have the same set of solutions. Since the last $r - 1$ equations are the same for the two sets, we need merely check the first one. Suppose we have a solution of (2.1); then $f_2 = c_2$ implies $kf_2 = kc_2$ and that, with $f_1 = c_1$, implies by addition $f_1 + kf_2 = c_1 + kc_2$, and hence any solution of (2.1) is a solution of (2.2). Conversely, any solution of

(2.2) satisfies $kf_2 = kc_2$, and thus

$$f_1 + kf_2 - kf_2 = c_1 + kc_2 - kc_2$$

that is,

$$f_1 = c_1$$

This completes the proof for this case, and any other case can be dealt with similarly.

It then follows from the theorem that, given any set of linear equations, one may perform any sequence of these three types of operations on the set of equations without changing the set of solutions. (It is not necessary that there be just as many equations as unknowns.)

2.2 MATRICES

We can save writing in dealing with equations of Sec. 2.1 by merely writing down in order the numbers as they occur. After all, it is only accidental that we used x, y, and z. We could just as well have used any other letters or no letters at all. In fact, we can keep track of where the numbers occur by writing them in a rectangular array. Any rectangular array of numbers is called a *matrix*. So we associate with the set of equations at the beginning of Sec. 2.1 the following matrix, where the first three columns constitute the matrix of coefficients and the last column designates the constants:

$$\begin{bmatrix} 2 & 7 & 9 & 16 \\ 1 & 2 & 3 & 5 \\ 3 & 8 & b & c \end{bmatrix}$$

We could then perform the same elementary operations on the rows of the matrix to get the following succession of matrices:

$$\begin{bmatrix} 0 & 3 & 3 & 6 \\ 1 & 2 & 3 & 5 \\ 0 & 2 & b-9 & c-15 \end{bmatrix} \quad \begin{bmatrix} 0 & 1 & 1 & 2 \\ 1 & 2 & 3 & 5 \\ 0 & 2 & b-9 & c-15 \end{bmatrix} \quad \begin{bmatrix} 0 & 1 & 1 & 2 \\ 1 & 2 & 3 & 5 \\ 0 & 0 & b-11 & c-19 \end{bmatrix}$$

It would have been simpler to have performed the operations on the matrices and in the end, when the computation was almost finished, translate back into equations. So we define analogously to the elementary operations on equations:

Elementary row operations on a matrix:

1. Add a multiple of one row to another; that is, add the multiple of each element of one row to the corresponding element in the other row.
2. Multiply a row by a nonzero number.
3. Interchange two rows.

In the first two cases we leave all but one row unchanged and in the third case change only two rows.

2.3 MATRICES AND VECTOR SPACES

Suppose a matrix, whether derived from a set of equations or not, has s rows and t columns. Then each row will be a t-tuple and thus a vector in V^t. This set of row vectors will span a vector space; this we call the *row space* of the matrix. There is a theorem for matrices analogous to Theorem 2.1. We state it and leave its proof as an exercise.

Theorem 2.2 Elementary operations on the rows of a matrix do not change its row space.

Thus the same process which we used to solve sets of linear equations can also be used on a set of vectors which span a space to simplify this set and/or to find a basis. Let us illustrate this for the following example:

Example Find a basis for the space spanned by the following set of vectors:

$$\alpha = (3,0,5,9) \qquad \beta = (4,7,0,-8) \qquad \gamma = (7,7,5,1)$$

Recall that a basis of a space is a linearly independent set of vectors which spans it. So we write the matrix with these vectors as rows:

$$\begin{bmatrix} 3 & 0 & 5 & 9 \\ 4 & 7 & 0 & -8 \\ 7 & 7 & 5 & 1 \end{bmatrix} \tag{2.3}$$

There are many possible sequences of elementary operations which will lead to the desired result. That which we choose is not the simplest one, but we are illustrating a systematic method rather than a tricky one. Since there is a 1 in the last place in the third row, we add 8 times the third row to the second and -9 times the third row to the first to get

$$\begin{bmatrix} -60 & -63 & -40 & 0 \\ 60 & 63 & 40 & 0 \\ 7 & 7 & 5 & 1 \end{bmatrix}$$

The vector now represented by the first row is the negative of that for the second row, and hence the row space is spanned by the following two linearly independent vectors:

$$(60,63,40,0) \qquad \text{and} \qquad (7,7,5,1)$$

It is easy to see that these are linearly independent and hence form a basis for the row space of the original matrix. Thus we have shown that the row space of the matrix (2.3) has dimension 2.

We call the dimension of the row space the *row rank*. More formally we have the

Definition The dimension of the row space of a matrix is called its *row rank*.

We have seen (Sec. 1.8) that the dimension of a vector space is independent of the particular basis we happen to choose. Hence, also, however we use the elementary row operations on a matrix, the number of linearly independent rows (vectors) we come out with in the end will not depend on the particular elementary row operations we used.

We have already seen (Sec. 1.7) that in the case of a set of linear equations the space spanned by the column vectors is also important. This we call the *column space* and its dimension the *column rank*.

Notice that we have used parentheses, ( ), for row vectors and brackets, [ ], for matrices. This is not to say that row vectors are not matrices but rather is a distinction of convenience. The square brackets for a matrix are much easier for the printer, and parentheses for row vectors are easier for us. This should be an acceptable compromise.

EXERCISES

1. Find all solutions for each set of equations below. If no solution exists, explain why not.

(*a*)
$$\begin{aligned} x + 2y + 3z + 4t &= 8 \\ 3x \qquad - 4z + 5t &= 25 \\ 3y + z - 2t &= 5 \end{aligned}$$

(*b*)
$$\begin{aligned} 2x + 5y + 7z &= 1 \\ x + 3y + 4z &= 2 \\ 4x + 2y + 6z &= -24 \end{aligned}$$

(*c*)
$$\begin{aligned} 2x + 5y + 7z &= 0 \\ x + 3y + 4z &= 0 \\ 4x + 2y + 6z &= 0 \end{aligned}$$

(*d*)
$$\begin{aligned} x + 2y &= 5 \\ 2x + 3y &= 7 \\ x + 3y &= 8 \\ 5x + 7y &= 19 \end{aligned}$$

(*e*)
$$\begin{aligned} x + 3y + z - t &= 5 \\ 2x + 6y + 2z + t &= 6 \\ -x - 3y + 3z + 2t &= 7 \end{aligned}$$

2. For each set of equations in Exercise 1, find the row rank of the matrix of the equations.

3. For each set of equations in Exercise 1, find the row rank of the matrix without the column of constants. For instance, for set (*a*) the matrix to be considered has three rows and four columns.

4. Prove Theorem 2.2.

5. Find a basis for each of the following sets of vectors:
(*a*) (1,2,3,4), (0,1,3,2), (1,−1,0,5), (2,2,6,11).
(*b*) (1,2,3,4,5), (5,4,3,2,1), (1,1,1,1,1).
(*c*) (1,−1,1,0), (0,1,−1,1), (1,0,1,−1), (−1,1,0,1).

6. Find the column rank of each of the matrices of the sets of equations (*a*) to (*c*) in Exercise 1 and compare it with the row rank.

7. Suppose a certain matrix has three rows and five columns, and the first three columns of this matrix have the form

$$\begin{bmatrix} a \\ 0 \\ 0 \end{bmatrix} \quad \begin{bmatrix} b \\ d \\ 0 \end{bmatrix} \quad \begin{bmatrix} c \\ e \\ f \end{bmatrix}$$

where a, d, and f are different from zero. Show that 3 is the row rank of the matrix. Would the same result hold if the given columns were *any* three columns of the matrix? Give reasons for your conclusions.

8. Prove that the column rank of a matrix is never greater than the number of rows.

9. The following table gives the number of units of vitamins A and B in 1 ounce of each of two types of food:

	A	B
Type I	5	1
Type II	3	1

Suppose a person wants to combine these two types of food to get c units of vitamin A and d units of vitamin B. For what values of c and d can this be done? (Note: c and d need not be integers.)

10. Suppose, in addition to the table in Exercise 9, there is a third type of food having in 1 ounce 1 unit of vitamin A and 4 of vitamin B. Would this increase the range of values of c and d?

2.4 HOMOGENEOUS EQUATIONS AND THE ECHELON FORM

A linear equation $a_1x_1 + a_2x_2 + \cdots + a_nx_n = c$ is called *homogeneous* if $c = 0$. A set of homogeneous equations is called a *homogeneous set*. If one or more of the equations of a set is not homogeneous, we call it a *nonhomogeneous set.* We shall see that there is a close connection between the set of solutions of a nonhomogeneous set of equations and that of the corresponding set of homogeneous equations, that is, those which are obtained by replacing all the constants by zeros. In this section we concern ourselves only with homogeneous equations, the development of a method of solution, and the general character of the solutions. Of course, any set of homogeneous linear equations has at least the solution in which all the variables are zero, the so-called *trivial solution.*

Let us first illustrate the process of obtaining a solution for the set of three homogeneous linear equations in five unknowns with the following matrix of coefficients:

$$\begin{bmatrix} 2 & 6 & 8 & 14 & 2 \\ 1 & 3 & 5 & 9 & -2 \\ -1 & -3 & -5 & -12 & 4 \end{bmatrix}$$

Notice that we have omitted the column of constants since they are all zero. There is no harm in doing this since elementary operations on the rows cannot alter such a column.

To solve this set we wish, by a sequence of elementary operations on the rows, to obtain a sequence of equations in which each has fewer unknowns than the previous one. One way to accomplish this is to add appropriate multiples of the first row to the others so that all but one of the elements of the first column are zero. Then we would add appropriate multiples of the second row to that below it, so that the last element in the second column is also zero. As the matrix stands, we would be adding $-\frac{1}{2}$ times the first row to the second. So to avoid fractions it is simpler first to interchange the first two rows to get

$$\begin{bmatrix} 1 & 3 & 5 & 9 & -2 \\ 2 & 6 & 8 & 14 & 2 \\ -1 & -3 & -5 & -12 & 4 \end{bmatrix}$$

Then subtract twice the first row from the second and add the first row to the last to get

$$\begin{bmatrix} 1 & 3 & 5 & 9 & -2 \\ 0 & 0 & -2 & -4 & 6 \\ 0 & 0 & 0 & -3 & 2 \end{bmatrix} \tag{2.4}$$

Inspection shows that the row rank of this matrix is 3. The set of homogeneous equations corresponding to this matrix is

$$\begin{aligned} x_1 + 3x_2 + 5x_3 + 9x_4 - 2x_5 &= 0 \\ -2x_3 - 4x_4 + 6x_5 &= 0 \\ -3x_4 + 2x_5 &= 0 \end{aligned}$$

We can now solve the last equation for x_4 in terms of x_5, the second for x_3 in terms of x_5, and the first for x_1 in terms of x_2 and x_5. Thus we obtain

$$(2.5) \qquad x_4 = \tfrac{2}{3}x_5 \qquad x_3 = \tfrac{5}{3}x_5 \qquad x_1 = -3x_2 - \tfrac{37}{3}x_5$$

So we can choose x_2 and x_5 as we please and then specify x_1, x_3, x_4 in accordance with (2.5). We say that we have "two arbitrary variables" in the set of solutions. As we have done it, the arbitrary variables are x_2 and x_5. We could have carried it out so that x_2 and x_4 were the arbitrary variables or, indeed, in a number of other ways. Notice that the number of arbitrary variables is equal to the number of variables or unknowns minus the row rank. We shall prove in Sec. 2.7 that this is no accident.

We say that the matrix (2.4) is in *row-echelon form.* In general we define it as follows:

Definition A matrix is in *row-echelon form* if the first (counting from the left) nonzero element in any row is to the left of the first nonzero element in every row below it.

This definition of row-echelon form is different from that which appears in some other texts. Here we think it better to continue the process only to the point where we can determine by inspection the row rank of the matrix, that is, the dimension of the row space. Not until we compute the multiplicative inverse of a matrix in Sec. 2.9 do we need to carry the process beyond the echelon form as defined above.

Now, in proving the following theorem, we describe a process by which any matrix may be taken into row-echelon form by elementary row operations.

Theorem 2.3 Every matrix may, by elementary row operations, be taken into row-echelon form.

Proof Let A denote the matrix. Consider the first column, looking from the left, whose elements are not all zero. Then interchange rows to make the element at the top of the column different from zero. By subtracting proper multiples of this first row from all the other rows, make the rest of that column zero. Then, leaving the first row as it is, carry through the same process on the rest of the matrix so that the first two rows have the property that all elements below the first nonzero element are zero. Now, leaving the first two rows, apply the process, row by row, to the rest of the matrix. Proceeding in this way will finally lead to a matrix in row-echelon form.

It is not hard to see that once a matrix is in row-echelon form, its row rank is the number of nonzero rows. We state this formally:

Theorem 2.4 The row rank of a matrix is the number of nonzero rows in its row-echelon form.

Proof Let $\rho_1, \rho_2, \ldots, \rho_s$ be the nonzero rows in order in the row-echelon form. In order to show that s is the dimension of the row space, we must show that the sequence of s rows is linearly independent, that is, that

$$a_1\rho_1 + a_2\rho_2 + \cdots + a_s\rho_s = \theta \tag{2.6}$$

implies that every a_i is zero. We suppose that this is not the case and reach a contradiction. If a_1 were different from zero, then the first row, ρ_1, would have to be a linear combination of the other rows. This is not possible, for suppose $\rho_1 = (0, 0, \ldots, 0, r, \ldots)$, where $r \neq 0$ is in the kth column. Now all the other row vectors have zeros in the first k columns, and hence ρ_1 is not a linear combination of the other rows; thus a_1 in (2.6) must be zero. The same argument can be used for the second row, ρ_2, to show that a_2 must be zero. We can continue through all the rows of the row-echelon form to complete the proof.

EXERCISES

1. Reduce each of the following matrices to row-echelon form:

$$(a)\ \begin{bmatrix} 1 & 3 \\ 2 & 4 \\ 1 & 5 \end{bmatrix} \qquad (b)\ \begin{bmatrix} 1 & -1 & 0 & 2 \\ 2 & 0 & 2 & 2 \\ 3 & 5 & 8 & -2 \end{bmatrix} \qquad (c)\ \begin{bmatrix} 0 & 1 & 3 & 2 \\ 1 & 0 & 1 & 3 \\ 5 & 6 & 5 & -1 \end{bmatrix}$$

2. Define a *column-echelon form* in a manner analogous to the row-echelon form. Reduce each of the matrices in Exercise 1 to column-echelon form.

3. Suppose a matrix is in row-echelon form. Show that elementary column operations do not change the row rank.

4. Consider each matrix of Exercise 1 to be the matrix of coefficients of a set of linear homogeneous equations. In each case note a relationship between the rank of the matrix and the number of arbitrary variables in the set of solutions.

5. In the example given using matrix (2.4), what are the possible pairs of arbitrary variables?

6. Prove that if $\mathcal{S}$ is a set of linear homogeneous equations for which the number of equations is less than the number of variables, then there is always a nontrivial solution.

7. Prove that for a matrix in row-echelon form the column rank is not greater than the row rank.

8. Prove that for a martix in row-echelon form the column rank is equal to the row rank.

2.5 DEFINITIONS AND OPERATIONS WITH MATRICES

Before proceeding further with the solutions of sets of linear equations, it will be helpful to deal more systematically with matrices. After this is done we shall return to further consideration of solutions of sets of linear equations. We shall also see that matrices have applications beside those to linear equations.

A matrix is a rectangular array, usually of numbers. In general the numbers will be in any field, but, if the reader prefers, he may initially regard them as complex numbers or real numbers. If the matrix has r rows and s columns, we shall refer to it as an *r by s matrix* and call r and s the *dimensions* of the matrix. If the matrix is square, the number of rows is called its *order*, and if we write of the order of a matrix it is understood to be square. We shall, in general, use capital letters for matrices. In designating the elements (entries) of a matrix, it is often convenient to use double subscripts, the first indicating the row and the second, the column. For instance, we write

$$A = \begin{bmatrix} a_{11} & a_{12} & a_{13} \\ a_{21} & a_{22} & a_{23} \end{bmatrix}$$

We designate this more compactly by $A = (a_{ij})$, for $i = 1, 2$, and $j = 1, 2, 3$.

It is convenient to be able to work with matrices as we work with numbers. Though there are some essential differences, we make our definitions so that, for matrices, as many of the properties of a field as possible will carry over. In fact, we shall find later in this section that, by judicious definitions of equality, sum, and product, the set of all square matrices of a fixed order will satisfy all the postulates of a field (see Appendix A) except

1. Commutativity of multiplication.
2. Existence of a multiplicative inverse.

This situation is often described by saying that the set of matrices of order n forms a *ring* with an identity element. We shall even find that for a large class of matrices a kind of division is also possible.

Notice that a vector is a matrix with one row or one column. We shall take our cue for the definitions of equality, addition, and scalar multiplication from the corresponding ones for vectors. So we have

Equality Two matrices are called *equal* if they both have the same dimensions and if corresponding elements (entries) are equal.

Addition If A and B are two matrices with the same dimensions, the matrix $A + B$ is that matrix obtained by adding corresponding elements of A and B.

Multiplication by a Scalar If c is a number and A a matrix, the matrix cA is obtained from A by multiplying each element of A by c.

Let us give examples of these and describe them in notational terms. The following pair of matrices are not equal:

$$\begin{bmatrix} 1 & 3 \\ 2 & 4 \end{bmatrix} \neq \begin{bmatrix} 1 & 2 \\ 3 & 4 \end{bmatrix}$$

They can be equal only if for *every* i and j the element in the ith row and jth column of one matrix is equal to the element in the corresponding position in the other. That is, using the notation earlier in this section, if

$$A = (a_{ij}) \qquad B = (b_{ij}) \qquad \text{for } 1 \leq i \leq r,\ 1 \leq j \leq s$$

then $A = B$ if and only if $a_{ij} = b_{ij}$, for all i and j in the ranges indicated.

For addition,

$$\begin{bmatrix} 1 & 2 & 3 \\ 4 & 5 & 6 \end{bmatrix} + \begin{bmatrix} 0 & 1 & -2 \\ 3 & -1 & 5 \end{bmatrix} = \begin{bmatrix} 1 & 3 & 1 \\ 7 & 4 & 11 \end{bmatrix}$$

Thus, in general, the element in the ith row and jth column of $A + B$ is $(a_{ij} + b_{ij})$. If the dimensions of the two matrices are not the same, addition is not defined for them, just as we cannot add two vectors with different numbers of components.

As for vectors, the definition of multiplication by a scalar is rather forced upon us by the definition of addition. Thus the matrix $B = A + A$ can be obtained from the matrix A by multiplying each element by 2, and we want $A + A$ to be equal to $2A$. So we define the matrix cA to be that whose elements are (ca_{ij}).

These definitions also force us to define the zero matrix to be that matrix whose elements are all zero. We shall designate it by (0) or sometimes by Z. Though we use the definite article *the*, it is of course understood that there is one zero matrix for any pair of dimensions. It has the following property:

$$(0) + A = A + (0) = A \qquad \text{for all matrices } A$$

Furthermore, we define $-A$ to be $(-1)A$ and see that $(-A) + A = A + (-A) = (0)$, and we write $A - B$ as equivalent to $A + (-B)$.

It follows immediately from the corresponding properties for numbers that

Addition of matrices is commutative and associative.

We may summarize the properties so far, exclusive of scalar multiplication, in the statement

The set of matrices of fixed dimensions form an additive group.

(See Appendix A.) The set also forms a vector space as we shall see later in this section.

Defining the product of two matrices is a somewhat more difficult matter. First we give two examples which point in a certain direction. As we proceed in the book, we shall see other more compelling reasons for defining the product of two matrices as we do.

Since we have not, to this point, multiplied vectors, this presents no clue. But we have used matrices for linear equations. Consider the set

$$\begin{aligned} 3x - 2y + z &= 5 \\ x + 3y - 5z &= 7 \end{aligned} \tag{2.7}$$

The matrix of coefficients is

$$\begin{bmatrix} 3 & -2 & 1 \\ 1 & 3 & -5 \end{bmatrix} = A$$

Now consider the following expression:

$$\begin{bmatrix} 3 & -2 & 1 \\ 1 & 3 & -5 \end{bmatrix} \begin{bmatrix} x \\ y \\ z \end{bmatrix} = \begin{bmatrix} 5 \\ 7 \end{bmatrix} \quad \text{or} \quad A \begin{bmatrix} x \\ y \\ z \end{bmatrix} = \begin{bmatrix} 5 \\ 7 \end{bmatrix} \tag{2.8}$$

The left-hand side of the equation looks like a kind of product of a matrix A with a column matrix whose elements are x, y, and z. We now show how we can define the product of matrices so that Equation (2.8) in matrices is equivalent to the set of Equations (2.7). We use a "row by column" multiplication. That is, to get the element in the first row and first column of the matrix product on the left of (2.8), we write the sum of the products of elements of the first row of A with the respective elements x, y, z of the first (and only) column of the matrix on its right to get $3x - 2y + z$. To find the element in the second row and first column of the product, we write the sum of the products of the elements of the second row of A by x, y, and z, respectively. Thus we define the matrix product as follows:

$$\begin{bmatrix} 3 & -2 & 1 \\ 1 & 3 & -5 \end{bmatrix} \begin{bmatrix} x \\ y \\ z \end{bmatrix} = \begin{bmatrix} 3x - 2y + z \\ x + 3y - 5z \end{bmatrix}$$

and equating the latter matrix to the matrix $\begin{bmatrix} 5 \\ 7 \end{bmatrix}$, we recover the set of Equations (2.7).

As a second example, suppose there are two warehouses W_1 and W_2 and three kinds of commodity C_1, C_2, and C_3 whose quantities in the respective warehouses are indicated by the following matrix:

$$\begin{array}{c} \\ W_1 \\ W_2 \end{array} \begin{array}{c} \begin{array}{ccc} C_1 & C_2 & C_3 \end{array} \\ \begin{bmatrix} q_{11} & q_{12} & q_{13} \\ q_{21} & q_{22} & q_{23} \end{bmatrix} \end{array}$$

That is, for example, there are q_{11} units of commodity C_1 in warehouse W_1. Denote this matrix by Q. Suppose that on a certain day D_1 the prices of the three commodities are p_{11}, p_{21}, p_{31}, respectively, and on another day D_2 they are p_{12}, p_{22}, p_{32}. We can tabulate these facts by means of a matrix P, the price matrix:

$$\begin{array}{c} \\ C_1 \\ C_2 \\ C_3 \end{array} \begin{array}{c} \begin{array}{cc} D_1 & D_2 \end{array} \\ \begin{bmatrix} p_{11} & p_{12} \\ p_{21} & p_{22} \\ p_{31} & p_{32} \end{bmatrix} \end{array} = P$$

where now the rows are associated with the kind of commodity and the columns with the day on which the prices are set. Now suppose we define the matrix product QP as we did in the previous example. That is, the element in the first row and first column of the product is defined to be the sum of the products of the elements of the first row of Q with corresponding elements of the first column of P. The element in the first row and second column of the product is defined to be the sum of the products of the elements in the first row of Q with corresponding elements of the second column of P. Thus if we define the product QP as follows,

$$\begin{bmatrix} q_{11} & q_{12} & q_{13} \\ q_{21} & q_{22} & q_{23} \end{bmatrix} \begin{bmatrix} p_{11} & p_{12} \\ p_{21} & p_{22} \\ p_{31} & p_{32} \end{bmatrix}$$
$$= \begin{bmatrix} q_{11}p_{11} + q_{12}p_{21} + q_{13}p_{31} & q_{11}p_{12} + q_{12}p_{22} + q_{13}p_{32} \\ q_{21}p_{11} + q_{22}p_{21} + q_{23}p_{31} & q_{21}p_{12} + q_{22}p_{22} + q_{23}p_{32} \end{bmatrix}$$

we see that the elements in the first row of the product are the total prices of the commodities in warehouse W_1 on the two days and that the second row of the product gives the corresponding data for warehouse W_2.

On the basis of these two examples and for other reasons which we shall see later, we make the following definition of the product of two matrices:

Definition If the number of columns in matrix A is the same as the number of rows in matrix B, then the *product* AB is the matrix obtained in the following way: To get the element in the ith row and jth column of AB, compute the sum of the products of the elements in the ith row of A and the corresponding elements in the jth column of B.

In notation, suppose A has r rows and s columns and B has s rows and t columns. Then the element in the ith row and jth column of AB is

$$a_{i1}b_{1j} + a_{i2}b_{2j} + \cdots + a_{is}b_{sj} = \sum_{k=1}^{s} a_{ik}b_{kj} \qquad \text{for } 1 \le i \le r,\ 1 \le j \le t$$

where $A = (a_{ij})$ and $B = (b_{ij})$. The product AB has r rows and t columns.

Notice that for such a product AB to be defined, there must be just as many elements in each column of B as there are in each row of A. That is, the number of rows of B is the same as the number of columns of A. For instance, if

$$A = \begin{bmatrix} 1 & 3 & 4 \\ 2 & -1 & 5 \end{bmatrix} \qquad B = \begin{bmatrix} 1 \\ 2 \\ 3 \end{bmatrix}$$

the product AB is defined but the product BA is not defined. Also, multiplication is not necessarily commutative even when the product in both directions is defined, as the following example shows:

$$\begin{bmatrix} 0 & 1 \\ 0 & 0 \end{bmatrix} \cdot \begin{bmatrix} 0 & 0 \\ 0 & 1 \end{bmatrix} = \begin{bmatrix} 0 & 1 \\ 0 & 0 \end{bmatrix} \quad \text{but} \quad \begin{bmatrix} 0 & 0 \\ 0 & 1 \end{bmatrix} \cdot \begin{bmatrix} 0 & 1 \\ 0 & 0 \end{bmatrix} = \begin{bmatrix} 0 & 0 \\ 0 & 0 \end{bmatrix}$$

The second equality shows also that the product of two matrices may be the zero matrix without either being the zero matrix.

It is unfortunate that these things happen, but there is no way of avoiding them if we define multiplication as we have done. We shall see that in spite of these failings, or perhaps because of them, this definition of multiplication is very useful.

We do have a multiplicative identity which we designate by I (in some texts E is used) and which is a square matrix with 1s along the diagonal from upper left to lower right and zeros everywhere else. One identity matrix is

$$\begin{bmatrix} 1 & 0 & 0 \\ 0 & 1 & 0 \\ 0 & 0 & 1 \end{bmatrix}$$

It is easily seen that if this is I, then, for any matrix with three rows,

$$IA = A$$

Here I is a *left identity*. If matrix A also has three columns, then $AI = A$ for the identity matrix I of order 3. Here I is a *right identity*. The left identity and right identity for a matrix A are the same if and only if A is square. If A had four columns, the product AI would have meaning only if I is of order 4, and in that case, $AI = A$.

Products of matrices are associative when the products are possible; that is,

$$(AB)C = A(BC)$$

One can prove this using the definition of product, but we shall have a better way to show it later (see Sec. 3.10); hence, until that point, we shall assume it.

However, we shall note now that if the product $(AB)C$ is defined, so is the product $A(BC)$; for let A be an r by s matrix. For AB to be defined, B must have s rows. Thus, if B has t columns, AB has r rows and t columns. Hence C must have t rows if $(AB)C$ is to be defined. Now, considering the product $A(BC)$, we see that BC is defined since B has t columns and C has t rows. Then since BC has the same number of rows as B, namely, s rows, and A has s columns, it follows that the product $A(BC)$ is defined. Thus if the product $(AB)C$ has meaning, so has $A(BC)$.

Notice that multiplying a matrix A on the left or right by the matrix cI has the effect of multiplying each element of A by c. That is,

$$(cI)A = A(cI) = cA$$

So multiplying any matrix by cI on the left or right is equivalent to multiplying by the scalar c. For this reason the matrix cI, for $c \neq 0$, is usually called a *scalar matrix*. All its elements are zero except for the number c along the diagonal. Notice that if $c = 0$, then cI is the zero matrix (0) which has the effect of multiplying each element of A by zero. Thus

$$(0)A = A(0) = (0)$$

It is not hard to show that the distributive property holds for matrices; that is,

$$A(B + C) = AB + AC \qquad \text{and} \qquad (B + C)A = BA + CA$$

Of course in both equations the dimensions of B and C must be the same, while in the first equation the number of rows of B and C must be the same as the number of columns of A, and in the second equation the number of columns of B and C must be equal to the number of rows of A. We leave the proof of the distributive property as an exercise.

If, in a set of matrices, addition *and* multiplication are to have meaning, then all the matrices of the set must be square and of the same order. With this restriction, we have an additive identity and a multiplicative identity, both of which are unique, and all the postulates of a field (see Appendix A) hold except the commutativity of multiplication and the existence of a multiplicative inverse. This can be briefly expressed by saying (see Appendix A) that the set of all (square) matrices of given order forms a ring with an identity element.

One can take another point of view with regard to the set of all matrices of order n. If we look back at the abstract definition of a vector space in Sec. 1.9 and take into account multiplication of matrices by scalars, it is not hard to see that the set of all matrices of order n is also a vector space. We leave this as an exercise. We can do more—actually find the dimension of this space. In fact we prove

Theorem 2.5 The dimension of the vector space consisting of all matrices of order n over a field is n^2.

Proof We accomplish the proof by showing that the space has a basis containing n^2 matrices. Let E_{ij} be the matrix of order n having a 1 at the intersection of the ith row and the jth column and all other elements zero. There are n^2 such matrices. Then if $A = (a_{ij})$ is any matrix of the set, it can be written as a linear combination of the E_{ij} as follows:

$$A = \Sigma a_{ij} E_{ij} \tag{2.9}$$

where the i and j range over the integers from 1 to n inclusive. Since the matrix A determines the coefficients in the linear combination and conversely, the representation (2.9) is unique and hence the E_{ij} are linearly independent. This completes the proof.

To clarify the above proof, take the case when $n = 2$. Then

$$E_{11} = \begin{bmatrix} 1 & 0 \\ 0 & 0 \end{bmatrix} \quad E_{12} = \begin{bmatrix} 0 & 1 \\ 0 & 0 \end{bmatrix} \quad E_{21} = \begin{bmatrix} 0 & 0 \\ 1 & 0 \end{bmatrix} \quad E_{22} = \begin{bmatrix} 0 & 0 \\ 0 & 1 \end{bmatrix}$$

and

$$a_{11}E_{11} + a_{12}E_{12} + a_{21}E_{21} + a_{22}E_{22} = \begin{bmatrix} a_{11} & a_{12} \\ a_{21} & a_{22} \end{bmatrix}$$

It should be noticed that we need not have taken the set of matrices to be square, since the product of vectors does not enter into the definition of a vector space. In fact, Theorem 2.5 is also true for the set of all matrices with r rows and s columns with n^2 replaced by rs. We leave this modification as an exercise.

EXERCISES

1. Given the following set of matrices:

$$A = \begin{bmatrix} 1 & 4 \\ 5 & 6 \\ 7 & 8 \end{bmatrix} \qquad B = \begin{bmatrix} 1 & 3 & -1 \\ 4 & 5 & 0 \end{bmatrix}$$

$$C = \begin{bmatrix} 7 & 3 & 4 \\ 2 & 5 & 6 \end{bmatrix} \qquad D = [1 \quad 5 \quad 4]$$

For which pairs is the sum defined? Where it is defined, find the sum.

2. For what pairs of matrices in Exercise 1 is the product defined in one order or the other? For these pairs find the products.

3. For the following matrix A find A^2, A^3, and $A + A^2 + A^3$.

$$\begin{bmatrix} 0 & 1 & 0 \\ 0 & 0 & 1 \\ 1 & 0 & 0 \end{bmatrix}$$

4. Find the square of the following matrix:

$$\frac{1}{3}\begin{bmatrix} 1 & 2 & -2 \\ 2 & 1 & 2 \\ -2 & 2 & 1 \end{bmatrix}$$

5. Prove that if A is a matrix for which the sum $A + A$ and the product A^2 are both defined, then A must be square.

6. Prove that if A and B are two matrices for which $A + B$ and AB are both defined, then A and B must be square.

7. If A and B are two matrices such that AB and BA are defined, must they be square? If so why; if not what must be true about their dimensions?

8. Let A be a square matrix (a_{ij}) with two rows. Find PA and AP for each of the following matrices P:

$$\begin{bmatrix} 0 & 1 \\ 1 & 0 \end{bmatrix} \quad \begin{bmatrix} 3 & 0 \\ 0 & 1 \end{bmatrix} \quad \begin{bmatrix} 1 & 4 \\ 0 & 1 \end{bmatrix}$$

9. For each of the matrices P in Exercise 8 find the matrix Q such that $QP = I$. For such a Q is $PQ = I$ also true?

10. If A is a square matrix, define its trace, written $\text{tr}(A)$, to be the sum of the elements on the principal diagonal; that is,

$$\text{tr}(A) = \sum_{i=1}^{n} a_{ii}$$

where $A = (a_{ij})$. Prove that $\text{tr}(AB) = \text{tr}(BA)$ for two n by n matrices A and B.

11. Let S be the set of matrices over a field which has r rows and s columns. Prove that S is a vector space and find its dimension.

12. Prove the distributive property for matrices. Why does $A(B + C) = AB + AC$ not imply $(B + C)A = BA + CA$?

13. If A and B are square matrices and

$$A(AB - BA) = (AB - BA)A$$

prove that $A^nB - BA^n = n(AB - BA)A^{n-1}$, first for $n = 2, 3$ and then for all positive integers. (A^n is the product of n A's. An argument by induction would seem appropriate.)

14. Prove that the set of matrices $\begin{bmatrix} a & b \\ -b & a \end{bmatrix}$ is isomorphic as a vector space to the set of vectors (a,b). If, in addition, a and b are real numbers, show that the correspondence

$$\begin{bmatrix} a & b \\ -b & a \end{bmatrix} \leftrightarrow a + bi$$

is preserved under multiplication; that is, the product of two such matrices corresponds to the product of the corresponding complex numbers.

15. Define $A * B$ to be $(AB + BA)/2$ (the Jordan product). Which of the following properties of matrices hold for such a "product"?

(*a*) Commutativity
(*b*) Associativity
(*c*) Existence of a multiplicative identity
(*d*) Distributivity

2.6 TRANSPOSES AND INVERSES OF MATRICES

It is often convenient to associate with a given matrix A its so-called *transpose*, defined as follows:

Definition The *transpose* of a matrix A is that matrix whose rows are the columns of A, in the same order. The transpose of A is denoted by A^T. (There are in other texts other notations for the transpose of a matrix. The most common are A^t and A'.)

This notation is convenient, for instance, when we are dealing with vectors as matrices. We usually consider a vector

$$\alpha = (a_1, a_2, \ldots, a_n)$$

as a matrix consisting of a single row. Then α^T is the same vector written in column form. A set of linear equations in $x_1, x_2, \ldots, x_n$ could then be written in the following form:

$$A\xi^T = \gamma^T$$

where A is the matrix of coefficients,

$$\xi = (x_1, x_2, \ldots, x_n) \qquad \text{and} \qquad \gamma = (c_1, c_2, \ldots, c_n)$$

You should verify that the same set of equations can also be written in the form

$$\xi A^T = \gamma$$

This illustrates an important property of transposes noted as

Theorem 2.6 If A and B are two matrices for which the product AB is defined, then $(AB)^T = B^T A^T$. That is, the transpose of a product is equal to the product of the transposes taken in reverse order.

Proof The proof is an exercise in the use of the summation sign. Let $A = (a_{ij})$ and $B = (b_{ij})$, and suppose A has r rows and s columns, while B has s rows and t columns. Then the element in the ith row and jth column of AB is

$$\sum_{k=1}^{s} a_{ik} b_{kj}$$

This is the element in the jth row and ith column of $(AB)^T$. Now if we designate the element in the kth row and ith column of A^T by a'_{ki}, we have $a'_{ki} = a_{ik}$. With a similar designation for B^T, we have $b'_{jk} = b_{kj}$. Hence the element in the jth row and ith column of $(AB)^T$ is

$$\sum_{k=1}^{s} a_{ik} b_{kj} = \sum_{k=1}^{s} a'_{ki} b'_{jk} = \sum_{k=1}^{s} b'_{jk} a'_{ki}$$

which is the element in the jth row and ith column of the product $B^T A^T$. This completes the proof.

An instance of this result is the change we made above from the equation $A\xi^T = \gamma^T$ to $\xi A^T = \gamma$. The latter is obtained from the former by taking the transpose of both sides of the equation and using the property

$$(\xi^T)^T = \xi$$

We have seen that we can add, subtract, and multiply matrices if their dimensions conform; in particular we can always perform these operations within a set of n by n square matrices. But division is another matter.

Suppose we recall what happens with numbers. To divide by 3, for instance, is equivalent to multiplying by $\frac{1}{3}$, where the number $\frac{1}{3}$ has the property that $3 \times \frac{1}{3} = 1$. So if we can find a matrix C such that $AC = I$, dividing by A would be equivalent to multiplying by C. The trouble is that even if A is not the zero matrix, such a C does not always exist. Let us explore briefly why this statement is true. Suppose there were a matrix B, different from the zero matrix, such that $BA = Z$, the zero matrix, as in an example in Sec. 2.5. Then the following sequence of equations shows that $AC = I$ leads to a contradiction:

$$AC = I \quad \text{and} \quad BA = Z$$

imply

$$B = BI = B(AC) = (BA)C = ZC = Z$$

On the other hand, suppose there is a matrix C such that $AC = I$; then for any matrix D of the proper dimensions, $X = CD$ is a solution of $AX = D$, since

$$D = ID = (AC)D = A(CD)$$

Note, however, that CD is not necessarily a solution of $XA = D$ since $(CD)A = D$ is not necessarily true.

If $AC = I$ we call C a *right inverse* of A. (Though it is a multiplicative inverse, we usually omit this adjective.) Again, there is no reason why CA should be I. If it happens that there is a matrix R such that $RA = I$, we call R a *left inverse* of A.

Now let A be a matrix with r rows and n columns, and let us explore the possibility of the existence of a matrix C such that $AC = I$. Since I is square and has the same number of rows as A, it must be of order r. This implies that C has n rows and r columns. Let ϵ_1^T denote the first column of I and γ_1^T the first column of C. Then $AC = I$ implies that γ_1^T is a solution of $AX = \epsilon_1^T$; that is, ϵ_1^T is a linear combination of the columns of A. Similarly $AC = I$ implies that every column of I is a linear combination of the columns of A. Thus every column of I is in the space spanned by the columns of A. Since the columns of I are linearly independent, the column rank of A is at least r. But the columns of A are r-tuples, and hence by Theorem 1.5 the column rank of A is not greater than r. Thus we have shown that if $AC = I$, then r is the column rank of A. This is part of the following theorem.

Theorem 2.7a If A has r rows and n columns, the matrix equation $AX = I$ is solvable if and only if the column rank of A is r; that is, A has a right inverse if and only if its column space has dimension equal to its number of rows.

To complete the proof we assume that the column rank of A is r and show that $AX = I$ is solvable. Under this assumption, the columns of A span a vector space of dimension r. The columns of A are in V^r and hence span V^r. But the columns of I are also in V^r and hence are linear combinations of the columns of A. That is, there are vectors γ_i such that $A\gamma_i^T = \epsilon_i^T$. Then, if we take C to be the matrix whose columns in order are γ_i^T, we have the matrix equation $AC = I$. This completes the proof of Theorem 2.7a.

To prove a companion theorem, notice that if A has r rows and n columns, then A^T has n rows and r columns. Then, applying Theorem 2.7a to A^T in place of A, we see that $A^TY = I$ has a solution if and only if the column rank of A^T is n, that is, if and only if the row rank of A is n. But $A^TY = I$ is equivalent to

$$I = I^T = (A^TY)^T = Y^TA$$

Thus we have proved

Theorem 2.7b If A has r rows and n columns, the matrix equation $YA = I$ is solvable if and only if the row rank of A is n; that is, A has a left inverse if and only if its row space has dimension equal to its number of columns.

Corollary If an r by n matrix has a right inverse, then $n \geq r$. If an r by n matrix has a left inverse, then $n \leq r$.

The first conclusion of this corollary follows since the existence of a right inverse, by Theorem 2.7a, implies that the dimension of the column space is r and this dimension cannot be greater than the number of columns, n. The second sentence follows similarly.

Under what conditions will a matrix have both a right and left inverse? The corollary implies that $r = n$; that is, it is necessary that the matrix be square. Then the two theorems above imply

Theorem 2.8 A matrix A has both a right and a left inverse if and only if both of the following conditions hold:

1. A is square.
2. Its column rank and its row rank are both equal to its order.

Actually we can strengthen Theorem 2.8 without much difficulty to read as follows:

Theorem 2.9 A matrix has both a right and left inverse if and only if

1. A is square, and
2. Its column rank *or* its row rank is equal to its order.

Proof Suppose n is the order of A and its column rank is n. Then, by Theorem 2.7a, $AX = I$ is solvable. Suppose the row rank of A were less than n. Then some nontrivial linear combination of the rows of A would be the zero vector; that is, for some vector $\alpha \neq \theta$,

$$\alpha A = \theta$$

But then, $$\theta X = \alpha A X = \alpha I = \alpha \qquad \text{and} \qquad \theta = \alpha$$

contrary to our choice of α. Hence the row rank must also be n.

The proof for the case in which the row rank is equal to the order goes in the same way.

It should be noted that if a matrix has a right inverse but not a left inverse, the right inverse need not be unique. For example, take

$$A = \begin{bmatrix} 1 & 0 & 1 \\ 0 & 1 & 0 \end{bmatrix} \qquad R = \begin{bmatrix} a & b \\ 0 & 1 \\ 1 - a & -b \end{bmatrix}$$

For all values of a and b, R is a right inverse of A. Similarly if a matrix has a left inverse but not a right inverse, the left inverse need not be unique. But if both the right and left inverses exist, they are not only the same but unique. This can be seen as follows. Recall that A must be square and suppose $AX = I$ and $CA = I$. Then

$$X = IX = (CA)X = C(AX) = CI = C$$

So $X = C$. If $AY = I$ it follows similarly that $Y = C$ and hence $X = Y$. Thus we have proved

Theorem 2.10 If A is a matrix which has both a right and a left inverse, the two are equal.

This gives us the right to make the following definition.

Definition *The inverse* of a matrix A, when it exists, is that matrix C such that $CA = AC = I$. We denote the inverse by A^{-1}. If a matrix has an inverse we call it *nonsingular*. If a square matrix is not nonsingular, we call it *singular*. (Note: A nonsingular matrix *must* be square.)

Using this terminology, we can recapitulate the results above in the following

Theorem 2.11 Each of the following is a necessary and sufficient condition that a matrix A be nonsingular:

1. A has both a right inverse and a left inverse.
2. A is square and its row rank or column rank is equal to its order.
3. A is square and it has a right inverse or a left inverse.

We leave as an exercise the proofs of the following for nonsingular matrices A and B.

1. $(AB)^{-1} = B^{-1}A^{-1}$
2. $(A^T)^{-1} = (A^{-1})^T$

Because of the second property we often write the transpose of the inverse (that is, the inverse of the transpose) of a matrix A as A^{-T}.

Now that we have the inverse, what do we do with it? We can use it to solve a set of linear equations, to be sure, for if the set of equations is written in the form $A\xi^T = \gamma^T$, the solution can be written as

$$\xi^T = A^{-1}\gamma^T$$

In fact, we shall develop in Sec. 2.9 a method for computing the inverse of a matrix. However, in solving a set of linear equations, it is more difficult to find the inverse than to solve by the methods we have used earlier in this chapter. Computing the inverse would be justified only if we had to solve a succession of sets of equations in which the set of constants alone varied. In practice and in theory what is usually of most importance is the *existence* of an inverse.

So we have, under certain circumstances, a kind of division of matrices. But a caution is in order. For numbers we write the solution of the equation $2x = 3$ as $x = \frac{3}{2}$. There is no ambiguity here since $3 \times \frac{1}{2} = \frac{1}{2} \times 3$. But for a matrix equation when A is nonsingular, *the* solution of $AX = B$ is $X = A^{-1}B$. To see this note first that $A^{-1}B$ is *a* solution since $A(A^{-1}B) = B$ and is *the* solution since if $AX = B$ for some X we can multiply both sides of the equation on the left by A^{-1} and have $A^{-1}(AX) = X = A^{-1}B$. But $A^{-1}B$ can be quite different from BA^{-1}. So instead of dividing, which implies commutativity, we multiply by the inverse, taking care on which side we multiply.

For future reference it is convenient to list here names of various kinds of matrices, noting first that the diagonal (or principal diagonal) of a matrix $A = (a_{ij})$ is the sequence $\{a_{ii}\}$. In each case below, A is a square matrix.

1. A is called a *diagonal* matrix if its only nonzero elements are on its diagonal.
2. A is called a *scalar* matrix if it is equal to cI for some number $c \neq 0$.
3. A is called *symmetric* if $A^T = A$.
4. A is called *skew symmetric* (or *skew*) if $A^T = -A$.
5. A is called *idempotent* if $A^2 = A$.
6. A is called *nilpotent* if, for some positive integer r, $A^r = (0)$.
7. A is called *involutory* if $A^2 = I$.

EXERCISES

1. Which of the following matrices have left inverses, which have right inverses, which have both, and which have neither? Justify your answers.

(*a*) $\begin{bmatrix} 1 & 2 & 5 \\ 2 & 4 & 7 \end{bmatrix}$

(*b*) $\begin{bmatrix} 1 & 2 & 8 \\ 2 & 4 & 16 \end{bmatrix}$

(*c*) $\begin{bmatrix} 1 & 3 \\ 2 & 4 \\ 5 & 6 \end{bmatrix}$

(*d*) $\begin{bmatrix} 1 & 2 & 3 \\ 4 & 5 & 6 \\ 7 & 8 & 9 \end{bmatrix}$

(*e*) $\begin{bmatrix} -1 & 1 & 1 \\ 1 & -1 & 1 \\ 1 & 1 & -1 \end{bmatrix}$

2. In Exercise 1, where a matrix has left or right inverses, find them all.

3. Show that the following matrix is involutory:

$$\frac{1}{3}\begin{bmatrix} 1 & 2 & -2 \\ 2 & 1 & 2 \\ -2 & 2 & 1 \end{bmatrix}$$

4. Can a nilpotent matrix be nonsingular? Explain.

5. Find the inverse of the matrix

$$B = \begin{bmatrix} 2 & 3 \\ 3 & 4 \end{bmatrix}$$

6. Use the result of Exercise 5 to solve each of the following matrix equations: $BX = C$ and $YB = C$ where $C = (1, -1)^T$. Are Y and X equal?

7. Find the inverse of the matrix

$$B = \begin{bmatrix} 2 & 5 \\ 1 & 6 \end{bmatrix}$$

For this matrix B, solve the equations of Exercise 6. In this case are X and Y equal?

8. If $AX = B$ and A is nonsingular, prove that $(AXA^{-1})A = B$.

9. Prove that if A and B are nonsingular matrices, then $(AB)^{-1} = B^{-1}A^{-1}$.

10. Prove that if A is a nonsingular matrix, then $(A^T)^{-1} = (A^{-1})^T$.

11. Prove that if A and B are nonsingular, then $(AB)^{-T} = A^{-T}B^{-T}$.

12. Prove $(ABC)^T = C^TB^TA^T$.

13. Let A be a square matrix. Show that $A + A^T$ is symmetric and that $A - A^T$ is skew symmetric. Hence show that if† $1 + 1 \neq 0$, any square matrix A can be written as the sum $S + R$, where S is a symmetric matrix and R a skew-symmetric matrix. Are S and R uniquely determined by A?

14. If A and B are symmetric matrices, prove that AB is symmetric if and only if $BA = AB$.

15. In Exercise 14 replace *symmetric* in succession by skew symmetric, idempotent, nilpotent, involutory. In each case show whether or not the conclusion is true.

16. Prove that if A and B are matrices for which both products AB and BA are defined, then A and B^T have the same dimensions and conversely.

17. Let A and P be square matrices of the same order with P nonsingular. Prove that PAP^T is symmetric if and only if A is.

† Of course this is always true for the familiar fields like those of the rational, real, or complex numbers, but there are fields in which $1 + 1 = 0$ (see Appendix A).

18. Is the conclusion of Exercise 17 true if the condition that P be nonsingular is omitted? Explain.

19. For which of the following replacements for the word *symmetric* does the statement of Exercise 17 hold: skew symmetric, idempotent, nilpotent, involutory?

20. If P is nonsingular and A is a matrix of the same order as P, prove that A is nilpotent if and only if PAP^{-1} is.

21. For which of the following replacements for the word *nilpotent* does the statement of Exercise 20 hold: symmetric, skew symmetric, idempotent, involutory.

22. Prove that if a matrix has a left inverse, then its transpose has a right inverse.

23. If $A^r - I = (0)$, for r an integer greater than 1, and if $A - I$ is nonsingular, prove that

$$A^{r-1} + A^{r-2} + \cdots + A = -I$$

Show that A has an inverse which is a polynomial in A.

24. If A is nonsingular, prove that $A^{-1} = (A^TA)^{-1}A^T$. Thus the inverse of A is expressed in terms of the inverse of the symmetric matrix A^TA.

25. If D is nonsingular, prove that

$$(D^{-1}AD)^k = D^{-1}A^kD$$

for every positive integer k.

26. Prove that if $AX = XA$ for all matrices X for which the products have meaning, then A is a scalar matrix.

27. Prove that if A is a square matrix, the following are equivalent statements:
(*a*) A is nonsingular.
(*b*) If ξ is a row vector such that $A\xi^T = \theta^T$, then $\xi = \theta$.

28. Is there a matrix A without an inverse such that for all matrices B of the same dimensions as A, both $AX = B$ and $YA = B$ are solvable? Give reasons for your answer.

29. Let A be a nonzero matrix which is not necessarily square. Show that $AX = I$ is solvable if and only if $CA = (0)$ implies $C = (0)$.

2.7 SOLUTION SPACES

There is another point of view with regard to the set of solutions of a set of linear equations which not only sheds light on the character of the set of solutions and its connection with arbitrary variables but also is useful in later developments.

Suppose $\mathcal{L}$ is a set of a linear homogeneous equations in n unknowns or variables. We have noted in special cases that the set of solutions, $\mathcal{S}$, is also a vector space. Let us be more formal about it, state this result as a theorem, and prove it.

Theorem 2.12 The set $\mathcal{S}$ of solutions of a set $\mathcal{L}$ of homogeneous equations is a vector space.

Proof Since the zero vector is a solution, $\mathcal{S}$ is not empty. Suppose first that $\mathcal{L}$ consists of the single equation

$$a_1x_1 + a_2x_2 + \cdots + a_nx_n = 0$$

Since every solution is an n-tuple, the set of solutions is a subset of V^n. We need therefore, by Sec. 1.10, merely show that $\mathcal{S}$ is closed under addition and scalar multiplication. Thus if $(c_1, c_2, \ldots, c_n)$ and $(d_1, d_2, \ldots, d_n)$ are solutions, substitution shows that $(c_1 + d_1, c_2 + d_2, \ldots, c_n + d_n)$ and $k(c_1, c_2, \ldots, c_n)$ are also solutions. This completes the proof for the case of just one equation.

Then if $\mathcal{L}$ contains s equations, the set of solutions will be the intersection of the sets of solutions of the individual equations. We know from Theorem 1.8 that the intersection of a set of vector subspaces is also a vector space. This completes the proof.

This gives us the right to the following definition:

Definition The set of solutions of a set of homogeneous linear equations is called the *solution space* of the set of equations.

We have seen that a set of linear equations may be written in the form

$$\gamma_1{}^Tx_1 + \gamma_2{}^Tx_2 + \cdots + \gamma_n{}^Tx_n = (0)$$

where the $\gamma_i{}^T$ are the columns of the matrix of coefficients. (We use the superscript T to emphasize the fact that we are using column vectors.) If the column rank is n, then we know that the columns are linearly independent, and hence the only solution is the trivial one in which all the x_i are zero. If the column rank is $n - 1$, then one of the columns is a linear combination of the others, and thus we have a nontrivial solution which is unique except for a scalar multiple; in other words, the dimension of the solution space is 1. It seems reasonable to suppose that if the column rank is $n - 2$, then the dimension of the solution space is 2, and so forth. But we need to pin this down more carefully in the proof of the following theorem.

Theorem 2.13 If r is the column rank of the matrix A of a set of linear homogeneous equations in n variables, then the dimension of the solution space is $n - r$.

Proof The student may prefer to first read this proof lightly and then reread it more carefully after following through the example given after the corollary. To make the notation easier, let us assume that the first r columns of A are linearly independent. Then note that $A\epsilon_i^T$ is the ith column of matrix A, where, as elsewhere, ϵ_i denotes the ith row of the identity matrix. If we let V_1 be the column space of A, and S its solution space, we have

$$V_1 = \text{sp}(A\epsilon_1^T, A\epsilon_2^T, \ldots, A\epsilon_r^T)$$

and S is the space spanned by the column vectors of V^n which are solutions of $AX^T = \theta$. Thus V_1 is of dimension r, and our task is to show that the dimension of S is $n - r$.

Now, every column of A is in V_1, and this can be expressed as

$$A\epsilon_j^T = \alpha_j^T$$

where $\qquad \alpha_j^T$ is in V_1 for $1 \leq j \leq n$

Also, if we let

$$V_2 = \text{sp}(\epsilon_1^T, \epsilon_2^T, \ldots, \epsilon_r^T)$$

we see that any vector in V_1, that is, every column of A, can be expressed in the form $A\beta^T$, where β^T is in V_2. Hence

$$A\epsilon_j^T = A\beta_j^T \qquad \text{or} \qquad A(\epsilon_j^T - \beta_j^T) = 0$$

This shows that

$$\epsilon_j^T - \beta_j^T \text{ is in } S \text{ for } 1 \leq j \leq n$$

in particular for all j between $r + 1$ and n inclusive. Now define $\gamma_j^T = \epsilon_j^T - \beta_j^T$, for $r + 1 \leq j \leq n$. Since there are $n - r$ such vectors γ_j^T, our proof consists, first, in showing that the set

$$\gamma_{r+1}^T, \gamma_{r+2}^T, \ldots, \gamma_n^T \tag{2.10}$$

is linearly independent and, second, that it spans S.

First, to show linear independence, suppose

$$\theta = \sum_{j=r+1}^{n} c_j\gamma_j = \sum_{j=r+1}^{n} c_j(\epsilon_j - \beta_j)$$

This implies

$$\sum_{j=r+1}^{n} c_j\epsilon_j = \sum_{j=r+1}^{n} c_j\beta_j$$

But the right side is in V_2, that is, is a linear combination of $\epsilon_1, \epsilon_2, \ldots, \epsilon_r$, where the left side is a linear combination of $\epsilon_{r+1}, \epsilon_{r+2}, \ldots, \epsilon_n$. Since the complete set of ϵ_i is linearly independent, we see that all the c_i must be zero. This establishes linear independence.

Second, to show that the set (2.10) spans S, we first consider the following sequence of vectors:

$$\epsilon_1, \epsilon_2, \ldots, \epsilon_r, \gamma_{r+1}, \gamma_{r+2}, \ldots, \gamma_n \tag{2.11}$$

This is a linearly independent sequence, for suppose

$$\begin{aligned} \theta &= \sum_{i=1}^{r} s_i\epsilon_i + \sum_{i=r+1}^{n} s_i\gamma_i = \sum_{i=1}^{r} s_i\epsilon_i + \sum_{i=r+1}^{n} s_i(\epsilon_i - \beta_i) \\ &= \left[\sum_{i=1}^{r} s_i\epsilon_i - \sum_{i=r+1}^{n} s_i\beta_i\right] + \sum_{i=r+1}^{n} s_i\epsilon_i \end{aligned} \tag{2.12}$$

The expression in square brackets is in V_2, that is,

$$\theta = \delta + \sum_{i=r+1}^{n} s_i\epsilon_i \qquad \text{for some } \delta \text{ in } V_2$$

that is, δ is a linear combination of ϵ_i for $1 \leq i \leq r$. This implies that $s_i = 0$ for $r + 1 \leq i \leq n$. Hence in (2.12) all the s_i are zero. This shows the linear independence of sequence (2.11). So now let ρ be any vector in S. We know that it can be expressed in the form

$$\rho = \sum_{i=1}^{r} d_i\epsilon_i + \sum_{i=r+1}^{n} d_i\gamma_i \tag{2.13}$$

Then

$$\theta = A\rho^T = \sum_{i=1}^{r} d_iA\epsilon_i{}^T + \sum_{i=r+1}^{n} d_iA\gamma_i{}^T \tag{2.14}$$

Since the $\gamma_i{}^T$ are in S, $A\gamma_i{}^T = \theta$, and (2.14) implies

$$\theta = \sum_{i=1}^{r} d_i(A\epsilon_i{}^T)$$

But the $A\epsilon_i{}^T$ are linearly independent, being the first r columns of A. Hence $d_1, d_2, \ldots, d_r$ are all zero. Then, referring back to Equation (2.13), we see that ρ is a linear combination of the γ_i, for $r + 1 \leq i \leq n$; that is, $\gamma_{r+1}^T, \gamma_{r+2}^T, \ldots, \gamma_n{}^T$ span S, and the theorem is proved.

Since now any solution can be expressed in the form

$$x_{r+1}\gamma_{r+1}^T + x_{r+2}\gamma_{r+2}^T + \cdots + x_n\gamma_n{}^T$$

with the x's arbitrary (compare Sec. 2.4), we express this by writing "$n - r$ is the number of *arbitrary variables*" in the solution. Using this term we can state the theorem in the following form:

Corollary If r is the column rank of a set of linear equations in n variables, then the number of arbitrary variables in the set of solutions is $n - r$.

To help in understanding the above proof, let us see what it amounts to for a numerical example, using the same notation. Consider the following set of equations:

$$\begin{aligned} 2x_1 - x_2 + 3x_3 + 4x_4 &= 0 \\ x_1 + 2x_2 + 4x_3 - 3x_4 &= 0 \end{aligned}$$

Then $r = 2$ and

$$A\epsilon_3^T = \begin{bmatrix} 3 \\ 4 \end{bmatrix} = 2\begin{bmatrix} 2 \\ 1 \end{bmatrix} + \begin{bmatrix} -1 \\ 2 \end{bmatrix} = 2A\epsilon_1^T + A\epsilon_2^T$$

$$A\epsilon_4^T = \begin{bmatrix} 4 \\ -3 \end{bmatrix} = \begin{bmatrix} 2 \\ 1 \end{bmatrix} - 2\begin{bmatrix} -1 \\ 2 \end{bmatrix} = A\epsilon_1^T - 2A\epsilon_2^T$$

Thus $\beta_3^T = 2\epsilon_1^T + \epsilon_2^T$ and $\beta_4^T = \epsilon_1^T - 2\epsilon_2^T$, and we have

$$\gamma_3^T = \epsilon_3^T - (2\epsilon_1^T + \epsilon_2^T) \qquad \gamma_4^T = \epsilon_4^T - (\epsilon_1^T - 2\epsilon_2^T)$$

Hence the general solution is

$$\begin{aligned} x_3\gamma_3^T + x_4\gamma_4^T &= x_3(\epsilon_3^T - 2\epsilon_1^T - \epsilon_2^T) + x_4(\epsilon_4^T - \epsilon_1^T + 2\epsilon_2^T) \\ &= (-2x_3 - x_4, -x_3 + 2x_4, x_3, x_4)^T \end{aligned}$$

You may have begun to suspect that the row rank and column rank of any matrix are the same. We showed this for nonsingular matrices in the proof of Theorem 2.8. With the above equipment this is not hard to prove.

Theorem 2.14 The column rank and the row rank of any matrix are equal.

Proof Let A be the matrix, c its column rank, r its row rank, and B the matrix obtained from A by converting it into row-echelon form and omitting the zero rows. The set of equations can be expressed in the form $A\xi^T = \theta^T$. Since converting to row-echelon form does not change the solution set, we see, from Theorem 2.13, that $n - c$ is not only the dimension of the solution set of $A\xi^T = \theta$ but also of $B\xi^T = \theta^T$. Hence, the column rank of B as well as of A is c. Since the columns of B are in V^r, it follows that

$$c \leq r$$

that is, the column rank of a matrix is not greater than its row rank.

The following sequence shows that the inequality holds in the other direction:

$$\begin{aligned} r = \text{row rank}(A) &= \text{column rank}(A^T) \leq \text{row rank}(A^T) \\ &= \text{column rank}(A) = c \end{aligned}$$

Hence $r = c$ and our proof is complete.

In the light of this theorem we can henceforth speak of *the rank* of a matrix without qualification as to whether it is for rows or for columns. Notice that thus the row space of any matrix is isomorphic to its column space, the dimensions being the same. But the two spaces need not be the same.

EXERCISES

1. For each set of equations below, find the rank of the matrix of coefficients, a basis for the solution space, and the general solution in terms of arbitrary variables.

(*a*)
$$\begin{aligned} 2x_1 + 3x_2 &= 0 \\ 3x_1 + 2x_2 &= 0 \end{aligned}$$

(*b*)
$$\begin{aligned} x_1 + 2x_2 - x_3 + x_4 &= 0 \\ x_1 - x_2 + x_3 - 2x_4 &= 0 \\ 2x_1 + x_2 - x_4 &= 0 \end{aligned}$$

(*c*)
$$\begin{aligned} x_1 + 5x_2 - x_3 &= 0 \\ -x_1 + 3x_2 + 2x_3 &= 0 \\ 8x_2 + x_3 &= 0 \\ 2x_1 + 2x_2 - 3x_3 &= 0 \end{aligned}$$

(*d*)
$$\begin{aligned} x_1 - x_2 + 4x_3 + 2x_4 + x_5 &= 0 \\ 2x_1 - 2x_2 + 5x_3 + 4x_4 &= 0 \\ 3x_1 - 3x_2 + 6x_3 + 6x_4 + 3x_5 &= 0 \end{aligned}$$

2. Prove that if a set of linear homogeneous equations has fewer equations than variables, there is always a nontrivial solution. In fact show that if s is the number of equations and n the number of variables, then the number of arbitrary variables is never less than $n - s$.

3. Suppose the solution set of s linear homogeneous equations in n variables is a vector space of positive dimension t. What does this imply about s and n?

4. The solution set of a set $\mathcal{S}$ of linear homogeneous equations has the following basis:

$$(1.2,1,0) \qquad (2,3,0,1)$$

What is the minimum number of equations which $\mathcal{S}$ can have? Find a set $\mathcal{S}$ in which the number of equations is a minimum. Is there more than one such set? If so, find them all.

5. Let A be the matrix of a set $\mathcal{S}$ of linear homogeneous equations and B a matrix whose columns are vectors of the solution set of $\mathcal{S}$. Show that $AB = (0)$.

6. Prove the converse of Theorem 2.13; that is, if A is a matrix of a set of linear homogeneous equations in n variables having a solution space of dimension $n - r$, then r is the column rank of A.

7. Let A be the matrix of a set $\mathcal{S}$ of n linear homogeneous equations in n variables. Denote by $\mathfrak{J}$ the set of linear homogeneous equations having A^T as its matrix. Do the solution spaces of $\mathcal{S}$ and $\mathfrak{J}$ have the same dimension? If so, are they the same? Justify your answers.

8. Answer the same question in Exercise 7 if A has r rows and n columns, with $n > r$.

2.8 SETS OF NONHOMOGENEOUS LINEAR EQUATIONS

We have spent much time and effort on the solution sets of homogeneous linear equations. Now we show that we have done most of the labor needed for the solution of sets of nonhomogeneous linear equations. Let us write the set in matrix form as follows:

$$A\xi^T = \gamma^T \qquad \text{for } \gamma \neq \theta \tag{2.15}$$

where γ^T is a column of constants; that is, $\gamma = (c_1, c_2, \ldots, c_n)$. As we have seen (Sec. 1.7), the set (2.15) is solvable if and only if the vector γ^T is in the space spanned by the columns of A. This means that there will be a solution if and only if the space spanned by the columns of A is the same as that spanned by the columns of the so-called *augmented matrix*

$$(A,\gamma^T) \tag{2.16}$$

with the column of the components of γ added. This is worth stating formally.

Theorem 2.15 The set of linear equations (2.15) is solvable if and only if the rank of the augmented matrix (2.16) is the same as that of A.

Suppose the set (2.15) has a solution; what can we say about the set of solutions? To answer this suppose that $\xi_0{}^T$ is a solution of (2.15) and let $\xi_1{}^T$ be any other solution. Then $A\xi_0{}^T = \gamma^T$, and $A\xi_1{}^T = \gamma^T$ implies $A(\xi_1{}^T - \xi_0{}^T) = \theta$; that is, $\xi_1{}^T - \xi_0{}^T$ is a solution of $A\xi^T = \theta$.

Conversely, let ξ_0 be a solution of (2.15) and σ^T a solution of $A\xi^T = \theta^T$. Then $A(\xi_0{}^T + \sigma^T) = A\xi_0{}^T = \gamma^T$. We have

Theorem 2.16 If $\xi_0{}^T$ is a solution of $A\xi^T = \gamma^T$, then the set of solutions of this matric equation is given by

$$\xi_0{}^T + \sigma^T$$

where the σ^T range over the solutions of $A\xi^T = \theta^T$.

Using this theorem and the Corollary of Theorem 2.13, we get the following:

Corollary If (2.15) has a solution, the number of arbitrary variables in the set of solutions is $n - r$, where r is the rank of A and n is the number of columns.

However, we should notice that the set of solutions of (2.15) does not form a vector space, for let $\xi_0{}^T + \sigma_1{}^T$ and $\xi_0{}^T + \sigma_2{}^T$ be two solutions of (2.15), where $\sigma_1{}^T$ and $\sigma_2{}^T$ are in the solution set S of the corresponding homogeneous equations. Then, if the solutions were a vector space, we would have

$$\xi_0{}^T + \sigma_1{}^T + \xi_0{}^T + \sigma_2{}^T = \xi_0{}^T + \sigma_3{}^T$$

for some vector $\sigma_3{}^T$ in S. But this would imply

$$\xi_0{}^T = \sigma_3{}^T - \sigma_2{}^T - \sigma_1{}^T$$

The right side of the equation is a vector in S, which is a contradiction since $\xi_0{}^T$ is not. In fact, $\xi_0{}^T$ is not a solution of $A\xi^T = \theta^T$, unless $\gamma^T = \theta^T$.

We say that the set of solutions of (2.15) is a *coset* of S. In general we have the

Definition If S is a subspace of V^n and α is a vector in V^n, then the set of vectors $\sigma + \alpha$, where σ is in S, is called a *coset* of S or a *linear manifold* and is written

$$S + \alpha$$

To see what this means geometrically, let us consider an example in three dimensions, leaving the two-dimensional case as an exercise. Each of the following equations

$$\begin{aligned} x + 2y + z &= 3 \\ 2x - y + z &= 4 \end{aligned} \tag{2.17}$$

represents a plane, and thus, since the planes are not parallel, the set of solutions of both will lie on a line. One solution is (1,0,2). To get all the solutions we look at the corresponding pair of homogeneous equations, and, solving

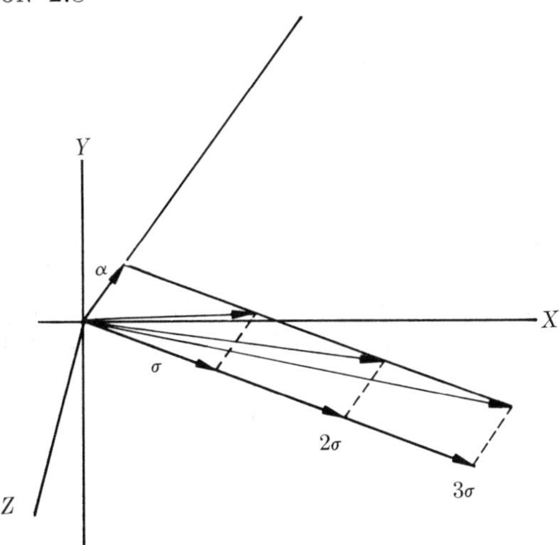

Figure 2.1

for x and z in terms of y, we have $x = 3y$ and $z = -5y$. Thus the solution space S is the set of vectors

$$(3a,a,-5a)$$

Note that it is one dimensional as we would expect. Thus the general solution of (2.17) is

$$(1,0,2) + (3a,a,-5a) = (1 + 3a,\ a,\ 2 - 5a) \tag{2.18}$$

In fact, the points on the line (2.17) have coordinates as indicated in (2.18). Actually, the form (2.18) is in many cases much more convenient to use than that of (2.17).

Reverting to our vector notation, let α denote the vector (1,0,2) and $\sigma = (3,1,-5)$. Then Fig. 2.1 shows the resultants of α and σ, α and 2σ, α and 3σ, and so forth. The arrows for the resultants all lie on the line represented by the set (2.18). The line (2.18) is not changed if α is replaced by any other point on the line. Thus, since $(4,1,-3)$ is on the line (2.18), then the points of that line can equally be represented by

$$(4,1,-3) + (3b,b,-5b) = (4 + 3b,\ 1 + b,\ -3 - 5b)$$

In fact, here $b = a - 1$. If, on the other hand, σ is kept the same and α is replaced by any point not on the line, we obtain a parallel line instead. Thus if, instead of α, we use the vector (1,2,3), the set

$$(1,2,3) + (3a,a,-5a) = (1 + 3a,\ 2 + a,\ 3 - 5a) \tag{2.19}$$

will be the points of a line parallel to line (2.18).

In general, let S be a proper subspace of V^n and α a vector of V^n. Then the set $S + \alpha$ is a coset or linear manifold. If β is a vector in V^n, then there are two possibilities:

1. The vector β is in $S + \alpha$; that is, $\alpha - \beta$ is in S. In this case $S + \alpha = S + \beta$; that is, the two cosets are the same.

2. The vector β is not in $S + \alpha$; that is, $\alpha - \beta$ is not in S. Then the cosets $S + \alpha$ and $S + \beta$ have no vectors in common since $\sigma_1 + \alpha = \sigma_2 + \beta$ for σ_i in S would imply $\alpha - \beta$ in S. So we have a definition and theorem.

Definition Let S be a subspace of V^n and let α and β be vectors of V^n. The cosets $S + \alpha$ and $S + \beta$ are called *parallel cosets*. We include equality in paralellism.

Theorem 2.17 Let S be a subspace of V^n and let α and β be two vectors of V^n. If $\alpha - \beta$ is in S, the parallel cosets $S + \alpha$ and $S + \beta$ are the same; if $\alpha - \beta$ is not in S, the cosets have no vector in common.

Recall that in the beginning of this book when we made the transition from a line segment with an arrow on one end to an n-tuple, we restricted ourselves to physical vectors emanating from the origin. The idea of a coset can free us from such a restriction. To see why this is so, let us confine ourselves to two dimensions and consider two points A and B whose pairs of coordinates are (a_1,a_2) and (b_1,b_2), respectively. Let S be the one-dimensional vector space $c(b_1,b_2)$ as c ranges over the set of real numbers. Then, as we saw in the example above, the coset

$$c(b_1,b_2) + (a_1,a_2)$$

is the set of all points on the line through A parallel to OB. These points can thus correspond to the set of physical vectors (that is, line segments with an arrow on one end) emanating from A in a direction parallel to OB. That is, in a sense, we have moved the "tails" of the vectors from the origin to the point A simply by forming the coset.

It should be pointed out that the results of this section are chiefly useful in determining the character of the sets of solutions of nonhomogeneous equations and the connections with homogeneous equations and vector spaces. When it comes to solving sets of linear equations, it is still best to revert to the elimination method, that is, to the reduction of the matrix of the set to echelon form.

We should note that in differential equations there is a situation very similar to that which we have found in this section on nonhomogeneous linear equations. Consider the following differential equation:

$$\frac{d^2y}{dx^2} - 3\frac{dy}{dx} + 2y = 20x^2 - 120x + 130 \tag{2.20}$$

First consider the solutions of the corresponding homogeneous equation, that is, in which the expression on the right is replaced by zero. It is known that since $z^2 - 3z + 2 = (z - 2)(z - 1)$, two solutions of the homogeneous differential equation are

$$e^{2x} \quad \text{and} \quad e^x$$

These are linearly independent over the set of real numbers. Thus the general solution of the homogeneous differential equation is

$$c_1e^{2x} + c_2e^x \tag{2.21}$$

In fact, the two exponential functions span a vector space. Then to get all the solutions of the nonhomogeneous equation, we add to (2.21) *any* solution of (2.20). One such solution is

$$f(x) = 10x^2 - 30x + 10 \tag{2.22}$$

Thus the general solution of (2.20) is

$$c_1e^{2x} + c_2e^x + 10x^2 - 30x + 10$$

There are two arbitrary variables, c_1 and c_2. Furthermore, $f(x)$ in (2.22) could be replaced by any other solution of the nonhomogeneous equation (2.20). So the general solution of (2.20) is a coset of the set (2.21). If in (2.20) the polynomial on the right of the equality sign were replaced by another polynomial, the solutions of the new differential equation would be a coset parallel to that of (2.20).

Finally, let us consider a connection between the above and an important idea in Euclidean three-dimensional geometry. A set of points, $\mathcal{H}$, in this geometry is called *convex* if for any two points in $\mathcal{H}$ it is true that the line segment determined by these two points is also in $\mathcal{H}$. Thus, if $\mathcal{H}$ is a plane, its being convex is equivalent to the property that if A and B are any two points of $\mathcal{H}$, the line segment $\overline{AB}$ is in $\mathcal{H}$. More generally, suppose $\mathcal{H}$ is the set of solutions of a set of linear nonhomogeneous equations in three variables and $\mathcal{S}$ is the set of solutions of the corresponding set of homogeneous equations. Then, if λ is in $\mathcal{H}$ and $\sigma \neq \theta$ is in $\mathcal{S}$, we see that λ and $\sigma + \lambda$ are in $\mathcal{H}$. Now $\{x\sigma + \lambda\}$, for x real, is the set of points on the line (one-dimensional linear manifold) determined by λ and $\sigma + \lambda$. All these are in $\mathcal{H}$. Thus not only are all the points of the line segment determined by the given points in $\mathcal{S}$, but all points of the complete line as well. We thus have a stronger property than convexity. This can be extended to any number of dimensions.

EXERCISES

1. Find all the solutions of each of the following sets of equations:

(*a*)
$$\begin{aligned} x + 2y + 3z &= 0 \\ 2x - y + z &= 5 \\ 3x - 2y + z &= 11 \end{aligned}$$

(*b*)
$$\begin{aligned} x + 2y + 3z &= -1 \\ 2x + 4y - z &= 5 \\ -x - 2y + 4z &= -6 \end{aligned}$$

(*c*)
$$\begin{aligned} x + 3y + t &= 5 \\ y + 2z - t &= 4 \\ 2x + 3y - z &= -7 \end{aligned}$$

(*d*)
$$\begin{aligned} x + 2y &= 7 \\ 3x + 5y &= 2 \\ 2x + 3y &= -5 \\ 4x + 7y &= 9 \end{aligned}$$

2. For each set of equations in Exercise 1, determine the set of solutions of the corresponding set of homogeneous equations, and show the relationship between this set and the solutions of the nonhomogeneous set.

3. For what values of c_1, c_2, and c_3 is the following set of equations solvable?

$$\begin{aligned} x + 2y + 3z &= c_1 \\ 2x - y + z &= c_2 \\ 3x - 2y + z &= c_3 \end{aligned}$$

4. Consider the equation $x + 2y + 3 = 0$, which represents a line in two dimensions. Express the coordinates of the points on the line as a coset. What is the geometrical meaning of this coset?

5. Solve the following set of equations for x and y in terms of x' and y':

$$\begin{aligned} \tfrac{4}{5}x - \tfrac{3}{5}y + 1 &= x' \\ -\tfrac{3}{5}x - \tfrac{4}{5}y + 2 &= y' \end{aligned}$$

What, in terms of x' and y', is the coset of solutions?

6. Let S be a vector subspace of V^n. Does the set of all cosets of S form a vector space? Justify your answer. If the answer is "yes," find the dimension of the space of cosets.

7. Express the condition that three points α, β, γ be collinear.

8. Express in terms of cosets or linear manifolds the following statement in Euclidean geometry:

If α, β, and γ are three noncollinear points, there is exactly one plane containing them.

9. Consider the following two sets of linear equations:

$$A\xi^T = \gamma^T \qquad \text{and} \qquad A\xi^T = \delta^T$$

where $\gamma \neq \delta$. Let S be the set of solutions of the set of homogeneous equations $A\xi^T = \theta^T$. Under what conditions will the sets of solutions of the given equations be the same cosets?

10. Given a square matrix A and let C be the set of vectors γ^T such that $A\xi^T = \gamma^T$ is solvable. Is C a vector space? What relation does it have to A?

11. Given a square matrix A of order n and let N be the set of vectors γ^T such that $A\xi^T = \gamma^T$ is *not* solvable. Does N constitute a vector space? Is N a coset? Explain.

12. Let A be a square matrix and suppose that for every column matrix $C \neq (0)$ having the same number of components as the order of A, it is true that $AX = C$ has no more than one solution. Then prove that for each such C, $AX = C$ does have a solution and it is unique.

13. Let S be a subspace of V^n and let $S + \alpha$ be a coset. Suppose A is a square matrix whose order is n. Consider the set T of all column vectors $A\xi + \alpha$, where ξ is a vector of S represented as a column vector. Prove that T is a coset. Define T in terms of A and S.

2.9 COMPUTATION OF INVERSE MATRICES

In this section we shall show how the elementary operations on matrices described in Sec. 2.2 can be accomplished by means of matrix multiplication and use this information to find a practical method for computing the inverse of a matrix when it has one.

First recall that the three so-called elementary operations on the rows of a matrix (or a set of equations) are

1. Adding a multiple of one row to another.
2. Multiplying a row by a nonzero number.
3. Interchanging two rows.

For each of these we define a matrix:

1. $I_{ij}(k)$ is the matrix obtained from the identity matrix by adding k times the ith row to the jth row.

2. $I_i(c)$ is the matrix obtained from the identity matrix by multiplying the ith row by c, for $c \neq 0$.

3. I_{ij} is the matrix obtained from the identity matrix by interchanging the ith and jth rows.

These are called *elementary matrices.* Using, for instance, the general matrix of order 3, it is easy to see that

1. $I_{ij}(k)A$ is the matrix obtained from A by adding k times the ith row to the jth row.
2. $I_i(c)A$ is the matrix obtained from A by multiplying the ith row by c.
3. $I_{ij}A$ is the matrix obtained from A by interchanging the ith and jth rows.

So we have what could be called the *golden rule of matrices:*

Do unto the identity matrix what you would do unto the matrix.

Though we need not operate on the columns of a matrix to get the inverse, it is interesting to see how we would do so. It can be seen that if we perform one of the elementary operations on the columns of the identity matrix and multiply A on the right by such a matrix, we perform the same operation on the columns of A. The corresponding matrices are, respectively,

$$I_{ij}^T(k) \qquad I_i^T(c) \qquad I_{ij}^T$$

Now let us consider a method of computing the inverse of a matrix. Let A be a square matrix of order n and write the following matrix with n rows and $2n$ columns, where I is the identity matrix of order n,

$$(A,I) \tag{2.23}$$

What we do is to operate on the rows of (2.23) in the same manner as was done to reduce A to row-echelon form, but carry the process further until, if possible, A is taken into the identity matrix. Then we shall find that I becomes the inverse of A. We shall illustrate this process below and then show that it can be carried through in general. At this point, suppose it can be done. Then, if E_i are the elementary matrices corresponding to the elementary row operations, we will have

$$E_kE_{k-1} \cdots E_1A = I$$

If we write this more briefly as $EA = I$, where E is the product of the elementary matrices, we see that, on the one hand, E is the inverse of A and, on the other,

$$E(A,I) = (EA,EI) = (I,E)$$

which shows that the last n columns display the inverse of A. Setting it up in this way avoids the necessity of keeping track of what the E's are.

Now we illustrate this process. If

$$A = \begin{bmatrix} 1 & 0 & 3 \\ 2 & 1 & 0 \\ 7 & 3 & 5 \end{bmatrix}$$

then

$$(A,I) = \begin{bmatrix} 1 & 0 & 3 & 1 & 0 & 0 \\ 2 & 1 & 0 & 0 & 1 & 0 \\ 7 & 3 & 5 & 0 & 0 & 1 \end{bmatrix}$$

Then we have the following sequence of matrices

$$\begin{bmatrix} 1 & 0 & 3 & 1 & 0 & 0 \\ 0 & 1 & -6 & -2 & 1 & 0 \\ 0 & 3 & -16 & -7 & 0 & 1 \end{bmatrix} \qquad \begin{bmatrix} 1 & 0 & 3 & 1 & 0 & 0 \\ 0 & 1 & -6 & -2 & 1 & 0 \\ 0 & 0 & 2 & -1 & -3 & 1 \end{bmatrix}$$

The matrix is now in echelon form. To transform the first three columns into the identity matrix requires the following steps:

$$\begin{bmatrix} 1 & 0 & 3 & 1 & 0 & 0 \\ 0 & 1 & -6 & -2 & 1 & 0 \\ 0 & 0 & 1 & -\frac{1}{2} & -\frac{3}{2} & \frac{1}{2} \end{bmatrix} \qquad \begin{bmatrix} 1 & 0 & 0 & \frac{5}{2} & \frac{9}{2} & -\frac{3}{2} \\ 0 & 1 & 0 & -5 & -8 & 3 \\ 0 & 0 & 1 & -\frac{1}{2} & -\frac{3}{2} & \frac{1}{2} \end{bmatrix}$$

Thus the inverse of A is the matrix composed of the last three columns in the last matrix of the sequence. This may easily be checked. Of course there are various paths from A to the identity matrix, but they all lead to the same inverse.

It remains to show why, by a sequence of elementary row operations, one may take a nonsingular matrix A into the identity transformation, that is, to prove the following theorem:

Theorem 2.18 If A is a nonsingular matrix, then there is a sequence of elementary row operations taking A into the identity matrix; that is, there is a sequence of elementary matrices E_i such that

$$E_k E_{k-1} \cdots E_1 A = I$$

Proof First we take A by a sequence of elementary transformations into row-echelon form as in Sec. 2.4. Now in the echelon form there can be no row of zeros since the rank of A is equal to the number of nonzero rows. Thus the echelon form will be

$$\begin{bmatrix} a_{11} & a_{12} & \cdots & a_{1n} \\ 0 & a_{22} & \cdots & a_{2n} \\ \cdots & \cdots & \cdots & \cdots \\ 0 & 0 & \cdots & a_{nn} \end{bmatrix}$$

Now no a_{ii} can be zero because if it were, by the definition of the echelon form, all the diagonal elements beyond it would be zero and hence a_{nn} would be zero; then we would have a row of zeros, which is a contradiction. To complete the process, first multiply the first row by a_{11}^{-1}, the second by a_{22}^{-1}, . . . , the last by a_{nn}^{-1}. Then by adding appropriate multiples of the last row to all the others, we can make the last column zero except for its element in the last row. By adding appropriate multiples of the $(n - 1)$st row to the others, we can do the same to the $(n - 1)$st column. Similarly we can make each column zero except for the 1 on the principal diagonal. We thus have finally the identity matrix.

Corollary Every nonsingular matrix can be expressed as a product of elementary matrices.

The proof is left as an exercise.

EXERCISES

1. Find the inverse of each of the following elementary matrices:

$$I_{ij} \qquad I_i(c) \qquad I_{ij}(k)$$

2. Prove the Corollary to Theorem 2.18.

3. Find the inverses of the following matrices:

$$(a)\ \begin{bmatrix} 2 & 0 & 1 \\ 1 & 2 & -3 \\ 4 & 5 & 7 \end{bmatrix}$$

$$(b)\ \begin{bmatrix} 3 & 1 & 0 \\ 2 & -1 & 1 \\ 6 & -1 & 2 \end{bmatrix}$$

$$(c)\ \begin{bmatrix} -1 & 2 & 3 \\ 1 & 4 & -1 \\ 2 & -2 & -5 \end{bmatrix}$$

4. Show that $I_{ij}^T = I_{ij}$. What matrix of type $I_{ij}(k)$ is equal to $I_{ij}^T(k)$?

5. If X_i denotes the ith row of X, prove $(AB)_i = A_iB$.

6. Show the results 1 to 3 just before the statement of the golden rule for $i = 1, j = 2$, and A a three-rowed square matrix, first for rows and second for columns.

7. Use the inverses of the matrices found in Exercise 3 to solve the following sets of equations:

(*a*)
$$\begin{aligned} 2x \qquad + z &= 5 \\ x + 2y - 3z &= 1 \\ 4x + 5y - 7z &= 2 \end{aligned}$$

(*b*)
$$\begin{aligned} 3x + y \qquad &= 2 \\ 2x - y + z &= 1 \\ 6x - y + 2z &= 0 \end{aligned}$$

(*c*)
$$\begin{aligned} -x + 2y + 3z &= 0 \\ x + 4y - z &= 1 \\ 2x - 2y - 5z &= 2 \end{aligned}$$

8. Find the inverse of the matrix

$$A = \begin{bmatrix} a & b \\ c & d \end{bmatrix} \qquad \text{where } ad - bc \neq 0$$

9. Find the conditions that the matrix A in Exercise 8 be its own inverse.

10. Prove that every square matrix of order 2 with integer elements can be represented as a product of powers of the matrices

$$\begin{bmatrix} 1 & 1 \\ 0 & 1 \end{bmatrix} \quad \begin{bmatrix} 1 & 0 \\ 1 & 1 \end{bmatrix} \quad \begin{bmatrix} a & 0 \\ 0 & 1 \end{bmatrix}$$

where a is an integer.

11. Starting with the matrix (A,I), can one also find the inverse of A by elementary column operations on (A,I)? If so, how? Could one use a combination of row and column operations?

12. Let S be the set of all square matrices of order 2 thought of as a vector space. Find a set of four idempotent matrices A_1, A_2, A_3, A_4 which forms a basis of S, that is, such that every matrix of S is a linear combination of the A's.

13. Show that a subspace of S in Exercise 12 is the set of matrices of the form

$$\begin{bmatrix} a & b \\ c & -a \end{bmatrix}$$

Find a basis of this subspace.

14. If V is the set of all square matrices of order 2 with complex elements, prove that there is a set of real matrices $\mathcal{S}$ such that every matrix of V is expressible uniquely as a linear combination with complex coefficients of the matrices of $\mathcal{S}$.

2.10 PARTITIONED MATRICES AND DIRECT SUMS

At times it is useful to partition a matrix by drawing or thinking of drawing one or more horizontal lines between rows and one or more vertical lines between columns as illustrated below:

$$A = \left[\begin{array}{c|c|c} A_{11} & A_{12} & A_{13} \\ \hline A_{21} & A_{22} & A_{23} \end{array}\right]$$

where the A_{ij} are matrices. Actually we do not usually draw the lines—thinking of them as having been drawn is sufficient—for the "imaginary lines" are merely a device for describing the requirement that the submatrices A_{ij} have the property that all in the same row have the same number of rows and all in the same column have the same number of columns. More formally, we say that a matrix A is *partitioned into submatrices* A_{ij} when it is written as a rectangular (or square) array of matrices with the understanding that any two matrices in the same row have the same number of rows and similarly for columns.

This idea of partitioning is intuitively very simple but notationally can become quite cumbersome. Hence we confine ourselves to the simplest case in what follows. (We could "leave the cumbersome details to the reader" as is often the custom, but if the reader is satisfied with the following we will not press the matter. If he is not satisfied, he can use a mathematical induction argument.) So let us consider two partitioned matrices A and B and explore what the partitioning must be, so that they can be multiplied in the customary way:

$$A = \begin{bmatrix} A_{11} & A_{12} \\ A_{21} & A_{22} \end{bmatrix} \qquad B = \begin{bmatrix} B_{11} & B_{12} \\ B_{21} & B_{22} \end{bmatrix} \tag{2.24}$$

For the product AB to have meaning, the number of columns of A must be the same as the number of rows of B. Formally we want to be able to express the product AB as follows:

$$AB = \begin{bmatrix} A_{11}B_{11} + A_{12}B_{21} & A_{11}B_{12} + A_{12}B_{22} \\ A_{21}B_{11} + A_{22}B_{21} & A_{21}B_{12} + A_{22}B_{22} \end{bmatrix} \tag{2.25}$$

For this to have meaning for the upper left-hand sum, the number of columns in A_{11} and A_{12} must be the same as the number of rows in B_{11} and B_{21}, respectively. If this is true then, from the conditions imposed for a partition, the other three sums on the right will also have meaning. We need a term for this as in the following definition:

Definition If the partitioned matrices (2.24) have the property that B_{11} has the same number of rows as A_{11} has columns and if B_{21} has the same number of rows as A_{12} has columns, then we say that the matrices are *partitioned conformably* for the product AB.

Notice that the partitioning can be conformable for the product AB without being conformable for the product BA even if both products have meaning. You are asked to show in an exercise that this is the case. If, however, the partitioning is conformable for both the product AB and for BA, we say that A and B are *partitioned conformably*.

We still have not shown that the equality indicated in (2.25) is true. For this we need some further notation. So let us choose letters for the number of rows and columns in the various submatrices as in the following table:

$$A: \begin{array}{c|c} r,\ c & r,\ c' \\ \hline r',\ c & r',\ c' \end{array} \qquad B: \begin{array}{c|c} c,\ d & c,\ d' \\ \hline c',\ d & c',\ d' \end{array} \tag{2.26}$$

meaning that A_{11} has r rows and c columns, A_{12} has r rows and c' columns, and so forth. The notation is chosen so that the matrices are partitioned conformably for the product AB. Now let us write the term in the ith row and jth column of the product AB

$$a_{i1}b_{1j} + \cdots + a_{ic}b_{cj} + a_{i,c+1}b_{c+1,j} + \cdots + a_{i,c+c'}b_{c+c',j} \tag{2.27}$$

Now for $1 \leq i \leq r$ and $1 \leq j \leq c$, the sum of the first c terms in (2.27) is the element in the ith row and jth column of the product $A_{11}B_{11}$, and the sum of the last c' terms is the element in the ith row and jth column of the product $A_{12}B_{21}$. For $1 \leq i \leq r$ and $c + 1 \leq j \leq c + c'$, the sum of the first c terms in (2.27) is the element in the ith row and jth column of the product $A_{11}B_{12}$, and the last c' terms give the elements of the product $A_{12}B_{22}$. For the second row on the right of (2.25), we let i range between $r + 1$ and $r + r'$ and repeat the argument above. Thus we have shown

Theorem 2.19 If the matrices are partitioned conformably for the product AB, then the product (2.25) holds.

Notice that if $d = r$ and $d' = r'$ in table (2.26) then A and B are partitioned conformably. Theorem 2.19 shows that if A and B are partitioned conformably, then the products by submatrices give the same results as the products of the matrices themselves in both directions.

It is quite apparent that there is no advantage in partitioning a matrix for multiplication unless it has some special form. So now we consider types of matrices for which partitioning is helpful. Suppose that in Equations (2.24) the matrices A_{12} and B_{12} are zero matrices. Then in the product (2.25), the matrix $A_{11}B_{12} + A_{12}B_{22}$ is also a zero matrix. So we have shown that if the matrices are partitioned conformably for the product AB and if both

A and B have the property that the submatrices in the upper right-hand corner are zero matrices, then the product has the same property.

If, further, in (2.24) both A_{12} and A_{21} are zero matrices, we say that A is the *direct sum* of A_{11} and A_{22} and write

$$\begin{bmatrix} A_{11} & (0) \\ (0) & A_{22} \end{bmatrix} = A_{11} \oplus A_{22}$$

It follows that if A and B are both direct sums partitioned conformably for AB, then

$$(A_1 \oplus A_2)(B_1 \oplus B_2) = A_1B_1 \oplus A_2B_2$$

More generally, if

$$A = A_1 \oplus A_2 \oplus \cdots \oplus A_k \quad \text{and} \quad B = B_1 \oplus B_2 \oplus \cdots \oplus B_k \tag{2.28}$$

where the partitioning is conformable for AB, then

$$AB = A_1B_1 \oplus A_2B_2 \oplus \cdots \oplus A_kB_k \tag{2.29}$$

In fact, if A and each A_i in (2.28) is square, then

$$A^r = A_1{}^r \oplus A_2{}^r \oplus \cdots \oplus A_k{}^r$$

for every positive integer r.

It should also be noticed that if A is a direct sum as in (2.28) with each A_i square, it can also be represented as a product of direct sums as follows: First write the identity matrix I as a direct sum of identity matrices I_i, where each I_i has the same order as A_i. Then denote by E_i the matrix obtained from this partition of I by replacing I_i by A_i. Then A can be written as

$$A = E_1E_2E_3 \cdots E_k$$

So far in this section we have dealt only with products of partitioned matrices, because most of the applications are to products. But one can add partitioned matrices in much the same way. Here, however, if the sum of matrices A and B in (2.24) is to have meaning for each i and j, it must be true that A_{ij} and B_{ij} have the same dimensions. Then for A and B as in (2.24),

$$A + B = \begin{bmatrix} A_{11} + B_{11} & A_{12} + B_{12} \\ A_{21} + B_{21} & A_{22} + B_{22} \end{bmatrix} \tag{2.30}$$

In particular,

$$(A_1 \oplus A_2) + (B_1 \oplus B_2) = (A_1 + B_1) \oplus (A_2 + B_2)$$

We leave as an exercise the proof of the following

Theorem 2.20 If two matrices A and B are partitioned as in (2.24), then the product (2.25) and the sum (2.30) have meaning and are true if and only if, for $i = 1$ and 2, A_{ii} and B_{ii} are square and of the same order. (A_{11} and A_{22} need not have the same order.)

EXERCISES

1. Let the matrices A and B be given as follows:

$$A = \begin{bmatrix} A_{11} & A_{12} \\ A_{21} & A_{22} \end{bmatrix}$$

where $A_{11} = \begin{bmatrix} 1 & 2 \\ 3 & 4 \end{bmatrix}$, $A_{12} = \begin{bmatrix} 3 \\ 4 \end{bmatrix}$, $A_{21} = [1 \quad -2]$, $A_{22} = (3)$

and
$$B = \begin{bmatrix} 1 & 3 & 5 & 7 \\ 0 & 2 & -3 & 0 \\ 6 & 1 & 0 & 2 \end{bmatrix}$$

Partition B so that A and B are partitioned conformably for AB. Find the product using partitions and check with the product AB.

2. In Exercise 1 can B be partitioned so that A and B are partitioned conformably for BA?

3. Let A be any matrix with two rows and three columns and B a matrix with three rows and two columns. Given any partition of A, can B be partitioned so that the partitioning is conformable for AB?

4. If for two matrices A and B the product AB has meaning, is there for every partitioning of A a partitioning of B so that it is conformable for AB? Reasons should be given for your answer, but a formal proof is not necessary.

5. Prove Theorem 2.20.

6. Give an example of partitions of two matrices A and B such that while both AB and BA have meaning and the partitioning is conformable for AB it is not for BA.

7. If in Exercise 6 A is a square matrix can such an example be given? Why or why not?

8. Let A be a matrix which is expressible as a direct sum of square matrices as in (2.28), and let $f(x)$ be any polynomial in x with coefficients in a field. Prove that

$$f(A) = f(A_1) \oplus f(A_2) \oplus \cdots \oplus f(A_k)$$

9. Suppose A and B are partitioned as in (2.24) and that the partitioning is conformable for AB. Further, suppose

$$A_{11} = aI_1 \quad \begin{matrix} A_{12} = (0) \\ \text{and} \end{matrix} \quad \begin{matrix} A_{21} = (0) \\ A_{22} = bI_2 \end{matrix} \quad \text{for } a \neq b$$

where I_1 and I_2 are identity matrices. Prove that $AB = BA$ if and only if both B_{12} and B_{21} are zero matrices.

3

TRANSFORMATIONS

3.1 FUNCTIONS, MAPPINGS, AND TRANSFORMATIONS

In mathematics, as elsewhere, we use different terms for the same thing depending on the context. A simple example of this outside of mathematics is the set of words: two, pair, brace, couple; all mean *two* in different connections. In such a sense, all the terms in the heading of this section mean the same thing. When we speak of a *function* we are apt to be in the field of analysis; the term *mapping* is definitely geometrical, while *transformation* is a little of both. Fundamentally these words all have the same meaning but since the last is most common in linear algebra, we shall use it most of the time. The reader who is more familiar with the term *function* might like to make the translation in the beginning.

Consider the transformation

$$x \overset{T}{\mapsto} x^2 \tag{3.1}$$

We call x^2 the *image of x under the transformation T* and say that the transformation *takes x into x^2*. (We customarily put a little tail on the arrow when we refer to the image of an element; we omit the tail when we refer to sets, as below.) Actually this is not complete, for we must specify the set of numbers on which T acts. If x is restricted to be a real number, the set of images are all real; then T takes the set of real numbers into the set of real numbers or, if we wish, takes the set of real numbers into the set of complex numbers. In the respective cases we write

$$\mathcal{R} \xrightarrow{T} \mathcal{R} \qquad \mathcal{R} \xrightarrow{T} \mathcal{C}$$

where $\mathcal{R}$ stands for the set of real numbers and $\mathcal{C}$ the set of complex numbers.

The principal requirement for a transformation, as for a function, is that the image of any element be unique. In general, if $\mathcal{S}$ and $\mathcal{S}'$ are two sets of numbers (or later vectors), we have the

Definition If $\mathcal{S}$ and $\mathcal{S}'$ are two sets of elements (numbers, vectors, what you will), we say that T is a *transformation* from $\mathcal{S}$ into $\mathcal{S}'$ if each element of $\mathcal{S}$, under T, has a unique image in $\mathcal{S}'$. The set $\mathcal{S}$ is called the *domain* of the transformation and $\mathcal{S}'$ its *codomain*. The set of images of the elements of $\mathcal{S}$ under the transformation T is called its *range*.

Various notations are used to represent the image of an element s of $\mathcal{S}$. Among them are

$$T(s) \qquad s^T \qquad sT \qquad Ts$$

We shall use the first and last, for the most part. Similarly we use $T(\mathcal{S})$ to designate the set of images of the elements of $\mathcal{S}$, that is, the range of T.

It should be emphasized that to define a transformation we must, first of all, specify its domain. That is, (3.1), where x is restricted to be a nonnegative real number, is quite different from the same correspondence when x is allowed to be any real number. Second, once the domain is specified, we must define what is to be the image of every element of the domain. Finally, we decide whether the codomain is to be the range, that is, the set of images, or something more. Then after these three steps we have defined the transformation. Notice that the range is always a subset of the codomain.

Then there are various kinds of transformations. A transformation can be *one to one*, that is, such that no two elements of the domain have the same image. Transformation (3.1) is one to one if the domain is the set of nonnegative real numbers but is not one to one if the domain is the set of reals; for in the latter case, x and $-x$ have the same image. If the transformation is one to one, every element of the range is the image of a unique element of the domain; that is, each element of the range determines a unique element of the domain. Thus we not only have a transformation T from the domain into the range but a transformation L from the range into the domain. We call L the *left* inverse since $L[T(a)] = a$ for each a in the domain. The domain of L is the range of T and the range of L is in the domain of T.

A transformation can be *onto*, namely, when its range and codomain are the same. This means that every element of the codomain is the image of an element of the domain. When the codomain is not necessarily the same as the range, we use the preposition *into*. That is, transformation (3.1) for a domain the set of real numbers or the set of nonnegative real numbers is *into* the set of real numbers as well as into the set of nonnegative real numbers, but it is *onto* only the set of nonnegative real numbers. A transformation which is onto has a *right inverse*, which may be seen as follows. Suppose b is in the range. Let $R(b)$ be any element of the domain of T whose image is b. Then $T[R(b)] = b$ for every b in the range. For instance, for (3.1) we could specify $R(x^2)$ to be the absolute value of x or the negative of its absolute value; in both cases the image $R(x^2)$ is x^2; thus R is the right inverse of T. The domain of T is the range of R, and the range of R is contained in the domain of T.

If a transformation is both one to one and onto, then both the right and left inverses exist. If this is the case, we shall show that they must be equal. First $L(b) = a$ for every b in the range of T, where a is defined by $T(a) = b$. Also $R(b) = a$ since $R(b)$ was defined to be *some* element of the domain of T whose image under T is b. But since T is one to one, there is no choice and that *some* element must be a. So we have that $R(b) = L(b)$ for all b in the range of T, and hence R and L must be the same transformation. We call them both *the* inverse of T as in the following

Definition Let T be a one-to-one transformation whose range and codomain are the same. Then the *inverse* of T, denoted by T^{-1}, is that transformation with the property that

$$T[T^{-1}(a)] = T^{-1}[T(a)] = a$$

for all a in the domain.

The paragraphs above show the existence of T^{-1}. That it is unique follows from the definition.

3.2 LINEAR TRANSFORMATIONS ON VECTOR SPACES

In this book we are concerned with transformations of a vector space into another (or the same) vector space. Thus, if V and V' are two vector spaces, then T is a transformation of V into V', written

$$V \xrightarrow{T} V'$$

when T assigns a unique vector in V' to each vector in V. If α is in V, we call $T(\alpha)$ the *image of α under T*. If W is the set of images of vectors of V, the transformation takes V *onto* W.

We assume throughout that the vector space is "over some field," that is, that the coefficients in the linear combinations of vectors are all in some field F. Usually we do not mention that the vector space is over a field unless we want to restrict the field in some way. But the field, sometimes called the *ground field,* is always there. When we use the word *number,* we mean an element of the ground field. When we are dealing with two or more vector spaces, we assume that the ground field is the same for all.

For our purposes not just any transformation will do. We require it to be linear. We want it not only to take vectors into vectors, but lines into lines, two-dimensional vector spaces into two-dimensional vector spaces, and so on for higher dimensions. To accomplish this we require it to be linear in the sense of Sec. 1.1; that is, we impose the following two requirements:

1. If $T(\alpha) = \alpha'$, then $T(c\alpha) = c\alpha'$.
2. If $T(\alpha) = \alpha'$ and $T(\beta) = \beta'$, then $T(\alpha + \beta) = \alpha' + \beta'$.

We can combine these two requirements into one as in the following

Definition A transformation T of a vector space V into a vector space V' is called *linear* if, for all vectors α and β of V and all numbers c and d of the field,

$$T(c\alpha + d\beta) = cT(\alpha) + dT(\beta) \tag{3.2}$$

Now $T(V)$, the set of images of the vectors of V, is itself a vector space as may be seen as follows. First, it is, by definition, a subset of the vector space V'. Thus, from Sec. 1.10, we need merely show that $T(V)$ is closed under addition and multiplication by a scalar. Now if α' and β' are in $T(V)$, there are vectors α and β of V such that $T(\alpha) = \alpha'$, $T(\beta) = \beta'$, and so, by (3.2),

$$\alpha' + \beta' = T(\alpha) + T(\beta) = T(\alpha + \beta)$$

and thus $\alpha' + \beta'$ is in $T(V)$. One can similarly show closure for multiplication by a scalar. The space $T(V)$ is called the *image space of V under T.*

There is another important consequence of the Definition. Let $\alpha_1, \alpha_2, \ldots, \alpha_n$ be a basis of V. Then, from (3.2),

$$T(c_1\alpha_1 + c_2\alpha_2 + \cdots + c_n\alpha_n) = c_1T(\alpha_1) + \cdots + c_nT(\alpha_n)$$

This shows that the set of images of a basis determines the images of all the vectors of V. In fact, one may choose the images of the basis as one pleases in V', and these together with the linearity of T determine the images of every vector of V. In other words, since we judge a transformation by what it does, if S and T are two linear transformations of V into V' and if for some basis $T(\alpha_i) = S(\alpha_i)$ for every element α_i of the basis of V, then S and T are the same transformation; that is, $S = T$. Notice that for any linear transformation T the image of the zero vector of V is the zero vector of V', since

$$T(\theta) = T(0 \cdot \alpha) = 0 \cdot T(\alpha) = \theta'$$

What can be said about the dimension of $T(V)$? Continuing with the same notation, we see that $T(V)$ is spanned by

$$T(\alpha_1),\ T(\alpha_2),\ \ldots,\ T(\alpha_n) \tag{3.3}$$

If the vectors of (3.3) are linearly independent, then V and $T(V)$ have the same dimension; otherwise $T(V)$ has smaller dimension than V. Thus the *dimension of the range of a linear transformation is never greater than that of the domain.*

Let us give some examples.

Example 1 Let $V = V^3$ and let $\epsilon_1, \epsilon_2, \epsilon_3$ be the canonical basis. Let $V' = V^2$ and let T be defined as follows:

$$T(\epsilon_1) = (1,1) \qquad T(\epsilon_2) = (1,0) \qquad T(\epsilon_3) = (0,1)$$

Then

$$\begin{aligned} T(a,b,c) = T(a\epsilon_1 + b\epsilon_2 + c\epsilon_3) &= a(1,1) + b(1,0) + c(0,1) \\ &= (a + b,\ a + c) \end{aligned}$$

Here the mapping is onto V^2.

Example 2 Let $V = V^2$ and $V' = V^3$ with T defined as follows:

$$T(\epsilon_1) = (1,1,0) \qquad \text{and} \qquad T(\epsilon_2) = (1,0,1)$$

Then $T(a,b) = a(1,1,0) + b(1,0,1) = (a + b,\ a,\ b)$. The transformation is into V^3 but onto a two-dimensional subspace of V^3.

Example 3 Let $\alpha_1, \alpha_2, \alpha_3$ be a basis of V and $\alpha_1', \alpha_2', \alpha_3'$ a basis of V'. Note that in this case we are not expressing the basis elements in terms of triples of numbers, though this could be done. Suppose T is defined by

$$T(\alpha_1) = \alpha_1' \qquad T(\alpha_2) = \alpha_2' \qquad T(\alpha_3) = \alpha_3'$$

This defines a transformation of V onto V'. Now we can define a linear transformation U of V' into V by

$$U(\alpha_1') = \alpha_1 \qquad U(\alpha_2') = \alpha_2 \qquad U(\alpha_3') = \alpha_3$$

This, being linear, determines the image of every element of V' to be an element of V. Since $U[T(\alpha_i)] = \alpha_i$ for $i = 1, 2, 3$, the linearity implies that $U[T(\alpha)] = \alpha$ for every α in V. Hence U is a left inverse of T. Furthermore $T[U(\alpha_i')] = \alpha_i'$ shows that U is a right inverse of the transformation. Hence we may write $U = T^{-1}$. You might like to try this out for $V = V^3$ with the canonical basis and V' the three-dimensional vector space spanned by the following three vectors:

$$(1,1,0,0) \qquad (0,0,1,1) \qquad (0,1,1,0)$$

With this example as background, we can now deal with the inverse in greater generality. We have seen that if a transformation is one to one and onto, it has an inverse. We can be much more explicit for linear transformations of vector spaces and state and prove the following

Theorem 3.1 If T is a linear transformation of a vector space V into a vector space V', then the inverse transformation T^{-1} exists if and only if $V' = T(V)$ and the dimensions of V and V' are equal. When the inverse exists, it is linear.

Proof Suppose T has an inverse T^{-1}. Then every vector of V' must be the image of a vector in V, and hence $V' = T(V)$. Furthermore, V, the image space of T^{-1}, cannot have greater dimension than its domain $T(V)$; that is, $\dim\,(V) \leq \dim\, T(V)$. Similarly, $\dim\,(V) \geq \dim\, T(V)$. This shows that the dimensions are equal.

Now suppose $V' = T(V)$ and V and V' have the same dimensions. Then for a given basis of V, the images of the basis, set (3.3), not only span $T(V)$ but are linearly independent, that is, are a basis of $T(V)$. Then let U be the linear transformation defined on the basis (3.3) of $T(V)$ by the following

$$U[T(\alpha_i)] = \alpha_i \qquad \text{for } i = 1, 2, \ldots, n \tag{3.4}$$

It is a transformation of $T(V)$ onto V, because every element of the basis of V is thus the image of an element of $T(V)$. Furthermore

$$T(\alpha_i) = \alpha_i'$$

implies

$$T[U(\alpha_i')] = T(\alpha_i) = \alpha_i'$$

and hence U is the inverse of T. This completes the proof.

Now we consider another example to illustrate the equality of two transformations which might appear to be different.

Example 4 Let α_1, α_2 be a basis of V^2. Another basis is $\alpha_1 + \alpha_2$, $\alpha_1 - \alpha_2$. So we can define a linear transformation T by the images of α_1, α_2 and another linear transformation S by the images of $\alpha_1 + \alpha_2$, $\alpha_1 - \alpha_2$ as follows (notice that we can choose these images as we please).

$$T(\alpha_1) = \beta_1 \qquad T(\alpha_2) = \beta_2$$
$$S(\alpha_1 + \alpha_2) = \beta_1 + \beta_2 \qquad S(\alpha_1 - \alpha_2) = \beta_1 - \beta_2$$

where β_1 and β_2 are a basis of V^2. Then

$$S[(\alpha_1 + \alpha_2) + (\alpha_1 - \alpha_2)] = \beta_1 + \beta_2 + \beta_1 - \beta_2 = 2\beta_1$$

implies that $S(2\alpha_1) = 2\beta_1$, and hence $S(\alpha_1) = \beta_1$. Similarly one can show that $S(\alpha_2) = \beta_2$. Thus S has the same effect on the basis α_1, α_2 of V as T has. This implies that $T(\alpha) = S(\alpha)$, for all α in V, showing that S and T are the same transformation.

Let us look at linear transformations geometrically. Start with V^3 and let each triple in V^3 be the coordinates of a point in three-dimensional Euclidean geometry. Let T be a linear transformation of V^3 *onto* itself. (We shall mention the effect of *into* transformations below.) Then the image of every point is a point; that is, $T(\alpha) = \alpha'$. If $\alpha \neq \theta$, then $\alpha' \neq \theta'$ since otherwise the mapping would not be one to one (θ' would be the image of θ and α). Since T is linear, $T(c\alpha) = cT(\alpha) = c\alpha'$. And since the points $c\alpha$, for all values of c, are the points on the line determined by θ and α, all images lie on the line determined by θ and α'. So lines through the origin are transformed onto lines through the origin. To deal with lines not through the origin, we should refer back to Sec. 2.8 when we showed that every line can be represented by a coset $S + \alpha$, where S is a one-dimensional vector space and α is a vector in V. Then if $T(S) = S'$ and $T(\alpha) = \alpha'$, we see that the images of all points of $S + \alpha$ lie in $S' + \alpha'$ and the dimensions are both 1. Hence a linear transformation which is onto takes lines onto lines; that is, the image of every line is a line. Similarly it can be shown that such a linear transformation takes planes into planes.

It can be shown along the above lines that if V^n represents an n-dimensional Euclidean geometry, any linear transformation of V^n onto itself takes any k-dimensional subspace or coset onto a k-dimensional subspace or coset. If the transformation T is not onto but only into, it will take a k-dimensional subspace or coset into one of the same or smaller dimension.

It should be noticed that not all transformations which take lines into lines are linear. The simplest example is that of a translation, say, in Euclidean plane geometry.

EXERCISES

1. Consider each function given below along with its domain D. It is a transformation onto what set? Name one set for which the transformation is into but not onto. Find the inverse function when it exists:

(*a*) D is the set of rational numbers and f takes x into x^3.

(*b*) D is the set of real numbers and f takes x into x^3.

(*c*) D is the set of positive real numbers and f takes x into $1/(x + 1)$.

(*d*) D is the set of points on a circle with a point A deleted; m is the line tangent to the circle at the other end of the diameter through A. The image of any point P different from A on the circle is the intersection of line AP with m.

(*e*) Let D be the set of pairs of coordinates (a,b) of points in the plane and let $T(a,b) = (a + 1, 0)$.

2. Omitting part (*d*), find which of the functions or transformations in Exercise 1 are linear.

3. In each case below find the image of every element of V^3 and find the dimension of the range. What would you regard as the codomain? In any of the cases does the transformation have an inverse?

(*a*) $T(\epsilon_1) = (1,0,1,0)$, $T(\epsilon_2) = (1,0,0,1)$, $T(\epsilon_3) = (1,0,1,1)$.

(*b*) $T(\epsilon_1) = (1,2)$, $T(\epsilon_2) = (1,-1)$, $T(\epsilon_3) = (0,-1)$.

(*c*) $T(\epsilon_1) = (1,0,1,0)$, $T(\epsilon_2) = (1,2,3,1)$, $T(\epsilon_3) = (0,1,0,1)$.

4. Let α_1, α_2 be a basis for a two-dimensional vector space. Define the following linear transformation of the space into itself:

$$T(\alpha_1) = a\alpha_1 + b\alpha_2$$
$$T(\alpha_2) = c\alpha_1 + d\alpha_2$$

Under what conditions on a, b, c, d is the mapping onto itself? If it is onto, find its inverse. If it is not onto, can it have an inverse?

5. Consider the transformation

$$T(x) = \frac{ax + b}{cx + d}$$

This is sometimes called a *linear* transformation. Is it one in our sense? Under what conditions has it an inverse?

6. Let $a + bi$ be a fixed complex number and consider $f(z)$ defined as follows for all complex numbers $z = x + yi$, with x and y real:

$$f(x + yi) = (a + bi)(x + yi) = u + vi$$

where $u = ax - yb$ and $v = bx + ay$. This is a transformation of the set of complex numbers into itself. Is it a linear transformation over the set of real numbers? Under what conditions has it an inverse?

7. For the transformation in Exercise 4, let the image of $x\alpha_1 + y\alpha_2$ be denoted by $x'\alpha_1 + y'\alpha_2$. Express x' and y' in terms of x and y and the letters a, b, c, and d.

8. Let

$$T(x_1,x_2,x_3,x_4) = (2x_1 + x_2 + x_3, x_1 - x_4)$$

Is T a linear transformation?

3.3 OPERATIONS WITH TRANSFORMATIONS

It is possible to define addition and multiplication of transformations so that they, as well as matrices, form a ring, that is, satisfy all the postulates of a field except commutativity of multiplication and the existence of an inverse. Once this is done, we can, with some exceptions, work with transformations just as we do with numbers.

If T and S are two linear transformations, it would be natural to define $(T + S)(\alpha)$ to be $T(\alpha) + S(\alpha)$. For this sum to have meaning, it is necessary that the domains of T and S be the same and that their ranges be the same. So formally we have the

Definition If T and S are two linear transformations of a vector space V into a vector space V', then we define $(T + S)(\alpha)$ to be $T(\alpha) + S(\alpha)$, for all α in V.

Since $T(\alpha)$ and $S(\alpha)$ are both in V', it makes sense to add them. With this definition it is easy to see that

Addition of transformations is commutative and associative.

Furthermore, if we define the zero transformation to be that which takes every vector of V into the null vector of V' and denote this by Z, we have

$$T + Z = Z + T = T$$

for all transformations T, where it understood that Z is also a transformation of V into V'.

We define scalar multiplication of a transformation in the obvious way:

Definition If c is a number and T a linear transformation taking V into V', then the transformation (cT) is defined by

$$(cT)(\alpha) = c(T\alpha) \qquad \text{for all } \alpha \text{ in } V$$

Also define (Tc) by

$$(cT)\alpha = (Tc)\alpha = c[T(\alpha)]$$

and

$$(-T)(\alpha) = (-1)T(\alpha)$$

Hence $[(-T) + T](\alpha) = [-(T\alpha) + T(\alpha)] = \theta$, and $(-T) + T$ is the zero transformation. In view of the other properties of transformations it is easy to see that the set of all transformations of V into V' forms an additive group (see Appendix A). In fact, the set of linear transformations of V into V' forms a vector space, as we shall prove later.

Now we proceed to define the product of two transformations. Here, at least from one point of view, it is natural to define product as iteration. That is, we define $(ST)(\alpha)$ to be $S(T\alpha)$. This requires that the range of T shall be contained in the domain of S. Thus we have the

Definition If T is a linear transformation of V into V' and S is a linear transformation of V' into V'', then $(ST)(\alpha)$ is defined to be $S(T\alpha)$, for all α in V.

So ST is a transformation of V into V''. Note that for such a product to have a meaning, the range of T must be contained in the domain of S.

Notice that in applying transformations we work from the right to the left just as we do with integration. Here associativity is no problem, since if R, S, and T are three linear transformations over matching vector spaces,

$$(RS)T = R(ST)$$

that is, for all α in V, $(RS)T(\alpha) = R(ST)(\alpha)$. In each case one finds the image of α under T, then the image of $T(\alpha)$ under S, and finally the image of $ST(\alpha)$ under R.

For example, let V, V', and V'' have respective bases α_1, α_2; β_1, β_2, β_3; and γ_1, γ_2, and define S and T by

$$\begin{array}{llll} T: & \beta_1 = a_1\alpha_1 + a_2\alpha_2 & \beta_2 = b_1\alpha_1 + b_2\alpha_2 & \beta_3 = c_1\alpha_1 + c_2\alpha_2 \\ S: & \gamma_1 = r_1\beta_1 + r_2\beta_2 + r_3\beta_3 & \gamma_2 = s_1\beta_1 + s_2\beta_2 + s_3\beta_3 & \end{array}$$

Then if we substitute the expressions for β_1, β_2, β_3 in T into those for γ_1 and γ_2 in S and collect the terms, we have the following transformation for ST:

$$\begin{aligned} \gamma_1 &= (r_1a_1 + r_2b_1 + r_3c_1)\alpha_1 + (r_1a_2 + r_2b_2 + r_3c_2)\alpha_2 \\ \gamma_2 &= (s_1a_1 + s_2b_1 + s_3c_1)\alpha_2 + (s_1a_2 + s_2b_2 + s_3c_2)\alpha_2 \end{aligned}$$

We define the identity transformation I to be that transformation which takes every vector into itself. There is one little difficulty here. Suppose T takes V into V'. Then $TI = T$, where the domain and range of I are both V. To write $IT = T$, the range and domain of I would have to be V'. In each case the domain and range of I are the same. Each space has its own identity transformation, and it is convenient to use the same symbol for all.

We leave as an exercise the showing of the distributive property; that is, if R, S, and T are transformations of V into itself, then

$$R(S + T) = RS + RT \qquad \text{and} \qquad (S + T)R = SR + TR$$

To summarize our results, we have

Theorem 3.2 The set of transformations of a vector space into itself forms a ring with an identity element.

We can also prove the following

Theorem 3.3 If V and V' are vector spaces over a field F, then the set of linear transformations of V into V' is a vector space over F.

Proof This follows from the definitions of the operations on transformations. We have already noted that the set of such transformations is an additive group. We have defined multiplication by a scalar and addition so that

$$(cT + dT')(\alpha) = cT(\alpha) + dT'(\alpha)$$

for all α in V. The other properties follow easily, and we do not go through them in all detail here.

The vector space of transformations referred to in Theorem 3.3 has a basis which we now proceed to develop, leading up to the next theorem.

Let $\alpha_1, \alpha_2, \ldots, \alpha_n$ be a basis of V and $\alpha_1', \alpha_2', \ldots, \alpha_s'$ a basis of V'. Then define ns linear transformations from V into V' as follows:

$$T_{11}(\alpha_1) = \alpha_1' \quad \text{and} \quad T_{11}(\alpha_i) = \theta \quad \text{for } 2 \leq i \leq n$$
$$T_{12}(\alpha_1) = \alpha_2' \quad \text{and} \quad T_{12}(\alpha_i) = \theta \quad \text{for } 2 \leq i \leq n$$

and, in general,

$$(3.5) \quad T_{pq}(\alpha_p) = \alpha_q' \quad \text{and} \quad T_{pq}(\alpha_i) = \theta \quad \text{for } i \neq p \text{ and } 1 \leq i \leq n$$

where p can be any integer from 1 to n inclusive and q any integer from 1 to s inclusive. So we have defined ns linear transformations of V into V'.

We need to show that every linear transformation of V into V' is a linear combination of the above. Suppose T is a linear transformation taking V into V'. It can be defined as follows:

$$(3.6) \qquad T(\alpha_i) = c_{i1}\alpha_1' + c_{i2}\alpha_2' + \cdots + c_{is}\alpha_s'$$

for $1 \leq i \leq n$.

But T can also be written as follows:

$$(3.7) \qquad T(\alpha_i) = c_{i1}T_{i1}(\alpha_i) + c_{i2}T_{i2}(\alpha_i) + \cdots + c_{is}T_{is}(\alpha_i)$$

Actually we could write (3.7) as a linear combination of T_{jk} for all j, that is,

$$T(\alpha_i) = \Sigma c_{jk}T_{jk}(\alpha_i) \qquad \text{where } 1 \leq j \leq n,\ 1 \leq k \leq s, \text{ and } c_{jk} = 0 \text{ for } j \neq i$$

But when $j \neq i$, the effect on α_i is nil, so we omit it.

Thus we have shown that every linear transformation of V into V' is a linear combination of the T_{pq} described in (3.5). We can further show that these are linearly independent. To this end, suppose there are numbers c_{ij} such that

$$\sum_{i,j} c_{ij}T_{ij} = Z$$

the zero transformation; that is,

$$\sum_{i,j} c_{ij}T_{ij}(\alpha_k) = \theta \qquad \text{for } 1 \leq k \leq n \tag{3.8}$$

Since we know by (3.5) that $T_{ij}(\alpha_k) = \theta$ unless $i = k$, the sum (3.8) reduces to

$$\sum_{j} c_{kj}T_{kj}(\alpha_k) = \theta \tag{3.9}$$

But, by definition, $T_{kj}(\alpha_k) = \alpha_j'$, and hence (3.9) becomes

$$\sum_{j} c_{kj}\alpha_j' = \theta \qquad \text{for each } k \tag{3.10}$$

Since the α_j' constitute a basis for V', they are linearly independent. This shows that, for (3.10) to hold, we must have $c_{kj} = 0$ for all j and, furthermore, for all k.

Thus we have shown

Theorem 3.4 The transformations (3.5) form a basis for the set of linear transformations from V into V'. This set is a vector space of dimension ns, where V and V' are of dimensions n and s, respectively.

It should be clear by this time that transformations behave very much like matrices. We have shown that the set of all transformations of V into V' forms a ring and they are a vector space. Indeed we shall see in Sec. 3.10 that, given bases of the spaces, the set of linear transformations of V into V', of dimensions n and s, respectively, is isomorphic to the set of n by s matrices. But for the most part it is usually much easier to deal with transformations as such rather than with the matrices with which they are associated, just as usually it is simpler to consider vectors as entities rather than as n-tuples. Indeed, just as we can get information about n-tuples by considering vectors, so we can inform ourselves of certain properties of matrices by deriving them from properties of transformations. Later in the chapter we shall explore these connections.

EXERCISES

1. Find a basis for the set of linear transformations taking V^2 into V^2 by specializing the discussion of the previous section.

2. Let α_1, α_2 be a basis for V; β_1, β_2 a basis for V'; and γ_1, γ_2 a basis for V''. Let transformations T of V into V' and S of V' into V'' be defined as follows:

$$\begin{aligned} T(\alpha_1) &= \beta_1 + \beta_2 & \qquad S(\beta_1) &= \gamma_1 - \gamma_2 \\ T(\alpha_2) &= 2\beta_1 + \beta_2 & \qquad S(\beta_2) &= 3\gamma_1 + \gamma_2 \end{aligned}$$

Find the transformation ST of V into V''.

3. Under what conditions on V, V', and V'' in Exercise 2 will the transformation TS be defined? If these conditions are imposed, what would one need to know in order to find TS?

4. Using the notation of Exercise 2, let transformations T and S be defined by

$$\begin{aligned} T(\alpha_1) &= \beta_1 + \beta_2 & \qquad S(\beta_1) &= \gamma_1 - \gamma_2 \\ T(\alpha_2) &= 2\beta_1 + 2\beta_2 & \qquad S(\beta_2) &= 3\gamma_1 + \gamma_2 \end{aligned}$$

Find the transformation ST of V into V''.

5. Answer the questions raised in Exercise 3, but now with reference to Exercise 4.

6. Show that in Exercise 4 the inverses of T and ST do not exist.

7. Let $T(x_1,x_2) = (x_1 + x_2,\ 2x_1 + x_2)$ and $S(x_1,x_2) = (x_1 - x_2,\ 3x_1 + x_2)$. Find $TS(x_1,x_2)$ and $ST(x_1,x_2)$. Compare your results with those of Exercise 2.

8. If in Exercise 2 the transformation T is as given but S is defined by

$$\begin{aligned} S(\beta_1) &= \gamma_1 - \gamma_2 \\ S(\beta_2) &= 2\gamma_1 - 2\gamma_2 \end{aligned}$$

find ST. By imposing conditions on V, V', and V'', can the transformation TS be also defined? If so, find it; if not, show why not.

9. Let R, S, and T be linear transformations of V^n into itself. Prove that $R(S + T) = RS + RT$.

10. Let $g(x)$ and $g'(x)$ be defined by

$$g(x) = \frac{ax + b}{cx + d} \qquad \text{and} \qquad g'(x) = \frac{a'x + b'}{c'x + d'}$$

Show that $g[g'(x)]$ is of the same form as g and g'. If g and g' correspond, respectively, to the matrices

$$A = \begin{bmatrix} a & b \\ c & d \end{bmatrix} \qquad \text{and} \qquad A' = \begin{bmatrix} a' & b' \\ c' & d' \end{bmatrix}$$

what is a corresponding matrix for $g[g'(x)]$?

11. If T is a linear transformation of V into V', under what conditions will the expression $T^2 + 3T + I$ have meaning?

12. If T is a transformation of V into itself, and $T^2 - I = Z$, the zero transformation, does it follow that $T - I = Z$ or $T + I = Z$? If so, why; if not, give an example to show it.

13. In Exercise 2, let $V = V' = V''$ and find the inverses of T and S. Then show that $(ST)^{-1}$ exists and is equal to $T^{-1}S^{-1}$.

14. Suppose that S and T are two linear transformations of a vector space V onto itself. Show that $(ST)^{-1}$ exists and is equal to $T^{-1}S^{-1}$.

3.4 RANK, NULLITY, INVERSES, AND KERNELS

We know from Sec. 3.2 that a linear transformation T of a vector space V of dimension n has a range (or image space) of dimension not greater than n. We begin with the

Definition The *rank* of a linear transformation is the dimension of its range or image space.

We know from the above remark that the rank of T is not greater than n, the dimension of its domain. Let us first consider the case in which the rank of T is exactly n and state the following theorem:

Theorem 3.5 If T is a linear transformation of V into itself, then the following three statements are equivalent; that is, each implies the others.

1. V and $T(V)$ have the same dimension n.
2. T has an inverse.
3. $T(\alpha) = T(\beta)$ implies $\alpha = \beta$.

Proof The equivalence of statements 1 and 2 is Theorem 3.1. Now to prove that statement 3 implies statement 1 we use the contrapositive; that is, we show that if statement 1 is false, so is statement 3. To this end suppose the vectors $T(\alpha_i)$, for α_i a basis of V, are linearly dependent. Then, for some c_i not all zero we have

$$\sum_{i=1}^{n} c_i T(\alpha_i) = \theta$$

that is,

$$T\left(\sum_{i=1}^{n} c_i\alpha_i\right) = \theta$$

Hence $c_1\alpha_1 + c_2\alpha_2 + \cdots + c_n\alpha_n$ is a nonzero vector whose image is the zero vector. This denies statement 3, since $T(\theta)$ is also θ.

We complete the proof by showing that statement 2 implies statement 3. This follows immediately by multiplying both sides of the equation $T(\alpha) = T(\beta)$ on the left by T^{-1}.

So we have shown that any one of statements 1 to 3 implies the other two.

Looking carefully at the above proof we can see that the vectors $T(\alpha_i)$ are linearly dependent if and only if there is some nonzero vector of V whose image is the null vector. This might lead us to guess that there is some connection between the rank of a transformation and the set of vectors whose images are the null vector. So we define a new term:

Definition The *kernel* of a linear transformation of a vector space V is the set of vectors of V whose images are the null vector. The kernel is also called the *null space* of the transformation.

We leave as an exercise the proof of the statement

The kernel of a linear transformation is a vector subspace of V.

This leads us to another definition:

Definition The *nullity* of a transformation is the dimension of its kernel.

We leave as an exercise the proof of the following, since it can be shown rather shortly from Theorem 3.5.

Theorem 3.6 Let V be an n-dimensional vector space, then the rank of the linear transformation T of V is n if and only if the kernel of T is the null space; that is, its nullity is zero.

Intuitively it might seem reasonable that as one increases the rank, one decreases the nullity. In fact, we have the following

Theorem 3.7 Let V be an n-dimensional vector space and T a linear transformation of V. If r is the rank of T, then its nullity is $n - r$.

Proof Let s be the dimension of the kernel of T and

$$\alpha_1, \alpha_2, \ldots, \alpha_s \tag{3.11}$$

a basis of K, the kernel of T. Now K is a subspace of V and hence by the completion theorem (Theorem 1.3) there are vectors γ_i such that

$$\alpha_1, \alpha_2, \ldots, \alpha_s, \gamma_1, \gamma_2, \ldots, \gamma_t \tag{3.12}$$

is a basis of V. Since $t = n - s$, we will have proved our result if we can show that the set

$$T(\gamma_1), T(\gamma_2), \ldots, T(\gamma_t) \tag{3.13}$$

is a basis of $T(V)$. To do this we need to demonstrate that the set (3.13) is linearly independent and spans $T(V)$. First, suppose

$$c_1T(\gamma_1) + c_2T(\gamma_2) + \cdots + c_tT(\gamma_t) = \theta \tag{3.14}$$

Then $$T(c_1\gamma_1 + c_2\gamma_2 + \cdots + c_t\gamma_t) = \theta$$

This implies that $c_1\gamma_1 + c_2\gamma_2 + \cdots + c_t\gamma_t$ is in the kernel of T and hence must be a linear combination of α_i. Since (3.12) is a linearly independent set, the c's in (3.14) must all be zero, and the set (3.13) is a linearly independent set.

Finally, we must show that the set (3.13) spans $T(V)$. Since the set (3.12) is a basis of V, every vector β of V is a linear combination of the vectors of (3.12); that is,

$$\beta = c_1\alpha_1 + c_2\alpha_2 + \cdots + c_s\alpha_s + d_1\gamma_1 + d_2\gamma_2 + \cdots + d_t\gamma_t$$

for some $c_1, \ldots, c_s$ and $d_1, \ldots, d_t$. Then

$$T(\beta) = T\left(\sum_i c_i\alpha_i\right) + T\left(\sum_i d_i\gamma_i\right)$$

But $T(\alpha_i) = \theta$ since each α_i is in the kernel of T. Hence

$$T(\beta) = \sum_i d_iT(\gamma_i)$$

This shows that every vector of $T(V)$ is a linear combination of the linearly independent set (3.13), and hence the rank of $T(V)$ is t. But since $t = n - s$, we have

$$s = n - r$$

where r is the rank of T and s is its nullity.

We thus have a kind of splitting of the space V. For let R be the space spanned by $\gamma_1, \gamma_2, \ldots, \gamma_t$ in (3.12), then

$$V = R \oplus K$$

that is, V is the direct sum of subspaces R and K (see Sec. 1.10).

Here is an example to illustrate Theorem 3.7 and its proof. Let the vector space be V^3 and define T as follows in terms of the canonical basis:

$$T(\epsilon_1) = \epsilon_1 + \epsilon_2 \qquad T(\epsilon_2) = -\epsilon_1 + \epsilon_3 \qquad T(\epsilon_3) = \epsilon_2 + \epsilon_3$$

We find the kernel by computing $T(a_1\epsilon_1 + a_2\epsilon_2 + a_3\epsilon_3) = (a_1 - a_2)\epsilon_1 + (a_1 + a_3)\epsilon_2 + (a_2 + a_3)\epsilon_3$. The image will be the null vector if and only if

$$a_1 = a_2 = -a_3$$

Hence the kernel $K = \text{sp}(\epsilon_1 + \epsilon_2 - \epsilon_3) = \text{sp}[(1,1,-1)]$, and, using the notation of the theorem, we take $\alpha_1 = (1,1,-1)$. We can complete the basis of V in various ways. One choice is $\gamma_1 = \epsilon_1$, $\gamma_2 = \epsilon_2$. Then

$$T(\gamma_1) = (1,1,0) \qquad \text{and} \qquad T(\gamma_2) = (-1,0,1)$$

These two images are linearly independent and they span the range space R of dimension 2. Here $V = R \oplus K$.

Notice that

$$T(\alpha) = T(\beta) \qquad \text{if and only if } T(\alpha - \beta) = \theta$$

This shows that two vectors have the same image if and only if their difference is in the kernel K. Then by use of either Theorem 3.5 or Theorem 3.6 we have

Theorem 3.8 A linear transformation T of a vector space V into $T(V)$ has an inverse if and only if its kernel is the null space.

Taking our cue from the corresponding definition for matrices in Sec. 2.6, we call a transformation *nonsingular* if it has an inverse; otherwise *singular*.

Finally, we prove a useful result on the ranks of products of transformations.

Theorem 3.9 Let T and S be two linear transformations of a vector space V into itself, let $r(T)$ denote the rank of T, and similarly for $r(S)$ and $r(TS)$. Then

$$r(TS) \leq r(T) \qquad \text{and} \qquad r(TS) \leq r(S)$$

Proof We know that $S(V) \subseteq V$, and hence $TS(V) \subseteq T(V)$ which implies the first inequality. Now notice that $S(\alpha) = \theta$ implies that $TS(\alpha) = \theta$, and hence the kernel of S is contained in the kernel of TS. This implies, using Theorem 3.7, that

$$n - r(S) \leq n - r(TS)$$

where $n = \dim(V)$. This implies the second inequality in the theorem.

3.5 APPLICATIONS

First let us look at a geometrical interpretation of the material in the previous section not only with a view to application but also to make the abstract ideas more concrete. We are familiar with the idea of a projection in Euclidean geometry. We can, for instance, in the plane, project every point onto the x axis by letting the point (a,b) correspond to the point $(a,0)$ on the x axis. This can be considered to be a transformation from V^2 onto V^1 and could be written as

$$T(1,0) = (1,0) \qquad \text{and} \qquad T(0,1) = (0,0)$$

that is,

$$T(a,b) = (a,0)$$

This is described as being the projection of the point (a,b) on the x axis "along the line $(0,b)$." The last phrase calls for some explanation. The set of points $(0,b)$ is the y axis. So the projection is parallel to the y axis. That is, to get the projection of any point (a,b) one draws a line through this point parallel to the y axis; where this line cuts the x axis is the projection of the point on the x axis. The points $(0,b)$ are the kernel of T since all these points project into the origin.

Similarly we could define another projection S by

$$S(1,0) = (1,0) \qquad \text{and} \qquad S(1,1) = (0,0)$$

that is,

$$S(a,b) = (a - b, 0)$$

In this case the projection is again on the x axis but along the line (b,b), that is, the line $x = y$. The points (b,b) are the kernel of S.

In three dimensions we could project onto the xy plane along the z axis by the transformation

$$T(a,b,c) = (a,b,0)$$

or onto the x axis along the yz plane by $T(a,b,c) = (a,0,0)$.

More generally, let α, β be a basis for V^2. Then the transformation T defined by

$$T(\alpha) = \alpha \qquad T(\beta) = \theta$$

is a projection of (α,β) on the line $\text{sp}(\alpha)$ along the line $\text{sp}(\beta)$. The kernel of the transformation in this case is $\text{sp}(\beta)$. With this as a background we can then define a projection as follows:

Definition A transformation T of a vector space V into itself is called a *projection* of V onto W along K if

1. K is the kernel of T.
2. $V = W \oplus K$.
3. $T(\omega) = \omega$ for all ω in W.

This means, of course, that we have to be a little careful about picking W. To illustrate this, let T be a transformation of V^3 into itself defined by

$$T(\epsilon_1) = \epsilon_1 \qquad T(\epsilon_2) = \epsilon_2 \qquad T(\epsilon_3) = \epsilon_1 + \epsilon_2 \tag{3.15}$$

Now the rank of $T(V)$ is 2, and hence by Theorem 3.7 the kernel of T has dimension 1. Since $T(1,1,-1) = \theta$, we see that the kernel K is spanned by $\alpha_1 = (1,1,-1)$. Now, following the proof of Theorem 3.7, we can take the basis to be

$$\alpha_1 \quad \epsilon_1 \quad \epsilon_2 \tag{3.16}$$

and $W = \text{sp}(\epsilon_1,\epsilon_2)$. In this case, (3.15) shows that T takes each vector of W into itself and hence T is a projection of V onto W along K.

On the other hand, we could just as well have completed the basis to yield

$$\alpha_1 \quad \epsilon_1 \quad \epsilon_3 \tag{3.17}$$

In this case $W = \text{sp}(\epsilon_1,\epsilon_3)$, and since $T(\epsilon_3) = \epsilon_1 + \epsilon_2$, which is not in W, we see that T is not a projection onto W.

The above shows that we must pick W carefully. But there are cases in which no amount of care will yield a W satisfying the requirements of the definition. This is only natural since not all transformations are projections, but an example should be helpful in showing what happens. So suppose the transformation T is defined by

$$T(\epsilon_1) = \epsilon_1 \qquad T(\epsilon_2) = 2\epsilon_2 \qquad T(\epsilon_3) = \epsilon_1 + \epsilon_2$$

First we should determine its kernel. A little calculation shows that

$$T(a\epsilon_1 + b\epsilon_2 + c\epsilon_3) = (a + c)\epsilon_1 + (2b + c)\epsilon_2$$

Hence $a\epsilon_1 + b\epsilon_2 + c\epsilon_3$ is in the kernel if and only if $a + c = 0$ and $2b + c = 0$. Thus, if we define $\beta = (-2,-1,2)$, we see that $K(T) = \text{sp}(\beta)$. Furthermore, W will be the set of vectors $a\epsilon_1 + b\epsilon_2 + c\epsilon_3$ for which

$$a\epsilon_1 + b\epsilon_2 + c\epsilon_3 = T(a\epsilon_1 + b\epsilon_2 + c\epsilon_3) = (a + c)\epsilon_1 + (2b + c)\epsilon_2$$

Hence, equating coefficients of the basis elements on both sides of the equation, we have $a = a + c$, $b = 2b + c$, $c = 0$. Hence $W = \text{sp}(\epsilon_1)$ and $\dim (W) + \dim (K) = 2 < 3$. This shows that $W + K \neq V$.

The following theorem shows that the purely algebraic equality $T^2 = T$, which concerns T alone, is a necessary and sufficient condition that T be a projection.

Theorem 3.10 Let T be a linear transformation of V into itself. Then T is a projection if and only if $T^2 = T$.

Partial Proof Suppose T is a projection. By definition, it must be along K, the kernel of T. Also, there must be a subspace W such that $V = W \oplus K$ and $T(\omega) = \omega$ for all ω in W. Then any vector α of V can be expressed uniquely in the form

$$\alpha = \omega + \beta \qquad \text{where } \omega \text{ is in } W \text{ and } \beta \text{ is in } K$$

Then $T(\alpha) = T(\omega) + T(\beta) = T(\omega)$ and $T^2(\alpha) = T[T(\omega)]$. But since $T(\omega) = \omega$, it follows that $T[T(\omega)] = \omega$. But $T(\alpha) = \omega$ which implies that $T^2(\alpha) = T(\alpha)$, for all α in V; that is, that $T^2 = T$.

We leave the rest of the proof as an exercise.

The idea of a kernel of a transformation can be applied to differentiation (compare with the end of Sec. 2.8). We have seen (Sec. 1.9) that a set of polynomials with real coefficients in a single variable of degree less than, say, $n + 1$ forms a vector space. It is not hard to see that taking the derivative is a linear transformation. To this end, let $f(x)$ and $g(x)$ be polynomials. Then, denoting the derivative with respect to x by use of a prime, we know that

$$\text{if } F(x) = f(x) + g(x), \text{ then } F'(x) = f'(x) + g'(x)$$

and

$$\text{if } G(x) = cf(x), \text{ then } G'(x) = cf'(x)$$

So differentiation is a linear transformation. What is its kernel? It is the set of polynomials whose image under differentiation is zero, that is, the set whose derivatives are zero. We know that this is the set of real numbers, that is, the set of constant functions. So, if V is the set of all polynomials in x of degree less than $n + 1$, then $K = \{c\}$, for c a real number, and

$$V = K + \text{sp}(x^n, x^{n-1}, \ldots, x)$$

As we know, two polynomials have the same derivative if and only if they differ by a constant, that is, an element of the kernel. Most of this example carries over equally well for the set of differentiable functions of x.

EXERCISES

Throughout this set of exercises, where there is no indication to the contrary, R, S, and T are linear transformations of V^n into itself.

1. For $n = 3$, find the kernel of T in each of the following cases:
 (*a*) $T(\epsilon_1) = \epsilon_1 + \epsilon_2$, $T(\epsilon_2) = \epsilon_2 + \epsilon_3$, $T(\epsilon_3) = \epsilon_3 + \epsilon_1$.
 (*b*) $T(\epsilon_1) = \epsilon_1 + \epsilon_2$, $T(\epsilon_2) = \epsilon_2 + \epsilon_3$, $T(\epsilon_3) = \epsilon_3 - \epsilon_1$.

2. Find the ranks of ST and TS for T defined as in Exercise 1*b* and

$$S(\epsilon_1) = \epsilon_1 + 2\epsilon_2 \qquad S(\epsilon_2) = \epsilon_3 - \epsilon_1 \qquad S(\epsilon_3) = \epsilon_3 + 2\epsilon_2$$

3. Let S be defined as in Exercise 2. Find a transformation R such that the rank of RS is less than each of the ranks of R and S.

4. Is it possible to find a transformation R in Exercise 3 so that both RS and SR have ranks less than S and R? If so, give an example; if not, prove your conclusion.

5. Taking the basis of V to be (3.17), find the space $T(W)$. Show that every vector of this space is left unchanged by T.

6. Is the following transformation over V^3 a projection?

$$T(\epsilon_1) = \epsilon_1 \qquad T(\epsilon_2) = 2\epsilon_2 \qquad T(\epsilon_3) = 2\epsilon_1 + \epsilon_2$$

7. Prove

$$r(S) + r(T) \leq r(ST) + n$$

where r denotes the rank and $n = \dim(V)$.

8. Prove that if $TS = Z$, the zero transformation, then the sum of the ranks of T and S is not greater than n, the dimension of V.

9. Using the notation of Exercise 7, prove that

$$r(S + T) \leq r(S) + r(T)$$

10. Prove that the kernel of a linear transformation is a vector space.

11. Prove Theorem 3.6.

12. Let K be the kernel of T and $V = W \oplus K$. Give an example illustrating each of the following:

(*a*) $T(W) \cap K \neq \theta$

(*b*) $T(W) \cap K = \theta$

13. Using the notation of Exercise 12, prove that $T(W) \subseteq W$ implies that the second alternative [part (*b*)] must hold.

14. Finish the proof of Theorem 3.10.

15. Show that if the ranks of T and T^2 are the same, then the range and kernel of T are disjoint.

16. Suppose that there is a linear transformation S such that $ST = I$; that is, $ST(\alpha) = \alpha$ for all α in V. Then prove that T is nonsingular and $TS = I$.

17. Let T be a transformation of the set of real numbers into their squares; that is, $T(x) = x^2$. Since $T(x) = 0$ implies $x = 0$, we could consider 0 to be the kernel of this transformation. But the transformation does not have an inverse. Why does this not deny Theorem 3.8?

18. Let† V be the space of all polynomials in x with real coefficients, and for any polynomial $\alpha = \sum_{i=0}^{n} a_i x^i$ define two transformations T and S as follows:

$$S(\alpha) = \sum_{i=0}^{n} i a_i x^{i-1}$$

$$T(\alpha) = \sum_{i=0}^{n} \frac{a_i x^{i+1}}{i+1}$$

Note that these are linear transformations associated with the derivative and integral, respectively. Prove that $ST(\alpha) = \alpha$ for all α, but that there is some β in V such that $TS(\beta) \neq \beta$. Why is this example not inconsistent with Exercise 16?

3.6 FIXED POINTS AND INVARIANT SUBSPACES

In characterizing a transformation, it is important to know what it leaves unchanged. First, there is an important property which is unchanged by a linear transformation. We noticed in Sec. 3.2 that a linear transformation from a vector space V into V' preserves collinearity; that is, if three points are collinear, then their images are also. More generally, if a set of vectors is in a k-dimensional subspace of V, then their images are in a k-dimensional subspace of V'.

Second, if T is a transformation of V into itself, one may also have vectors or sets of vectors which are taken into themselves or onto themselves. If S is a subspace of V and $T(S) \subseteq S$, we call S an *invariant subspace,* while if $T(S) = S$, we call S *fixed under* T. In many instances we can tell what kind of a transformation T is by what it leaves fixed or invariant.

For instance, in plane Euclidean geometry where we confine ourselves to rigid motion transformations, that is, Euclidean transformations, a rotation is a Euclidean transformation which leaves exactly one point fixed, and every Euclidean transformation which leaves exactly one point fixed is a rotation, A translation is characterized by the property that it leaves fixed all lines parallel to a given line (that is, all lines in the direction of the translation) but leaves no point fixed. A reflection has the property that it leaves every point of a given line fixed and leaves fixed all lines perpendicular to the given line.

Notice that there are two kinds of "fixity." A translation takes certain lines into themselves without leaving fixed any point of these lines. But a

† Nering, *Linear Algebra and Matrix Theory,* p. 36, listed in the bibliography.

reflection leaves every point of a line fixed. So, many kinds of invariance are possible.

We now list two results with examples, leaving the proofs as exercises.

1. If W and W' are two invariant subspaces of T, then $W + W'$ and $W \cap W'$ are also invariant subspaces of T.

For example, if each of two lines through the origin is left invariant by T, then the plane determined by these two lines is also invariant. Also if each of two planes through the origin is left invariant under T, then their line of intersection is also left invariant.

2. If each vector of a basis of a vector space is invariant under T, then every vector of the space is invariant under T.

For example, if each of the points defined by three linearly independent triples of real numbers is left invariant under T, then every point defined by a triple is left invariant under T.

3.7 POLYNOMIALS IN A LINEAR TRANSFORMATION T

It turns out to be true that certain polynomials in a transformation T give us information about its invariant subspaces. Let us first look at this for the case of a projection. After that we shall explore these relationships in general.

In Sec. 3.5 we noted that a transformation T is a projection on a certain subspace if and only if $T^2 = T$, that is, if and only if T "satisfies" the equation

$$T^2 - T = Z \tag{3.18}$$

where Z is the zero transformation. We prove the following theorem, which casts new light on our previous results as well as prepares for what is to come in the more general case.

Theorem 3.11 Suppose T is a linear transformation of V^n into itself which satisfies Equation (3.18). Let W_1 be the kernel of T and W_2 the kernel of $T - I$. Then

1. $V = W_1 \oplus W_2$
2. $W_1 = (T - I)(V)$
3. $W_2 = T(V)$

Proof Since $T - (T - I) = I$, we have that for any vector α in V^n

$$T(\alpha) - (T - I)(\alpha) = \alpha \tag{3.19}$$

Now (3.18) shows that $(T - I)T(\alpha) = \theta$, and hence $T\alpha$ is in the kernel of $T - I$. Similarly $(T - I)(\alpha)$ is in the kernel of T. Thus (3.19) expresses α as a sum of an element in the kernel of $T - I$ and an element of the kernel of T. Thus $V = W_2 + W_1$. Furthermore, if β is in $W_1 \cap W_2$, that is, $(T - I)(\beta) = \theta$ and $T(\beta) = \theta$, Equation (3.19) with β in place of α shows that $\beta = \theta$. Hence $W_1 \cap W_2 = \theta$, and the sum $W_1 + W_2$ is direct.

Thus, by Exercise 14 of Sec. 1.10, n is the sum of the dimensions of W_1 and W_2. Hence, if $r(T)$ and $r(T - I)$ denote the ranks of T and $(T - I)$, respectively, we have

$$n - r(T) + n - r(T - I) = n$$

that is,

$$r(T) + r(T - I) = n \tag{3.20}$$

Now $T(V)$ is in the kernel of $T - I$ since $(T - I)T(V) = \theta$. But (3.20) shows that $T(V)$ has the same dimension as the kernel of $T - I$. It follows that $T(V)$ *is* the kernel of $T - I$. Similarly $(T - I)(V)$ is both the codomain of $T - I$ and the kernel of T. This completes the proof.

We shall show in Sec. 3.9 that not only is there a splitting of the vector space but also, in a certain sense, of the transformation.

Notice that now we have a complete description of the effect of T on V, for we may take $\alpha_1, \alpha_2, \ldots, \alpha_r$ to be the basis of the kernel of $T - I$ and $\alpha_{r+1}, \alpha_{r+2}, \ldots, \alpha_n$ to be the basis of the kernel of T, while α_i, for $1 \leq i \leq n$, is a basis of V. Then using this basis of V, we have

$$\begin{aligned} (T - I)(\alpha_i) &= \theta \qquad \text{for } 1 \leq i \leq r \\ T(\alpha_i) &= \theta \qquad \text{for } r + 1 \leq i \leq n \end{aligned}$$

that is,

$$T(\alpha_i) = \begin{cases} \alpha_i & \text{for } 1 \leq i \leq r \\ \theta & \text{for } r + 1 \leq i \leq n \end{cases}$$

Thus W_2 is a subspace of V in which every vector is fixed, and the images under T of all vectors of W_1 are the null vector. The transformation is a projection of V onto W_2 along W_1.

Now we consider the general situation. Suppose $f(x)$ is a polynomial in x with coefficients in some field, say, the real numbers or the complex numbers. We might write it as follows:

$$f(x) = a_s x^s + a_{s-1}x^{s-1} + a_{s-2}x^{s-2} + \cdots + a_1 x + a_0$$

We call a_s the *leading coefficient*. Now if T is a linear transformation of a vector space into itself, we can form the same polynomial in T as follows:

$$f(T) = a_s T^s + a_{s-1}T^{s-1} + a_{s-2}T^{s-2} + \cdots + a_1 T + a_0 I$$

We say that T "satisfies the equation" $f(T) = Z$ when the equation is true. Notice that for this, T *must* take V into itself, since for, say, T^2 to have a meaning the range and domain of T must be the same.

Now we know that T satisfies some polynomial equation since we showed in Sec. 3.3 that all transformations of V^n into itself form a vector space of dimension n^2. Thus the following set is a linearly dependent set:

$$I, T, T^2, \ldots, T^{n^2}$$

That is, T satisfies a polynomial equation of degree n^2 or less. We shall show in the next section that, as a matter of fact, every transformation T of V^n into V^n satisfies a polynomial equation of degree at most n, but we do not need this for the present.

Since there is *some* polynomial equation which T satisfies, there must be one of least positive degree. Furthermore, we may without loss assume that the leading coefficient of this minimum equation is 1. Thus we have the

Definition Given a transformation T of V^n into itself, *the minimum polynomial* of T is the polynomial $m(x)$ of least positive degree and leading coefficient 1 such that $m(T) = Z$, the zero transformation.

It is not hard to show that this minimum polynomial is unique. But before doing this we need to look a little more carefully into the behavior of polynomials in a transformation. There was no difficulty in replacing x by T above, but what about multiplying and adding? Let us illustrate this for two polynomials:

$$f(x) = ax^2 + bx + c \qquad \text{and} \qquad g(x) = dx^3 + ex^2 + hx + k$$

Now

$$f(x) + g(x) = dx^3 + (a + e)x^2 + (b + h)x + (c + k)$$

Because we have defined addition of transformations and multiplication by a scalar so that the manipulations are just the same as with x, we have

$$f(T) + g(T) = dT^3 + (a + e)T^2 + (b + h)T + (c + k)I$$

In fact, if S were another transformation of V into itself, we could compute $f(T) + g(S)$ just as we compute $f(x) + g(y)$.

When it comes to multiplication we have to be a little more careful, but there is really no trouble here if we are concerned with only a single transformation. We write

$$f(x)g(x) = adx^5 + (ae + bd)x^4 + (cd + be + ah)x^3 + (ec + bh + ka)x^2 + (bk + ch)x + ck$$

In computing this product we assumed, among other things, that $ax^2dx^3 = adx^5$. This same equality holds for the transformation T

$$aT^2dT^3 = adT^5$$

since the transformation is commutative with the coefficients. So there is no trouble here either, and we can get $f(T)g(T)$ by replacing x by T in the above sum and ck by ckI. It is even true that $f(T)g(T) = g(T)f(T)$ since any power of T is commutative with any power of T. But of course if we had two transformations there would be no reason why $f(T)g(S)$ should be equal to $g(S)f(T)$. From all this it should be clear without formal proof that we can deal with addition and multiplication of polynomials in a transformation T just as with polynomials in x. In fact, we set up our definitions of operations on T so that this would be true. However, it should be noticed that we have to be careful about factoring. In the example at the beginning of this section, $T^2 - T = Z$ did not imply that either $T = Z$ or $T = I$ since, for example, $T(T - I) = Z$ does not imply $T - I = Z$ unless T has an inverse. Now we can prove

Theorem 3.12 If $f(x)$ is a polynomial such that $f(T) = Z$ and if $m(x)$ is the minimum polynomial of T, then $m(x)$ is a factor of $f(x)$.

Proof We can divide $f(x)$ by $m(x)$ and get a quotient $q(x)$ and a remainder $r(x)$ which is either zero or of lower degree than $m(x)$; that is,

$$f(x) = q(x)m(x) + r(x)$$

Replacing x by T we have

$$f(T) = q(T)m(T) + r(T)$$

But $f(T) = Z = m(T)$ implies $r(T) = Z$. Since $m(x)$ is the polynomial of least positive degree such that $m(T) = Z$, we see that $r(x)$ must be zero, and hence $m(x)$ is a factor of $f(x)$.

Corollary The minimum polynomial of a linear transformation T is unique.

To show this, suppose there were two minimum polynomials $m(x)$ and $m'(x)$. By the theorem, each must divide the other. Then since their leading coefficients are both 1, the polynomials must be the same.

For example, suppose a transformation T of V^3 into itself is defined as follows in terms of a basis $\alpha_1, \alpha_2, \alpha_3$:

$$T(\alpha_1) = \alpha_2 \qquad T(\alpha_2) = \alpha_3 \qquad T(\alpha_3) = \theta$$

Then $T^3(\alpha_i) = \theta$, for $i = 1, 2, 3$. Thus if $f(x) = x^3$, we have $f(T) = Z$. So the minimum polynomial must be a divisor of x^3 and thus must be one of x^3,

x^2, x. But $T^2(\alpha_1) = \alpha_3$ shows that x^2 is not the minimum polynomial and thus x is not (why?). Hence x^3 is the minimum polynomial.

One can also extend this idea of a minimum polynomial as follows:

Definition If α is a nonzero vector of V, then the minimum polynomial of T with respect to α is that polynomial $m_\alpha(x)$ of least positive degree and with leading coefficient 1 such that

$$m_\alpha(T)(\alpha) = \theta$$

One can prove the following result for such a minimum polynomial in the same way as we proved Theorem 3.12. The proof is left as an exercise.

Theorem 3.12a If $f(x)$ is a polynomial such that $f(T)(\alpha) = \theta$ and if $m_\alpha(x)$ is the minimum polynomial of T with respect to α, then $m_\alpha(x)$ is a factor of $f(x)$.

Corollary The minimum polynomial of T with respect to α is unique. It is a factor of the minimum polynomial of T.

Now we have the equipment to show the usefulness of the minimum polynomial of T in finding its fixed points. It is contained in the following theorem:

Theorem 3.13 If $m(x)$ is the minimum polynomial of a linear transformation T, then there is a nonzero vector α such that

$$T(\alpha) = c\alpha \tag{3.21}$$

for some number c, if and only if $x - c$ is a factor of $m(x)$.

Proof First suppose $x - c$ is a factor of $m(x)$. Then write $m(x) = (x - c)m'(x)$, where $m'(x)$ is a polynomial of degree one less than that of $m(x)$. Now consider

$$m(T) = (T - cI)m'(T) = Z$$

If $T - cI$ were nonsingular we could multiply the second equation by its inverse on the left and get $m'(T) = Z$. This would deny the supposition that $m(x)$ is the minimum polynomial of T. But $T - cI$ singular implies that its kernel is of dimension at least 1 and hence that there is a nonzero vector α such that $(T - cI)(\alpha) = \theta$; that is, $T(\alpha) = c\alpha$.

Conversely, suppose that there is a nonzero vector α such that $T(\alpha) = c\alpha$. This implies that $(T - cI)(\alpha) = \theta$ and thus that the minimum polynomial of T with respect to α is $x - c$. By the Corollary of Theorem 3.12a, this must be a factor of $m(x)$. This completes the proof.

An immediate consequence of this theorem is that in order to get the fixed points, or, indeed, any points which are taken into scalar multiples of themselves by the transformation T, one finds the zeros of the minimum polynomial of T. That is, the zeros of the minimum polynomial yield invariant subspaces of dimension 1. In a later chapter we shall deal with invariant subspaces of any dimension.

For instance, if T is a projection, $x^2 - x$ is its minimum polynomial. The zeros of this polynomial are 0 and 1. Then $\text{sp}(\alpha)$ will be an invariant subspace of T of dimension 1 if $T(\alpha) = 1 \cdot \alpha$ or $T(\alpha) = 0 \cdot \alpha$, and these are the only invariant subspaces of dimension 1.

In fact, the zeros of the minimum polynomial of a transformation play a very important role in determining its properties. So we have a name for them.

Definition The zeros of the minimum polynomial of a linear transformation T are called its *characteristic roots*, and the corresponding vectors, its *characteristic vectors*. The set of characteristic roots is called the *spectrum of* T.

For instance in the example of a projection above, the characteristic roots are 1 and 0, and the vectors α are its characteristic vectors.

Elsewhere characteristic roots are often referred to as *latent roots, roots,* or *eigenvalues;* and characteristic vectors are often called *eigenvectors.* The use of the word *spectrum* in this connection is probably because, in a certain sense, the characteristic roots break the transformation into component parts just as the colors of the spectrum are the component parts of white light.

We shall show in the next section that the minimum polynomial of a transformation of a vector space V of dimension n has degree not greater than n. This implies that the number of distinct characteristic roots cannot be greater than the dimension of the space.

EXERCISES

1. In each case below, $f(x)$ is a polynomial such that $f(T) = Z$ for a transformation T of V into itself. List the possibilities for the minimum polynomial of T, and for each of these find the spectrum of T:

(*a*) $f(x) = x^5$ (*b*) $f(x) = x^3 - 1$

(*c*) $f(x) = (x - 1)^3$ (*d*) $f(x) = (x - 1)^2(x - 2)$

2. Let a transformation T of V into itself be defined for a basis $\alpha_1, \alpha_2, \alpha_3$ as follows:

$$T(\alpha_1) = -\alpha_2 \qquad T(\alpha_2) = \alpha_1 + 2\alpha_2 \qquad T(\alpha_3) = \alpha_3$$

What is the spectrum of T? Find, in terms of the basis, the set of characteristic vectors.

3. Suppose, for a transformation T of V into V, that $T(\alpha) = 2\alpha$ and $T(\beta) = 3\beta$ for vectors α and β of V. For what numbers r and s is $r\alpha + s\beta$ a characteristic vector of T?

4. Prove that if α, β, and $\alpha + \beta$ are characteristic vectors of a transformation of T into itself, with α and β vectors of V, then $r\alpha + s\beta$ is a characteristic vector of T for all numbers r and s.

5. Suppose T is a linear transformation of V^n into itself and that $T^2 - I = Z$, where I is the identity matrix and Z the zero matrix. State and prove the theorem corresponding to Theorem 3.11 for this case.

6. Prove Theorem 3.12a and its Corollary.

7. Let $x^3 - x$ be the minimum polynomial of a linear transformation T of V^3 into itself. Prove that there are three numbers c for which there exists a vector α such that $T(\alpha) = c\alpha$. Prove that the three vectors so defined are linearly independent.

8. Prove the two numbered statements at the end of Sec. 3.6.

9. Let $m_\alpha(x)$ and $m_\beta(x)$ be the minimum polynomials of a transformation T with respect to nonzero vectors α and β, respectively. Prove that the least common multiple of these two minimum polynomials is a factor of the minimum polynomial of T.

10. Is there a linear transformation T of V^2 into itself over the field of real numbers such that $T^2 = -I$? If so, find one; if not, show why not.

11. Let a linear transformation T be defined by the following equations with respect to a basis α_1, α_2 of V^2:

$$T(\alpha_1) = a\alpha_1 + b\alpha_2 \qquad T(\alpha_2) = c\alpha_1 + d\alpha_2$$

Let $f(x) = x^2 - (a + d)x + ad - bc$. Show that $f(T) = Z$.

12. For Exercise 11, show that $f(x)$ is the minimum polynomial of T if and only if $T = eI$ is false for every number e.

13. If T is a linear transformation of V into itself, prove that its kernel is an invariant subspace.

14. Prove that if c is a characteristic root of the transformation T of V into itself, then c^k is a characteristic root of T^k.

15. Let $f(x)$ be a polynomial in x. Prove that if c is a characteristic root of T as defined in Exercise 14, then $f(c)$ is a characteristic root of $f(T)$.

16. Prove that the minimum polynomial of a transformation T is the least common multiple of its minimum polynomials with respect to all the vectors of V.

17. Let $f(T) = Z$ and $g(T) = Z$ for polynomials $f(x)$ and $g(x)$. Suppose $d(x)$ is the g.c.d. (greatest common divisor: see Appendix C) of $f(x)$ and $g(x)$. Prove that $d(x)$ is divisible by the minimum polynomial of T and hence that $d(T) = Z$.

3.8 THE MAXIMUM DEGREE OF THE MINIMUM POLYNOMIAL

We shall show in this section that if T is a linear transformation of V^n into itself, then it satisfies a polynomial equation of degree less than or equal to n. In other words, its minimum polynomial cannot be of degree greater than n. Before doing this, however, let us look at a particular example to explore how such a polynomial may be found.

Suppose $n = 4$ and $\alpha_1, \alpha_2, \alpha_3, \alpha_4$ is a basis of V. Let the following define the transformation T:

$$T(\alpha_1) = \alpha_2 \qquad T(\alpha_2) = \alpha_1 + \alpha_2 \qquad T(\alpha_3) = \alpha_4 + \alpha_1 \qquad T(\alpha_4) = \alpha_3 + \alpha_2$$

Notice first that, since $T(\alpha_1) = \alpha_2$ and $T(\alpha_2) = \alpha_1 + \alpha_2$, it follows that $W = \text{sp}(\alpha_1,\alpha_2)$ is an invariant subspace under T. Furthermore, while α_1 and $T(\alpha_1)$ are linearly independent, α_1, $T(\alpha_1)$, and $T^2(\alpha_1)$ are not as is shown by

$$T^2(\alpha_1) = \alpha_1 + \alpha_2 = \alpha_1 + T(\alpha_1)$$

Thus

$$(T^2 - T - I)(\alpha_1) = \theta$$

and therefore

$$[T^2(T\alpha_1) - T(T\alpha_1) - T\alpha_1] = (T^2 - T - I)(\alpha_2) = \theta$$

Thus $(T^2 - T - I)(\alpha)$ for all α in W, and, by Theorem 3.12a, $g(x) = x^2 - x - 1$ must be a factor of the minimum polynomial of T.

Let W' be the range of $g(T)$. If we can find another polynomial $f(x)$ such that $f(T)(\beta) = \theta$, for all vectors β in W', it will follow that $f(T)g(T)(\alpha) = \theta$, for all α in V, since $g(T)(\alpha)$ is in W' and $f(T)$ takes W' into θ. So first we calculate the range of $g(T)$.

$$\begin{aligned} g(T)(\alpha_3) &= -\alpha_1 + 2\alpha_2 - \alpha_4 \\ g(T)(\alpha_4) &= 2\alpha_1 - \alpha_3 = T(-\alpha_1 + 2\alpha_2 - \alpha_4) \end{aligned} \tag{3.22}$$

Since $-\alpha_1 + 2\alpha_2 - \alpha_4 = \beta_1$ and $2\alpha_1 - \alpha_3 = \beta_2$ are linearly independent and $g(T)(\alpha_1) = g(T)(\alpha_2) = \theta$, we see that the dimension of W' is 2, and we should be able to find a polynomial $f(x)$ of degree 2 or less such that $f(T)(\beta) = \theta$, for all β in W'. A little calculation shows

$$T^2(\beta_1) = \beta_1 \qquad \text{and} \qquad T^2(T\beta_1) - T\beta_1 = T^2(\beta_2) - \beta_2 = \theta$$

Thus $(T^2 - I)(\beta) = \theta$, for all β in W', and we can take $f(x)$ to be $x^2 - 1$.

Thus if $F(x) = (x^2 - 1)(x^2 - x - 1)$, it follows that

$$F(T)(\alpha) = \theta \qquad \text{for all } \alpha \text{ in } V$$

Furthermore, $F(x)$ must be the minimum polynomial since it is divisible by both $f(x)$ and $g(x)$, polynomials which have no common factor of positive degree.

Now we state the general theorem and prove it along the lines of the example just completed.

Theorem 3.14 Let T be a linear transformation of V^n into itself. There is a polynomial $h(x)$ of positive degree no greater than n such that $h(T) = Z$, where Z is the zero transformation.

Proof We prove this by induction on n. First, if $n = 1$, then $V = \text{sp}(\alpha)$, for some nonzero vector α. Then $T(\alpha) = c\alpha$ for some number c; we can choose $h(x)$ to be $x - c$ and see that $(T - cI)(\alpha) = \theta$, that is, $T - cI = Z$.

Now suppose the theorem holds for all spaces of dimension less than n. Following the pattern of the example above, choose some vector α, not the null vector, in V^n and let

$$\alpha,\ T(\alpha),\ T^2(\alpha),\ \ldots,\ T^{s-1}(\alpha) \tag{3.23}$$

be a linearly independent set of vectors with $s \geq 1$, while $T^s(\alpha)$ is a linear combination of the vectors of (3.23). (In the example, $s = 2$ and $\alpha = \alpha_1$.) Since the vectors in (3.23) are linearly independent and contained in V^n, we know that $s \leq n$. Let W be the space spanned by the vectors of (3.23). We know that its dimension is s, which is positive. Since, for some numbers c_i,

$$T^s(\alpha) - c_{s-1}T^{s-1}(\alpha) - c_{s-2}T^{s-2}(\alpha) - \cdots - c_1T(\alpha) - c_0\alpha = \theta$$

we let

$$g(x) = x^s - c_{s-1}x^{s-1} - \cdots - c_1x - c_0$$

and see that

$$g(T)(\alpha) = \theta \qquad T^jg(T)(\alpha) = g(T)[T^j(\alpha)] = \theta$$

for all j. Hence $g(T)(\alpha) = \theta$, for all α in W.

Now the kernel of $g(T)$ is of dimension at least s, and hence its range has dimension $m \leq n - s < n$ by Theorem 3.7. Still following the pattern of the example, we call W' the range of $g(T)$. That is, W' consists of all vectors ω' such that $\omega' = g(T)(\alpha)$, for some α in V^n. But

$$T(\omega') = g(T)[T(\alpha)]$$

for ω' in W', and, since $T(\alpha)$ is in V, $T(W') \subseteq W'$; that is, T takes the space W' into itself. From the example, the next step is to find a polynomial $f(x)$ such that $f(T)(\omega') = \theta'$, for all ω' in W'. But here we know of the existence of such an $f(x)$ from the induction hypothesis, since $\dim(W') < \dim(V)$. So there is a polynomial $f(x)$ such that $f(T)(\omega') = \theta$, for all ω' in W', and the degree of $f(x)$ is not greater than m.

Hence, for α any vector in V^n, we have

$$f(T)g(T)(\alpha) = f(T)[g(T)(\alpha)] = \theta$$

since $g(T)(\alpha)$ is in W', the image space of $g(T)$, and $f(T)$ takes each vector of W' into the null vector. Thus we can choose $h(x) = f(x)g(x)$ and see that its degree is not greater than

$$s + m \leq s + (n - s) \leq n$$

This completes the proof.

It should be noted that $h(x)$ is not necessarily the minimum polynomial. However, if it turns out that $f(x)$ and $g(x)$ have no common factor of positive degree and if $f(x)$ is the minimum polynomial for W', then their product must be the minimum polynomial. The proof of this statement is left as an exercise. Notice that, by the construction, $g(x)$ must be the minimum polynomial of W, since otherwise the set (3.23) would be linearly dependent. Furthermore, though this method of computing the minimum polynomial depends on the basis, we know from the Corollary of Theorem 3.12 that the minimum polynomial itself is independent of the basis.

So now with the minimum polynomial at hand, we have the means of finding the characteristic roots and hence the fixed subspaces of dimension 1. This, in its turn, gives us much information about the transformation.

3.9 TRANSFORMATIONS CHARACTERIZED BY THEIR CHARACTERISTIC VECTORS

We noted in Sec. 3.6 that fixed points (indeed, more generally, characteristic vectors) help to characterize a transformation. In Euclidean geometry, any Euclidean transformation with exactly one fixed point must be a rotation. But, on the other hand, a translation is not the only Euclidean transformation which has no fixed points. It turns out to be true that a large class of linear transformations are determined by their characteristic vectors. In this section we shall show that indeed this happens whenever the minimum polynomial can be expressed as a product of distinct linear factors. The converse is also true, and since it is easier to prove, we consider it first. We have

Theorem 3.15 Let T be a linear transformation of a vector space V into itself. If there is a set of characteristic vectors which span V, then the minimum polynomial of T is a product of distinct linear factors.

Proof Let $\alpha_1, \alpha_2, \ldots, \alpha_n$ be a linearly independent set of characteristic vectors which span V, that is, a basis consisting of characteristic vectors. Let $c_1, c_2, \ldots, c_k$ be the distinct characteristic roots. Now for any α_i there is a c_j such that $T(\alpha_i) = c_j\alpha_i$. This means that $x - c_j$ is the minimum polynomial of T with respect to α_i. Hence by the Corollary of Theorem 3.12a, $x - c_j$ is a factor of the minimum polynomial of T. Thus $m(x)$, the minimum polynomial of T, must be divisible by all $x - c_j$ and hence by

$$h(x) = (x - c_1)(x - c_2) \cdots (x - c_k)$$

It remains to show that $h(x) = m(x)$, that is, that $h(T)(\alpha) = \theta$, for all α in V. To simplify notation, suppose $T(\alpha_1) = c_1\alpha_1$. Then

$$h(T)(\alpha_1) = (T - c_2I)(T - c_3I) \cdots (T - c_kI)(T - c_1I)(\alpha_1) = \theta$$

Since each characteristic vector has a corresponding characteristic root and since the linear polynomials in T are commutative, we see that $h(T)(\alpha_i) = \theta$, for $1 \leq i \leq n$, and thus $h(T)(\alpha) = \theta$, for all α. This completes the proof by showing that $h(x) = m(x)$.

Before stating the other principal theorem of this section, we need another one. This is an extension of Theorem 3.11, and the proofs are closely analogous. In fact, it is suggested that the reader look back at the proof of Theorem 3.11 and see if, using that as a model, he can devise a proof of the following:

Theorem 3.16 Let $m(x)$ be the minimum polynomial of a linear transformation T of a vector space V into itself and let

$$m(x) = (x - c)g(x) \tag{3.24}$$

where $g(c) \neq 0$. Let W_1 be the kernel of $T - cI$ and W_2 the kernel of $g(T)$. Then

$$V = W_1 \oplus W_2$$

and W_1 is the range of $g(T)$, while W_2 is the range of $T - cI$. (Notice that by the definition of W_1 and W_2 they must be invariant under T.)

Proof We can divide $g(x)$ by $x - c$ to get a quotient $q(x)$ and a remainder d, which is in the ground field and is not zero since $d = g(c) \neq 0$. Then $g(x) = q(x)(x - c) + d$ may be written

$$-(x - c)q(x) + g(x) = d \neq 0 \tag{3.25}$$

For any vector α in V, (3.25) shows that

$$-(T - cI)q(T)(\alpha) + g(T)(\alpha) = d\alpha \tag{3.26}$$

or

$$(3.27) \qquad (T - cI)h(T)\alpha + eg(T)(\alpha) = \alpha$$

where $h(x) = -q(x)/d$ and $e = 1/d$. Since

$$g(T)[T - cI]h(T)(\alpha) = m(T)h(T)(\alpha) = h(T)m(T)(\alpha) = \theta$$

for all vectors α, we see that $(T - cI)h(T)(\alpha)$, call it ω_2, is in the kernel of $g(T)$; that is, ω_2 is in W_2. Also $(T - cI)[eg(T)(\alpha)] = em(T)(\alpha) = \theta$ implies that $g(T)(\alpha)$, call it ω_1, is in W_1, the kernel of $T - cI$. Thus (3.27) can be written $\omega_2 + \omega_1 = \alpha$, with ω_i in W_i, no matter what α is. This shows that $V = W_1 + W_2$.

To show that the sum is direct we must show that W_1 and W_2 are disjoint. To this end, let α' be a vector in $W_1 \cap W_2$. Then look at (3.27) with α' in place of α and see that α' in W_1 implies that

$$(T - cI)h(T)(\alpha') = h(T)(T - cI)(\alpha') = \theta$$

while α' in W_2 implies that $eg(T)(\alpha') = \theta$. Hence the left side of (3.27), with α' in place of α, is the null vector. This implies that the right side, namely, α', is also the null vector. This shows that W_1 and W_2 are disjoint.

If $r[g(t)]$ and $r(T - cI)$ denote the ranks of the respective transformations, the fact that the sum of the dimensions of W_1 and W_2 is n can be expressed as follows:

$$n - r[g(T)] + n - r(T - cI) = n$$

which implies that

$$r[g(T)] + r(T - cI) = n$$

Hence $r(T - cI) = n - r[g(T)] = \dim(W_2)$. Also $g(T)(T - cI) = Z$ implies that $(T - cI)(V) \subseteq W_2$. These together imply that $(T - cI)(V) = W_2$. Similarly, $g(T)(V) = W_1$. This completes the proof.

We have just shown that corresponding to a factorization of the minimum function we have what can be called a *splitting* of the vector space. We have concomitantly a splitting of the transformation as well. Let T_1 be the *restriction* of T to the space W_1 and T_2 the *restriction* of T to W_2. That is, T_1 is a transformation of W_1 into itself, which has the same effect on every vector of W_1 as T has; and similarly for T_2. Then, since $T(\alpha) = c\alpha$, for all α in W_1, $T_1 = cI$ on W_1. Also $g(T_2) = Z$ on W_2. Here we write $T = T_1 \oplus T_2$ and mean the following: To get the image of a vector α under the transformation T, first express α in the form $\omega_1 + \omega_2$, where ω_1 is in W_1 and ω_2 is in W_2; then apply T_1 to ω_1 and T_2 to ω_2 and, finally, add the resulting vectors. In symbols,

$$(T_1 \oplus T_2)(\omega_1 + \omega_2) = T_1(\omega_1) + T_2(\omega_2)$$

The transformation T is thus a kind of direct sum of two transformations. Notice that this is quite a different situation from the sum of two transformations. When we write $T_1 + T_2$, we think of both T's being applied to the same vector space; for $T_1 \oplus T_2$ to have meaning we must first of all have a splitting of the vector space V into the direct sum of two disjoint vector spaces, and each of these subspaces must be invariant under T. Thus we have shown part of the following

Theorem 3.17 Under the hypotheses of Theorem 3.16, there is a splitting also of the transformation T into a direct sum $T_1 \oplus T_2$, where T_1 is the restriction of T to W_1 and T_2 is the restriction to W_2. The minimum polynomial of T_1 is $x - c$, and the minimum polynomial of T_2 is $g(x)$.

Proof We need only prove the statement in the last sentence of the theorem. Since $T_1 - cI = Z$ on W_1, the minimum polynomial of T_1 is $x - c$. Suppose $m'(x) \neq g(x)$ were the minimum polynomial of T_2. Then $m'(x)$ would have to be a factor of $g(x)$, by Theorem 3.12, and hence of lower degree. We shall show that $(T - cI)m'(T) = Z$, and since $(x - c)m'(x)$ is of lower degree than $m(x)$, we arrive at a contradiction. To do this, write $\alpha = \omega_1 + \omega_2$, where ω_i is in W_i, for $i = 1,2$. Then $(T - cI)\omega_1 = (T_1 - cI)\omega_1 = \theta$ and $m'(T)\omega_2 = m'(T_2)\omega_2 = \theta$. Thus $(T - cI)m'(T)(\alpha) = \theta$, for every element of V, which shows that indeed $(T - cI)m'(T) = Z$. But $m'(x)$ cannot be of lower degree than $g(x)$ since $(x - c)g(x)$ is the minimum polynomial. This completes the proof.

With this preparation, the proof of our principal theorem is short.

Theorem 3.18 If the minimum polynomial of a linear transformation T of V into itself is a product of distinct linear factors, then there is a set of characteristic vectors which spans V. Furthermore, if the minimum polynomial is

$$m(x) = (x - c_1)(x - c_2) \cdots (x - c_r)$$

where the c's are distinct, then

$$V = W_1 \oplus W_2 \oplus \cdots \oplus W_r$$

where W_i is the kernel of $T - c_iI$, and

$$T = T_1 \oplus T_2 \oplus \cdots \oplus T_r$$

where T_i is T restricted to W_i.

Proof We prove the theorem by induction on the number of factors of $m(x)$. If $m(x) = x - c$, then $(T - cI)(\alpha) = \theta$, for all α in V, and every vector of V will be a characteristic vector of T, thus establishing our result. Now assume the theorem for $r - 1$ distinct linear factors, and let $m(x) = (x - c)g(x)$, where $g(x)$ is a product of distinct linear factors, none of which is equal to $x - c$. Then, by Theorems 3.16 and 3.17, we have a splitting of V such that W_0 is the kernel of $T - cI$, W' is the kernel of $g(T)$, and

$$V = W_0 \oplus W' \qquad T = T_0 \oplus T'$$

Then any basis of W_0 will consist of characteristic vectors of T associated with the characteristic root c. Now $g(x)$, the minimum polynomial of T', has fewer linear factors than $m(x)$. Thus we know by the hypothesis of the induction that W' is spanned by a set of characteristic vectors of T'. Hence V is spanned by a set of characteristic vectors of T. The splitting of W' and T' follows also from the induction hypothesis. This completes the proof.

Let us see what this means in terms of a specific example. Suppose V is of dimension 4 and the minimum polynomial of T is

$$m(x) = (x - 1)(x + 1)(x - 2)$$

Then if W_1 is the kernel of $T - I$, W_2 is the kernel of $T + I$, and W_3 is the kernel of $T - 2I$, we know that

$$V = W_1 \oplus W_2 \oplus W_3$$

Since 4, the dimension of V, is the sum of the dimensions of the W's, we know that one must have dimension 2 and the other two, dimension 1. Suppose that W_1 has dimension 2 and W_2 and W_3 have dimension 1. Then W_1 has a basis of two vectors: α_1 and α_2. For each of these, $(T - I)(\alpha_i) = \theta$ implies $T(\alpha_i) = \alpha_i$; that is, α_1 and α_2 are characteristic vectors associated with the characteristic root 1. Similarly W_2 and W_3 will be spanned by the single vectors α_3 and α_4, respectively. Thus T can be defined as follows:

$$T(\alpha_1) = \alpha_1 \qquad T(\alpha_2) = \alpha_2 \qquad T(\alpha_3) = -\alpha_3 \qquad T(\alpha_4) = 2\alpha_4$$

The vectors α_1, α_2 are linearly independent since we chose them that way. Also α_3 is not a linear combination of α_1 and α_2 since W_1 and W_2 are disjoint. Furthermore, α_4 is not a linear combination of α_1, α_2, α_3 since W_3 is disjoint from $W_1 \oplus W_2$. So the α_i are linearly independent and span the space V.

We can combine Theorem 3.15 with part of Theorem 3.18 to give

Theorem 3.19 Let T be a linear transformation of a vector space V into itself. The minimum polynomial of T is a product of distinct linear factors if and only if there is a set of characteristic vectors of T which span V.

We should look a little more carefully at the subspaces W_i occurring in Theorem 3.18. For instance, W_1 is the kernel of $T - c_1I$, that is, is the set of all vectors α such that $T(\alpha) = c_1\alpha$. The dimension of W_1 is the nullity of $T - c_1I$. So a basis of W_1 contributes as many vectors to the basis of V

as its dimension. The same is true for each W_i. Furthermore, since the W_i are disjoint in pairs, that is, $W_i \cap W_j = \theta$ for $i \neq j$, no vector in one of them can be a linear combination of vectors in the others. That is, if we choose a set $\mathcal{S}$ of vectors, which is the union of sets of basis vectors of the several W_i, then $\mathcal{S}$ is a linearly independent set and is a basis of V.

There is a rather curious theorem whose proof we leave as an exercise.

Theorem 3.20 Let $m(x)$ be the minimum polynomial of a transformation T of a vector space V into itself. Let $h(x)$ be any polynomial whose coefficients are in the ground field, and let $g(x)$ be the greatest common divisor of $m(x)$ and $h(x)$. Then the ranks of $h(T)$ and $g(T)$ are the same.

EXERCISES

In these exercises, T is a linear transformation of a vector space V of dimension n into itself.

1. Let $V = V^2$ have a basis α_1, α_2 and let $T(\alpha_1) = -\alpha_2$, $T(\alpha_2) = \alpha_1 + 2\alpha_2$. Find the minimum polynomial of T and show that V has no basis consisting of characteristic vectors.

2. With the notation of Exercise 1, let $T(\alpha_1) = -2\alpha_2$, $T(\alpha_2) = \alpha_1 + \alpha_2$. Does this T have a basis of characteristic vectors? If so, find such a basis.

3. For Exercise 1 show that there is a basis β_1, β_2 of V such that $T(\beta_1) = \beta_1$ and $(T - I)(\beta_2) = \beta_2$.

4. Suppose for $V = V^3$ and a basis α_i we have $T(\alpha_1) = \alpha_2$, $T(\alpha_2) = \alpha_3$, and $T(\alpha_3) = \theta$. Show that there is a set of characteristic vectors spanning the space and find such a set. What is the minimum polynomial?

5. Suppose that $n = 4$ and that the minimum polynomial of T is $x^2 - 1$. What are all the possibilities for bases of V consisting of characteristic vectors?

6. Find the minimum polynomial of the transformation T, defined below, in terms of a basis of V^3.

$$T(\alpha_1) = \alpha_2 \qquad T(\alpha_2) = \theta \qquad T(\alpha_3) = \alpha_1 + \alpha_2 - \alpha_3$$

Is there a basis consisting of characteristic vectors? If so, find it.

7. Answer the same questions as those in Exercise 6 for the following transformation of V^3 into itself:

$$T(\alpha_1) = \alpha_2 \qquad T(\alpha_2) = \alpha_1 \qquad T(\alpha_3) = \alpha_1 - \alpha_2 + \alpha_3$$

8. Suppose $n = 3$ and the minimum polynomial of T is $x^2(x - 1)$. Show that there is a basis α_1, α_2, α_3 of V such that $T^2(\alpha_1) = T^2(\alpha_2) = \theta$ and $T(\alpha_3) = \alpha_3$.

9. Suppose the minimum polynomial of T over V^3 is $(x - 1)(x + 1)^2$. Prove that V^3 is spanned by the vectors in the kernels of $T - I$ and $(T + I)^2$.

10. Suppose $T(\alpha) = c\alpha$ and $T(\beta) = d\beta$, with $c \neq d$. Prove that α and β are linearly independent.

11. Generalize the result of Exercise 10 for k distinct characteristic roots.

12. Prove that if the minimum polynomial of T is of degree n and if it is a product of distinct linear factors, then V is the direct sum of n vector spaces, each of dimension 1.

13. Suppose T has two characteristic roots a and b and that the following is a basis of the space V:

$$\alpha_1, \alpha_2, \ldots, \alpha_r, \beta_1, \beta_2, \ldots, \beta_s$$

where $T(\alpha_i) = a\alpha_i$ and $T(\beta_i) = b\beta_i$. Prove that $(x - a)(x - b)$ is the minimum polynomial of T.

14. Carry through the proof for the case like Exercise 13, where now T has three characteristic roots and V is spanned by a set of characteristic vectors.

15. Suppose W and W' are two subspaces of V, and $g(x)$ and $g'(x)$ are two polynomials whose greatest common divisor is 1, such that $g(T)W = \theta$, $g'(T)W' = \theta$. Prove that the product $g(x)g'(x)$ is a factor of the minimum polynomial of T. Also prove that if the sum of the degrees of $g(x)$ and $g'(x)$ is n, then their product *is* the minimum polynomial of T except perhaps for a constant factor.

16. Let S be a nonsingular transformation of V into itself and T a transformation of V into itself. Prove that T and STS^{-1} have the same minimum polynomial and the same characteristic roots. Do they have the same characteristic vectors?

17. Find a linear transformation T of V^2 into itself whose minimum polynomial is $(x - 1)^2$. In this case is V^2 spanned by a set of characteristic vectors of T?

18. Prove Theorem 3.20. (Hint: Use Theorem C.2p in Appendix C.)

3.10 LINEAR TRANSFORMATIONS AND MATRICES

So far in this chapter we have made no mention of matrices because it is, in general, easier to deal with transformations as concepts independent of the idea of a matrix. But there is an intimate connection, and, as is usual in such circumstances, both theories benefit from establishing this relationship.

Suppose V is a vector space of dimension n with a basis $\alpha_1, \alpha_2, \ldots, \alpha_n$, and W is a vector space of dimension s with a basis $\beta_1, \beta_2, \ldots, \beta_s$. Then a transformation T taking V into W will express the image of each basis vector of V as a linear combination of the basis vectors of W. Specifically,

$$
\begin{aligned}
T(\alpha_1) &= t_{11}\beta_1 + t_{21}\beta_2 + \cdots + t_{s1}\beta_s \\
T(\alpha_2) &= t_{12}\beta_1 + t_{22}\beta_2 + \cdots + t_{s2}\beta_s \\
&\cdots\cdots\cdots\cdots\cdots\cdots \\
T(\alpha_n) &= t_{1n}\beta_1 + t_{2n}\beta_2 + \cdots + t_{sn}\beta_s
\end{aligned}
\tag{3.28}
$$

Notice that we have reversed the subscripts on the t's so that the second subscript corresponds to the row number and the first to the column number. The reason for this will soon be apparent.

We have seen (Sec. 1.9) that any vector can be represented as a row (or column) matrix. That is, if

$$\xi = x_1\alpha_1 + x_2\alpha_2 + \cdots + x_n\alpha_n$$

is any vector of V, we may represent it by the row matrix

$$(x_1, x_2, \ldots, x_n)$$

where the x_i are the components of ξ relative to the given basis. If the image of ξ under T is η, it can be written

$$\eta = y_1\beta_1 + y_2\beta_2 + \cdots + y_s\beta_s$$

and the row matrix of its components is $(y_1, y_2, \ldots, y_s)$. So now we seek a relationship in terms of the t_{ij} between the components of η and the components of ξ.

Using (3.28) above, and collecting coefficients of the β's on the right, we have

$$
\begin{aligned}
T(x_1\alpha_1 + x_2\alpha_2 + \cdots + x_n\alpha_n) = {} & (t_{11}x_1 + t_{12}x_2 + \cdots + t_{1n}x_n)\beta_1 \\
& + (t_{21}x_1 + t_{22}x_2 + \cdots + t_{2n}x_n)\beta_2 + \cdots \\
& + (t_{s1}x_1 + t_{s2}x_2 + \cdots + t_{sn}x_n)\beta_s
\end{aligned}
$$

Since y_i is the coefficient of β_i in η, we have

$$
\begin{aligned}
y_1 &= t_{11}x_1 + t_{12}x_2 + \cdots + t_{1n}x_n \\
y_2 &= t_{21}x_1 + t_{22}x_2 + \cdots + t_{2n}x_n \\
&\cdots\cdots\cdots\cdots\cdots\cdots \\
y_s &= t_{s1}x_1 + t_{s2}x_2 + \cdots + t_{sn}x_n
\end{aligned}
\tag{3.29}
$$

Using the definition of multiplication of matrices, this may be expressed briefly as

$$\eta^T = M(T)\xi^T \tag{3.30}$$

where $M(T) = (t_{ij})$ is the matrix of the set of Equations (3.29). We call $M(T)$ the *matrix of the transformation* T. Notice that $M(T)$ is the transpose of the matrix of coefficients in the set of Equations (3.28). In fact, we reversed the subscripts in Equations (3.28) so that in Equations (3.29) the first subscripts of the t's indicate the row, and the second the column. Also we are really using the symbols ξ and η in two different senses: initially as vectors which are linear combinations of the basis vectors and, in (3.30), as n-tuples. To avoid this we could write $M(\eta)$ instead of η in (3.30), but the context as well as the superscript T should make it clear that n-tuples are intended; that is, if part of an equation is clearly a matrix, the rest must be matrices as well, or scalars.

Thus, given bases of V and W, there is a unique matrix M associated with the linear transformation T from V into W. It is important to note that $M(T)$ depends not only on T but on the bases chosen for V and W. We sometimes emphasize this by writing $M(T,\mathcal{B}_1,\mathcal{B}_2)$, where $\mathcal{B}_1$ and $\mathcal{B}_2$ designate, respectively, the bases of V and W relative to which the matrix is found.

Conversely, given any matrix with s rows and n columns, then, for any bases of V and W, there will be a linear transformation of V into W defined by this matrix. So we have a one-to-one correspondence between matrices and transformations, once the bases have been decided upon. This extends, moreover, to the operations on matrices and transformations. We list these correspondences below, prove some of them, and leave the others for exercises:

1. $M(T_1 + T_2) = M(T_1) + M(T_2)$
2. $cM(T) = M(cT)$
3. $M(T_2 \cdot T_1) = M(T_2)M(T_1)$
4. $M(Z) = (0)$
5. $M(I) = I$
6. $[M(T)]^{-1} = M(T^{-1})$

where in correspondence 5 we use the same letter for the transformation I, on the left, and the matrix I, on the right. Also in correspondences 5 and 6 the *transformation must be of V into itself.* In this case we assume that the *bases for the domain and range are the same.* We develop this a little further below.

It is understood that in correspondences 1, 2, and 4 the bases of V and W are fixed throughout and the matrices are determined with reference to the same sets of bases. The proofs of these are left as exercises.

The others require more discussion. Let us begin with correspondence 3. Here we need to exercise a little care, first about the order of multiplication and second about the spaces involved. Recall (Sec. 3.3) that when we write $T_2T_1(\alpha)$ we mean that T_1 is first applied to α, and then T_2 is applied to the result. That is, the transformation on the right is applied first. Second, if T_1 takes V of dimension n into W of dimension s, then T_2 must be a transformation of W into a space, say, U, of dimension which we may call u. Then, having chosen bases for all three spaces, we see that $M(T_1)$ has s rows and n columns, while $M(T_2)$ has u rows and s columns. Hence the product $M(T_2)M(T_1)$ exists. Let

$$\begin{aligned} \xi &= x_1\alpha_1 + x_2\alpha_2 + \cdots + x_n\alpha_n \\ \eta &= y_1\beta_1 + y_2\beta_2 + \cdots + y_s\beta_s \\ \zeta &= z_1\gamma_1 + z_2\gamma_2 + \cdots + z_u\gamma_u \end{aligned}$$

where ξ is a vector in V, η its image in W, and ζ the image of η in U. Let $M(T_1) = (t_{ij})$ and $M(T_2) = (t'_{ij})$. Then

$$y_i = \sum_{j=1}^{n} t_{ij}x_j \qquad \text{and} \qquad z_k = \sum_{i=1}^{s} t'_{ki}y_i$$

where $1 \leq i \leq s$ and $1 \leq k \leq u$. Combining these we have

$$z_k = \sum_{i=1}^{s} t'_{ki} \sum_{j=1}^{n} t_{ij}x_j = \sum_{j=1}^{n} \left(\sum_{i=1}^{s} t'_{ki}t_{ij} \right) x_j$$

Thus the coefficient of x_j corresponding to z_k is

$$\sum_{i=1}^{s} t'_{ki}t_{ij}$$

which is just the element in the kth row and jth column of the matrix product $M(T_2)M(T_1)$. Thus the matrix of z_k in terms of x_j is the product of the matrix of z_k in terms of y_i on the left by that of y_i in terms of x_j on the right.

It is worthwhile to follow this through in matrix form, even though this does not constitute a proof since, up to this time, we have not shown that products of matrices are associative. In matrix form it is quite simple:

$$\eta^T = M(T_1)\xi^T \qquad \zeta^T = M(T_2)\eta^T$$

imply, by substituting the first equation in the second, that

$$\zeta^T = M(T_2)M(T_1)\xi^T$$

Notice that the order T_2T_1 is the same as the order $M(T_2)M(T_1)$. In both cases the transformation T_1 and the matrix $M(T_1)$ are applied first.

Before looking at correspondence 5, let us notice an immediate consequence for matrices of correspondence 3 for transformations, namely, the associative property for multiplication of matrices. To see this first recall the associative property for transformations:

$$(T_3T_2)T_1 = T_3(T_2T_1) \tag{3.31}$$

Then, from correspondence 3, we have

$$\begin{aligned} M[(T_3T_2)T_1] &= M(T_3T_2)M(T_1) = [M(T_3) \cdot M(T_2)]M(T_1) \\ M[T_3(T_2T_1)] &= M(T_3)M(T_2T_1) = M(T_3)[M(T_2) \cdot M(T_1)] \end{aligned}$$

But (3.31) shows that

$$M[(T_3T_2)T_1] = M[T_3(T_2T_1)]$$

Hence multiplication of matrices is associative.

For correspondence 5, let T_I denote the identity transformation. This has the property that it takes each vector into itself. This cannot happen unless $T_I(V) = V$, that is, unless the domain and range are the same. Then if we choose α_i to be a basis for both V and $T_I(V)$, we have

$$T_I(\alpha_i) = \alpha_i \qquad \text{for } 1 \leq i \leq n$$

If (t_{ij}) is the matrix of this transformation with respect to the basis, we see that the t_{ij} are determined by $\sum_{j=1}^{n} t_{ij}\alpha_j = \alpha_i$, and hence

$$t_{ij} = \begin{cases} 1 & \text{if } i = j \\ 0 & \text{if } i \neq j \end{cases} \tag{3.32}$$

Then the matrix of T_I is the identity matrix. It is convenient in this connection to define the so-called *Kronecker delta*, δ_{ij}. This symbol is defined to be 1 if $i = j$ and 0 if $i \neq j$. Thus condition (3.32) can be written $t_{ij} = \delta_{ij}$.

Notice that to get the identity matrix for T_I we not only have to assume that $T_I(V) = V$ but we must choose the bases of the domain and range to be the same. For instance, suppose in V^2 we take α_1, α_2 to be a basis for V and $\beta_1 = \alpha_1 + \alpha_2$, $\beta_2 = \alpha_1 - \alpha_2$ to be a basis for $T_I(V)$. Then, since $\alpha_1 = \frac{1}{2}(\beta_1 + \beta_2)$ and $\alpha_2 = \frac{1}{2}(\beta_1 - \beta_2)$, the matrix of T_I is

$$\begin{bmatrix} \frac{1}{2} & \frac{1}{2} \\ \frac{1}{2} & -\frac{1}{2} \end{bmatrix}$$

So, in general, we have the understanding that whenever $T(V) \subseteq V$, we express the images of the vectors of V as linear combinations of the basis of V.

With regard to correspondence 6, we make the same assumptions as above; that is, T is a linear transformation of V into itself and the images of the elements of V are linear combinations of the basis elements chosen for V. Then correspondence 6 is established by the following sequence of equations:

$$M(T^{-1})M(T) = M(T^{-1}T) = M(I) = I$$

and $M(T^{-1})$ is the inverse of $M(T)$.

We have seen that the operations on matrices and transformations correspond. This is not strange since we set them up that way. It is true that the notion of rank corresponds as well. We state this as a theorem:

Theorem 3.21 If T is a linear transformation of V into W and $M(T)$ is its matrix relative to some bases of V and W, then the rank of T is equal to the rank of $M(T)$. Also the nullities of T and $M(T)$ are equal.

Proof Recall (Sec. 3.4) that the rank of T is the dimension of its range, or image space, and that the rank of a matrix is the dimension of its column space, which is the same as the dimension of its row space (Sec. 2.7). It is best to go back to Equations (3.28). Suppose that a linear combination of the images of $\alpha_1, \alpha_2, \ldots, \alpha_n$ is the null vector as follows:

$$\sum_{i=1}^{n} c_i T(\alpha_i) = \theta$$

Then

$$\sum_{i=1}^{n} \sum_{j=1}^{s} c_i t_{ji} \beta_j = \theta$$

But the β's are linearly independent, and hence the coefficient of each β_j must be zero; that is, if we let $\gamma_i{}^T$ denote the ith column of $M(T)$, we have

$$\sum_{i=1}^{n} c_i \gamma_i{}^T = \theta$$

This means that if any linear combination of the elements $T(\alpha_i)$ of the range of T is the null vector, then the same linear combination of the columns of $M(T)$ will also be the null vector. Thus, if any subset of $\{T\alpha_i\}$ is linearly independent, so is the corresponding subset of γ_i. The same holds for linear dependence. So

$$\text{Rank } T = \text{column rank } M(T) = \text{rank } M(T)$$

and

$$\text{Nullity } T = \text{nullity } M(T)$$

Thus Theorem 3.7 implies Theorem 2.13. We can also deduce from Theorem 3.9 that the rank of the product of two matrices cannot be greater than the rank of either of them.

Recall that in Sec. 2.10 we discussed direct sums of matrices, while in Sec. 3.9 we described a direct sum of transformations. It is not hard to see that, by a proper choice of basis, the matrix of a direct sum of transformations becomes the direct sum of the matrices of the transformations. For this, refer to Theorem 3.17 where $V = W_1 \oplus W_2$ and $T = T_1 \oplus T_2$, with T_i the restriction of T to W_i, for $i = 1,2$. Then if α_i, with $1 \leq i \leq r$, is a basis of W_1 and α_j, with $r + 1 \leq j \leq n$, is a basis of W_2, the complete set of α's is a basis of V, and the transformation T with respect to this basis can be expressed as follows:

$$T(\alpha_i) = \sum_{k=1}^{r} t_{ki}\alpha_k + \sum_{u=r+1}^{n} 0 \cdot \alpha_u \qquad \text{for } 1 \leq i \leq r$$

$$T(\alpha_j) = \sum_{k=1}^{r} 0 \cdot \alpha_k + \sum_{u=r+1}^{n} t_{uj}\alpha_u \qquad \text{for } r + 1 \leq j \leq n$$

Then $M_1 = (t_{ik})$ and $M_2 = (t_{ju})$, where i, j, k, u have the ranges indicated above, are the matrices of T_1 and T_2, respectively, and $M_1 \oplus M_2$ is the matrix of T. This can easily be extended to the more general situation described in Theorem 3.18.

3.11 CHANGE OF BASIS

We have noted that the matrix of a transformation depends on the bases of the two spaces involved. If one changes the basis, this should not change the transformation itself, though it will change the form in which it is expressed. In this section we shall explore the effect on the matrix of a change in basis. But first we should remark on two different points of view with regard to a change of basis. To be specific, consider a translation in analytic Euclidean geometry:

$$x' = x + h \qquad y' = y + k$$

where (x',y') is the image of (x,y). One point of view is that one "moves" every point h units to the right and k units upward. This could be characterized vaguely by the word *alibi*—a change of location. Another point of view would be that we leave the points where they are but merely move the axes of reference h units downward and k units to the left. This is an *alias*—a change in name. Sometimes one point of view is more useful and sometimes the other, but the algebra involved is the same for both.

Now let $\mathcal{A} = (\alpha_1, \alpha_2, \ldots, \alpha_n)$ and $\mathcal{A}' = (\alpha_1', \ldots, \alpha_n')$ be two bases for V. Then a vector of V will have one set of components $(x_1, x_2, \ldots, x_n)$ with respect to the basis $\mathcal{A}$ and another set $(x_1', x_2', \ldots, x_n')$ with respect to the basis $\mathcal{A}'$. Here we are using the alias point of view by giving the same vector two different names with respect to the two bases. Then, as we saw in (3.30), there is a matrix R such that

$$\xi'^T = R\xi^T$$

where ξ' and ξ are the n-tuples consisting of the x_i and x_i', respectively.

Similarly if $\mathcal{B} = (\beta_1, \beta_2, \ldots, \beta_s)$ and $\mathcal{B}' = (\beta_1', \beta_2', \ldots, \beta_s')$ are two bases of W, and η' and η the corresponding s-tuples relative to the two bases, we have

$$\eta'^T = S\eta^T$$

Furthermore R and S are nonsingular matrices since their ranks are n and s, respectively. Now R is a matrix of a transformation of V onto V and S is a matrix of a transformation of W onto W.

If T is a linear transformation of V into W, it has a matrix M with respect to the bases $\mathcal{A}$ and $\mathcal{B}$, respectively. (We here take the alibi point of view and consider it moving the vectors of V into vectors of W. In fact, if $V \neq W$, we *must* take this point of view.) Then, in terms of the components,

$$\eta^T = M\xi^T \tag{3.33}$$

But $\xi'^T = R\xi^T$ and $\eta'^T = S\eta^T$ imply $\xi^T = R^{-1}\xi'^T$ and $\eta^T = S^{-1}\eta'^T$; substitution of these into (3.33) yields

$$(\eta')^T = (SMR^{-1})(\xi')^T \tag{3.34}$$

Thus SMR^{-1} is the matrix of the transformation T with reference to the bases $\mathcal{A}'$ and $\mathcal{B}'$. We state this formally as

Theorem 3.22 If M is the matrix of a transformation T of V into W with respect to bases $\mathcal{A}$ and $\mathcal{B}$ of V and W, respectively, then for any other pair of bases $\mathcal{A}'$ and $\mathcal{B}'$ of V and W, the matrix of T is

$$SMR^{-1}$$

where R is the matrix of the transformation taking $\mathcal{A}$ onto $\mathcal{A}'$ and S that taking $\mathcal{B}$ onto $\mathcal{B}'$.

Corollary If, in the notation of the theorem, $W = V$ and $\mathcal{A} = \mathcal{B}$, $\mathcal{A}' = \mathcal{B}'$, it follows that the matrix of the transformation relative to the basis $\mathcal{A}'$ is RMR^{-1}. (This follows since $S = R$.)

The point in changing bases is that often it is possible to simplify the matrix of a transformation by judicious choice of a basis. For instance, suppose T is a linear transformation of V of dimension n into W of dimension r. As in the Proof of Theorem 3.7 we let

$$\gamma_1, \gamma_2, \ldots, \gamma_t \qquad \alpha_1, \alpha_2, \ldots, \alpha_s \qquad \text{where } t + s = n \tag{3.35}$$

be a basis of V, where the α's are the basis of its kernel. Then choose

$$\beta_i = T(\gamma_i) \qquad \text{for } 1 \leq i \leq t$$

and complete the basis of W by $\beta_{t+1}, \beta_{t+2}, \ldots, \beta_r$. Then we have

$$\begin{aligned} T(\gamma_i) &= \beta_i && \text{for } 1 \leq i \leq t \\ T(\alpha_i) &= \theta && \text{for } 1 \leq i \leq s \end{aligned}$$

and the matrix $M(T)$ with r rows and n columns takes the form

$$\begin{bmatrix} I & 0_1 \\ 0_2 & 0_3 \end{bmatrix} \tag{3.36}$$

where I is the identity matrix of order t, and the 0_i are zero matrices such that 0_1 has t rows and $n - t$ columns, 0_2 has $r - t$ rows and t columns, and 0_3 has $r - t$ rows and $n - t$ columns.

It should be noticed that when T is a linear transformation of V into itself, it is natural to express both the vectors of the domain and the range in terms of the same basis $\mathcal{A}$. (See the correspondence between the identity matrix and identity transformation in Sec. 3.10.) In fact, if T^2 and higher powers of T are to have meaning, the domain and range must be the same (that is, T must be square), and if $M(T^2) = [M(T)]^2$ the same basis must be used for the domain and range. Then, as noted above, if R is the matrix of the transformation taking the basis $\mathcal{A}$ into the basis $\mathcal{A}'$, the matrix M is replaced by RMR^{-1}. In this case it is not always possible to choose R so that RMR^{-1} is a diagonal matrix as you are asked to show in Exercise 6 below. But there is an important class of transformations which, for a proper choice of basis, have diagonal matrices. We showed in Theorem 3.19 that if the minimum polynomial of a transformation T is a product of distinct linear factors, then there is a set of characteristic vectors of T which span the space. So we can use Theorem 3.19 to prove the following:

Theorem 3.23 A transformation T of a vector space V into itself has a minimum polynomial which is a product of distinct linear factors if and only if one may choose a basis of V so that the matrix of T with respect to this basis is diagonal.

Proof From Theorem 3.19 we know that T has a minimum polynomial which is a product of distinct linear factors if and only if it has a set of characteristic vectors which span V. So first we must prove that if there is a set of characteristic vectors which span V, then we can choose a basis of V so that the matrix of T with respect to this basis is diagonal. The basis we choose is the spanning set of characteristic vectors, α_i, for $1 \leq i \leq n$. Then $T(\alpha_i) = c_i\alpha_i$, which shows that for this choice the matrix of T is the diagonal matrix whose elements along the diagonal are $c_1, c_2, \ldots, c_n$.

Second, suppose the matrix of T with respect to a basis $\beta_1, \beta_2, \ldots, \beta_n$ is diagonal. Then $T(\beta_i) = c_i\beta_i$, where the c_i are the elements of the diagonal matrix. Thus the β_i are characteristic vectors. This completes the proof.

EXERCISES

1. Let the matrix of a transformation T with respect to the canonical basis of V^2 be

$$M = \begin{bmatrix} 0 & -2 \\ 1 & 1 \end{bmatrix}$$

Find the minimum polynomial and characteristic vectors of T. Let C be the matrix whose columns are the linearly independent characteristic vectors of T. Show that C is nonsingular and $C^{-1}MC$ is a diagonal matrix.

2. Do Exercise 1 with M replaced by

$$\begin{bmatrix} 1 & 2 \\ 2 & 4 \end{bmatrix}$$

3. Let the matrix of T with respect to the canonical basis of V^2 be

$$M = \begin{bmatrix} -1 & -4 \\ 1 & 3 \end{bmatrix}$$

Find a characteristic vector α of M and a characteristic vector β of $M - I$ where α and β are linearly independent. Then let C be the matrix whose columns are α and β and find $C^{-1}MC$.

4. If M in Exercise 3 is the matrix of T with respect to the canonical basis, what is the matrix of the same transformation with respect to the basis $\epsilon_1 + \epsilon_2$, $\epsilon_1 - \epsilon_2$?

5. Do Exercise 4 for the matrix in Exercise 2.

6. Suppose that the matrix of a linear transformation T with respect to some basis of V^2 is

$$M = \begin{bmatrix} 0 & -1 \\ 1 & 2 \end{bmatrix}$$

Prove that there is no nonsingular matrix C such that $C^{-1}MC = D$ is a diagonal matrix.

7. Let a transformation T of V^3 into itself have the following matrix with respect to the canonical basis $\epsilon_1, \epsilon_2, \epsilon_3$.

$$M(T) = \begin{bmatrix} 1 & 1 & -\frac{2}{3} \\ -1 & 0 & \frac{4}{3} \\ 1 & 2 & -1 \end{bmatrix}$$

Compute $\beta_1 = T(\epsilon_1)$, $\beta_2 = T^2(\epsilon_1)$, $\beta_3 = T^3(\epsilon_1)$ and show that the β_i are linearly independent. Find the matrix of T with respect to the basis β_i.

8. In Exercise 7 replace the matrix of T by

$$M(T) = \begin{bmatrix} 7 & 4 & -4 \\ 4 & -8 & -1 \\ -4 & -1 & -8 \end{bmatrix}$$

Are the vectors β_i linearly independent?

9. Using the matrix given in Exercise 8, compute $\gamma_1 = T(\epsilon_2)$, $\gamma_2 = T^2(\epsilon_2)$, $\gamma_3 = T^3(\epsilon_2)$. Are the γ_i linearly independent? If so, find the matrix of T with respect to the basis γ_i.

10. Let T be a linear transformation of V^3 into W. Let $\alpha_1, \alpha_2, \alpha_3$ be a basis of V and β_1, β_2 a basis of W. Define T by

$$T(\alpha_1) = \beta_1 \qquad T(\alpha_2) = \theta \qquad T(\alpha_3) = \beta_1 + \beta_2$$

Find the matrix of T with respect to these two bases.

11. Let α_1, α_2 and β_1, β_2 be two bases for V^2 and suppose

$$T(\alpha_1) = \alpha_1 + \alpha_2 \qquad T(\alpha_2) = 2\alpha_1 + 3\alpha_2$$

is the transformation T expressed in terms of the basis α_1, α_2. What is its matrix expressed in terms of the basis β_1, β_2 if

$$\alpha_1 = \beta_1 - \beta_2 \qquad \alpha_2 = 2\beta_1 + 2\beta_2$$

12. Prove correspondences 1, 2, and 4 of Sec. 3.10.

13. In matrix (3.36), there is the tacit assumption that $n \geq t$ and $r \geq t$. Why must these inequalities hold?

14. Let $\alpha_1, \alpha_2, \alpha_3$ be a basis of a vector space V of dimension 3 over the field of complex numbers. Find the characteristic roots of the following transformation and the corresponding characteristic vectors. Find a basis of V for which the matrix of T is a diagonal matrix:

$$T(\alpha_1) = \alpha_2 \qquad T(\alpha_2) = \alpha_3 \qquad T(\alpha_3) = \alpha_1$$

3.12 SIMILAR TRANSFORMATIONS AND THE TRACE

Suppose T is a linear transformation of a vector space V into itself. We observed at the end of Sec. 3.8 that the minimum polynomial of T is independent of the particular basis chosen. That is, we can speak of the *minimum polynomial of* T without reference to the basis. The characteristic roots of T are also independent of the basis since if $T(\alpha) = c\alpha$, the equality will still hold, no matter what name is given to the vector α, that is, no matter what basis is used to express it. Since we will need matrices to deal with another important invariant, we shall now state the above result and prove it using matrices.

Theorem 3.24 The characteristic roots of a transformation are independent of the basis used, provided that the basis for the domain is the same as that for the range.

Proof From the Corollary of Theorem 3.22 we see that if the same basis is used for the range as for the domain, a change of basis replaces the matrix M, of T relative to one basis, with RMR^{-1} relative to another basis, where R is a nonsingular matrix. Then, $M\alpha = c\alpha$ implies

$$MR^{-1}R\alpha = c\alpha \qquad \text{and} \qquad RMR^{-1}(R\alpha) = c(R\alpha)$$

Thus the characteristic vector has the name $R\alpha$, in terms of the new basis, but the characteristic root remains the same.

An interesting consequence of this, whose proof we leave as an exercise, is the following

Theorem 3.25 If $M - cI$ is a singular matrix, so is $RMR^{-1} - cI$.

There is another useful invariant of a transformation. It is most conveniently defined by means of a matrix but turns out to be independent of the basis chosen. Recall that the trace of a square matrix is the sum of the elements along its principal diagonal (Sec. 2.5, Exercise 10). We make the following corresponding definition:

Definition If T is a linear transformation of V into itself and if $M(T)$ is its matrix with reference to a certain basis, then we call the *trace* of T, written $\text{tr}(T)$, the trace of the matrix $M(T)$.

The wording of the definition assumes something we must prove, namely, the following

Theorem 3.26 The trace of a linear transformation of a vector space into itself is independent of the basis chosen (provided, as always, that the domain and range have the same basis).

Proof From Exercise 10 of Sec. 2.5 we see that $\text{tr}(AB) = \text{tr}(BA)$ for any two matrices. Hence

$$\text{tr}(CMC^{-1}) = \text{tr}(C^{-1}CM) = \text{tr}(M)$$

There is one application of the trace which we state as a theorem, sketching the proof and leaving the completion as an exercise.

Theorem 3.27 If a transformation T is idempotent (that is, $T^2 = T$), then its trace is equal to its rank provided that the characteristic of the ground field is zero (see Appendix A).

A Sketch of the Proof Recall that we showed in Sec. 3.5 that such a transformation is a projection. Now $x^2 - x$ is the minimum polynomial of such a T unless $T = 0$ or $T = I$, and the theorem is easily verified for these two cases. Since, for all other cases, the minimum polynomial is a product of two distinct linear factors, we can apply Theorem 3.23 to see that there is a basis consisting of characteristic vectors associated with the roots 1 and 0. From this point the rest of the proof is easy.

There is a convenient term which is defined as follows:

Definition Two matrices M and M' are said to be *similar* if there is a nonsingular matrix C such that

$$M' = CMC^{-1}$$

Notice that similar matrices must be square.

Using this term we can restate Theorem 3.23 as follows:

Theorem 3.28 If M is the matrix with respect to a basis of a linear transformation T of V into itself, and if the minimum polynomial of T has distinct linear factors, then M is similar to a diagonal matrix.

We can even find a transformation C which takes M into a diagonal matrix. Suppose $T(\alpha_i) = c_i\alpha_i$ for a basis α_i of V. If we think of the α_i as column vectors, then $M\alpha_i = c_i\alpha_i$. Choose C^{-1} to be the matrix whose columns are the α_i in order. Then

$$MC^{-1} = M(\alpha_1, \alpha_2, \ldots, \alpha_n) = (c_1\alpha_1, c_2\alpha_2, \ldots, c_n\alpha_n)$$

and

$$C^{-1}D = (c_1\alpha_1, c_2\alpha_2, \ldots, c_n\alpha_n)$$

where D is the diagonal matrix whose diagonal elements in order are the c_i. This shows that $CMC^{-1} = D$.

We leave as an exercise the proof of

Theorem 3.29 Similarity is an *equivalence relationship;* that is,

1. Any matrix is similar to itself.
2. If M is similar to M', then M' is similar to M.
3. If M is similar to M' and M' to M'', then M is similar to M''.

EXERCISES

1. In each case below, the minimum polynomial is given. Check that it is really the minimum polynomial and find a matrix C such that CMC^{-1} is diagonal:

(*a*) $M = \begin{bmatrix} 1 & 4 \\ 1 & 1 \end{bmatrix}$, $m(x) = (x + 1)(x - 3)$

(*b*) $M = \begin{bmatrix} 7 & -5 & -2 \\ 5 & -4 & -1 \\ 2 & -1 & -1 \end{bmatrix}$, $m(x) = x(x + 1)(x - 3)$

(c) $\begin{bmatrix} 7 & 4 & -4 \\ 4 & -8 & -1 \\ -4 & -1 & -8 \end{bmatrix}$, $m(x) = x^2 - 81$

(d) $\begin{bmatrix} 1 & -1 & 3 \\ 0 & 0 & 1 \\ 0 & 0 & -1 \end{bmatrix}$, $m(x) = x^3 - x$

2. Prove Theorem 3.25.

3. Complete the proof of Theorem 3.27.

4. Prove Theorem 3.29.

5. Let T and S be two linear transformations of V^n into itself. Prove first that $\mathrm{tr}(T + S) = \mathrm{tr}(T) + \mathrm{tr}(S)$ and, second, that $\mathrm{tr}(cT) = c[\mathrm{tr}(T)]$ for every number c.

6. Let $\mathcal{S}$ denote the set of all linear transformations of V^n into itself over a field F. Let f be a transformation of the elements (transformations) of $\mathcal{S}$ defined by $f(T) = \mathrm{tr}(T)$ for all T in $\mathcal{S}$. Show that f is a linear transformation of $\mathcal{S}$ into the field F.

7. Prove that if T and T' are two linear transformations of V^2 into itself, and if T has two distinct characteristic roots which are equal to those of T', then T and T' are similar.

8. Prove that the following two matrices are similar:

$$\begin{bmatrix} \cos\varphi & -\sin\varphi \\ \sin\varphi & \cos\varphi \end{bmatrix} \quad \text{and} \quad \begin{bmatrix} \cos\varphi + i\sin\varphi & 0 \\ 0 & \cos\varphi - i\sin\varphi \end{bmatrix}$$

where $i = \sqrt{-1}$.

9. Prove that the following two matrices are similar:

$$\begin{bmatrix} \cos\varphi & -\sin\varphi \\ -\sin\varphi & -\cos\varphi \end{bmatrix} \quad \text{and} \quad \begin{bmatrix} 1 & 0 \\ 0 & -1 \end{bmatrix}$$

10. Prove that if T is a transformation such that $T^2 = rT$, where $r \neq 0$, then the trace of T is r times its rank.

11. Suppose† A, B, and C are three idempotent matrices of the same order and that $A = B + C$. Prove that the rank of A is the sum of the ranks of B and C. (Hint: Use Theorem 3.27.) Prove that, under the conditions given, $BC = CB = (0)$. Here assume that the field is that of the real numbers.

12. Prove the converse of Theorem 3.28.

† See D. Z. Djokovic, Note on Two Problems on Matrices, *Am. Math. Monthly*, vol. 76, pp. 652–654, 1960.

3.13 AFFINE TRANSFORMATIONS

Recall that in Sec. 2.8, in connection with solutions of nonhomogeneous linear equations, we discussed cosets, that is, sets of vectors of the form $W + \alpha_0$, where W is a vector space and α_0 is a fixed vector. We shall see that a translation is a transformation which is analogous to a coset. In two dimensions we express a translation as follows:

$$T\begin{bmatrix} a_1 \\ a_2 \end{bmatrix} = \begin{bmatrix} a_1 + h \\ a_2 + k \end{bmatrix} = \begin{bmatrix} a_1 \\ a_2 \end{bmatrix} + \begin{bmatrix} h \\ k \end{bmatrix}$$

So in n dimensions we would define a translation as follows:

Definition A transformation T_{α_0} of V into itself is called a *translation* if, for some vector α_0 in V,

$$T_{\alpha_0}(\alpha) = \alpha + \alpha_0 \qquad \text{for all } \alpha \text{ in } V$$

(It is convenient to allow the possibility that α_0 is the null vector.)

It is not hard to prove the following results, which we leave as exercises:

1. $T_\alpha T_\beta = T_{\alpha+\beta}$.
2. $T_\alpha^{-1} = T_{-\alpha}$.
3. The set of translations of a vector space into itself forms a multiplicative group.

Notice that a translation is not a linear transformation. But it is convenient to combine it with a linear transformation in the following way:

Definition If L is a linear transformation and T a translation, then TL is called an *affine transformation.*

One can proceed to discuss such transformations without referring to matrices, but we shall see quickly that certain properties can be obtained easily from a matrix representation. The correspondence which works is the following:

$$T \leftrightarrow \begin{bmatrix} I & \alpha_0{}^T \\ \theta & 1 \end{bmatrix} \qquad L \leftrightarrow \begin{bmatrix} M(L) & \theta^T \\ \theta & 1 \end{bmatrix} \qquad \alpha \leftrightarrow \begin{bmatrix} \alpha^T \\ 1 \end{bmatrix}$$

where I and $M(L)$ are square matrices whose order is n, the dimension of the vector space, $M(L)$ is the matrix of L with respect to a basis, and θ, α_0, and α

are row vectors in V^n. To show why this works, we illustrate it for V^2. Let $M(L) = (m_{ij})$. Then $L\xi = \xi'$ is represented as follows:

$$\begin{aligned} m_{11}x_1 + m_{12}x_2 &= x_1' \\ m_{21}x_1 + m_{22}x_2 &= x_2' \end{aligned}$$

That is,
$$\begin{bmatrix} m_{11} & m_{12} \\ m_{21} & m_{22} \end{bmatrix} \begin{bmatrix} x_1 \\ x_2 \end{bmatrix} = \begin{bmatrix} x_1' \\ x_2' \end{bmatrix}$$

Equivalently we can also write this in the following form:

$$\begin{bmatrix} m_{11} & m_{12} & 0 \\ m_{21} & m_{22} & 0 \\ 0 & 0 & 1 \end{bmatrix} \begin{bmatrix} x_1 \\ x_2 \\ 1 \end{bmatrix} = \begin{bmatrix} x_1' \\ x_2' \\ 1 \end{bmatrix}$$

Now let T be the translation $x_1' + a_1 = x_1''$, $x_2' + a_2 = x_2''$. This can be written, following the pattern above,

$$T \begin{bmatrix} x_1' \\ x_2' \\ 1 \end{bmatrix} = \begin{bmatrix} 1 & 0 & a_1 \\ 0 & 1 & a_2 \\ 0 & 0 & 1 \end{bmatrix} \begin{bmatrix} x_1' \\ x_2' \\ 1 \end{bmatrix} = \begin{bmatrix} x_1' + a_1 \\ x_2' + a_2 \\ 1 \end{bmatrix} = \begin{bmatrix} x_1'' \\ x_2'' \\ 1 \end{bmatrix}$$

Then, whether we multiply matrices or substitute, we come out with

$$TL \begin{bmatrix} x_1 \\ x_2 \end{bmatrix} = \begin{bmatrix} m_{11}x_1 + m_{12}x_2 + a_1 \\ m_{21}x_1 + m_{22}x_2 + a_2 \end{bmatrix} = \begin{bmatrix} x_1'' \\ x_2'' \end{bmatrix}$$

You should check to see that this is so.

Then, multiplication of matrices shows us that the matrix of an affine transformation will be of the form

$$A = \begin{bmatrix} M & \alpha_0{}^T \\ \theta & 1 \end{bmatrix}$$

This can be shown in the same way for V^n. Given a basis of V^n there is a one-to-one correspondence between such matrices and affine transformations. The product of two such matrices is again one of the same type. If M is nonsingular, we call A nonsingular, and its inverse will also be a matrix of an affine transformation.

It should be noted that the set of affine transformations is a vector space, for first,

$$c[TL(\alpha)] = cL(\alpha) + cT(\theta) = L_1(\alpha) + T_1(\theta)$$

where
$$T_1(\theta) = cT(\theta) \qquad \text{and} \qquad L_1(\alpha) = cL(\alpha)$$

Second,
$$\begin{aligned} T_1L_1(\alpha) + T_2L_2(\alpha) &= L_1(\alpha) + T_1(\theta) + L_2(\alpha) + T_2(\theta) \\ &= (L_1 + L_2)(\alpha) + (T_1 + T_2)(\theta) \\ &= (T_1 + T_2)(L_1 + L_2)(\alpha) \end{aligned}$$

An affine transformation takes cosets into cosets, as we now proceed to show. Note first that this should seem natural since a translation preserves collinearity. So consider the coset $W + \beta$, where W is a subspace of V^n and β an element of V^n. Then

$$T_\alpha L(W + \beta) = T_\alpha[L(W) + L(\beta)] = L(W) + L(\beta) + \alpha$$

which is also a coset. Furthermore, if L is nonsingular, W and $L(W)$ have the same dimension. Hence a nonsingular affine transformation will take any coset of dimension r into another of the same dimension.

EXERCISES

1. Let $V = V^2$ and T be the translation $T(\alpha) = \alpha + \epsilon_1$ for all α in V^2. Let $L(\epsilon_1) = \epsilon_1 + \epsilon_2$ and $L(\epsilon_2) = \epsilon_1 - \epsilon_2$. Find $TL(\epsilon_i)$ and $LT(\epsilon_i)$ for $i = 1, 2$.

2. Define $L(\epsilon_i)$ as in Exercise 1 and $T(\alpha) = \alpha + \beta$ for some fixed β in V^2 and all α in V. Can β be chosen so that $TL = LT$? If so, why? If not, why not?

3. Do Exercise 1 with T as given but $L(\epsilon_1) = \epsilon_2$ and $L(\epsilon_2) = \theta$.

4. Define T as in Exercise 2 and L as in Exercise 3. Answer the question in Exercise 2 for these transformations.

5. Consider the affine transformation whose matrix is

$$A = \begin{bmatrix} \cos\varphi & -\sin\varphi & h \\ \sin\varphi & \cos\varphi & k \\ 0 & 0 & 1 \end{bmatrix}$$

Express A as a product LT, where L is the matrix of a linear transformation and T is a translation. Show that A can also be expressed in the form $T'L$, where T' is another translation. Find the relationship between T and T'.

6. Prove the three properties of a translation listed just after the first definition in the section.

7. Let L be a linear transformation of V^n into itself and T a translation. Show that there is a translation T' such that $LT = T'L$.

8. Using the notation of Exercise 7, is it always possible to find a translation T' such that $LT' = TL$?

9. By use of Exercise 7 or by other means, without using matrices but using the definition of an affine transformation, prove that the product of two affine transformations is an affine transformation.

10. Is the sum of two affine transformations of V^n into itself an affine transformation? Establish your result by proof or counterexample.

11. Prove that the multiplicative group of translations of a vector space V^n into itself is isomorphic to the additive group of the vectors of V^n.

3.14 APPLICATIONS TO PROJECTIVE GEOMETRY

Recall that in Sec. 1.11 we showed how an analytic projective geometry of $n - 1$ dimensions could be derived from a vector space of n dimensions by identifying a one-dimensional vector space with a point in projective geometry, a two-dimensional vector space with a line in projective geometry, and so forth. In other words, a point in projective geometry is the set of all multiples of a nonzero vector, a line in projective geometry (that is, a projective line) is the set of all linear combinations of two vectors, and so forth. Here we confine ourselves to the field of real numbers. (That is, the n-tuples are of real numbers and the coefficients of the linear combinations are real.) Recall also, as in Desargues' theorem (Theorem 1.14), that while we cannot find the "sum" of two points by adding the corresponding vectors, yet all linear combinations of two linearly independent vectors constitute a line in projective geometry which is independent of the particular vectors that we use to represent the points. In this section we show how what we have been learning about transformations applies to projective geometry.

First, consider a one-dimensional projective geometry P. This is obtained from V^2 by eliminating the zero vector and identifying the projective points with the subspaces $\text{sp}(\alpha)$, for $\alpha \neq \theta$ in V^2. Now if T is a linear transformation of V^2 into itself and if (x_1',x_2') is the image of (x_1,x_2) under T, we may write $T(x_1,x_2) = (x_1',x_2')$. Since the transformation is linear we have

$$\begin{aligned} x_1' &= ax_1 + bx_2 \\ x_2' &= cx_1 + dx_2 \end{aligned} \tag{3.37}$$

that is,

$$T(x_1,x_2) = (ax_1 + bx_2,\, cx_1 + dx_2) \tag{3.38}$$

In the projective geometry P, the transformation T_P corresponding to T takes the point Q with coordinates (x_1,x_2) into Q' with coordinates (x_1',x_2'); that is,

$$T_P(Q) = Q' \tag{3.39}$$

Suppose S is another linear transformation of V^2 into itself, where now $S(x_1,x_2) = e(ax_1 + bx_2, cx_1 + dx_2)$ for any number e different from zero. This is a different linear transformation for different numbers e, but the point Q' in the projective geometry is the same, no matter what e is. So we can write T_P as the following *set* of linear transformations:

$$(3.40) \qquad T_P: \quad \begin{aligned} ex_1' &= ax_1 + bx_2 \\ ex_2' &= cx_1 + dx_2 \end{aligned} \qquad \text{where } e \neq 0$$

Of course we could also replace the representation (x_1,x_2) of Q by (rx_1,rx_2) for any number r different from zero, but this would merely change e without changing the set T_P.

We should notice that if $T_P(Q)$ is to be a point for all points Q, we must provide that $T_P(Q)$ shall not be the null vector. This is equivalent to saying that the matrix of coefficients in (3.37) is nonsingular. If $ad - bc \neq 0$, we call a transformation T_P over P as defined in (3.40) a *collineation* for reasons which we shall give later in this section. We need such a term since T_P is not a linear transformation, though it is closely allied to one.

We have found that we can get information about a transformation by finding its fixed points. So in the collineation (3.40) we replace x_i' by x_i and have

$$(3.41) \qquad \begin{aligned} 0 &= (a - e)x_1 + bx_2 \\ 0 &= cx_1 + (d - e)x_2 \end{aligned}$$

This has a nontrivial solution if and only if the matrix of coefficients in (3.41) is singular, that is,

$$(3.42) \qquad (a - e)(d - e) - bc = 0$$

that is, if and only if e is a solution of the quadratic equation

$$(3.43) \qquad z^2 - (a + d)z + ad - bc = 0$$

Notice that from the construction e must be a characteristic root of the transformation T, and any nonzero solution of (3.41) is a corresponding characteristic vector. Thus (3.43) is either the minimum polynomial of T or a multiple of it. So the fixed points of T_P correspond to the characteristic vectors of T. Thus we have the following possibilities for fixed points of T_P:

1. Equation (3.43) has two distinct real roots. In this case there are two fixed points.
2. Equation (3.43) has exactly one real root, that is, is a square of a linear polynomial. We see below that here either the collineation is the identity or it has exactly one fixed point.
3. The roots of (3.43) are imaginary. Here there are no real fixed points.

To establish the result announced in case 2, notice that the number of linearly independent characteristic vectors (that is, the number of distinct points in P) associated with e depends on the rank of the matrix (see the text in Sec. 3.9, after Theorem 3.19):

$$\begin{bmatrix} a - e & b \\ c & d - e \end{bmatrix} \tag{3.44}$$

Since e is a characteristic root either (3.44) is of rank and nullity 1, in which case there is exactly one fixed point, or (3.44) is of rank zero. In the latter case,

$$a = e \qquad b = 0 \qquad c = 0 \qquad d = e$$

and the transformation is the identity collineation

$$ex'_1 = ax_1 \qquad ex'_2 = ax_2 \tag{3.45}$$

If the nullity of (3.44) is 1, then all characteristic vectors are scalar multiples of a single one; that is, there is exactly one fixed point.

Now let P be the projective geometry of dimension $n - 1$ obtained as indicated above from V^n. If S is any subspace of V^n, there corresponds a projective subspace (that is, a subgeometry) P_S of dimension 1 less than that of S. Since any nonsingular transformation on V^n takes S into a subspace of the same dimension, the corresponding T_P will take P_S into a projective subspace of the same dimension. In particular, any collineation takes a one-dimensional projective space, that is, a projective line, into a projective line. This is the reason for the term *collineation*. The general collineation takes the form

$$ex'_i = \sum_{j=1}^{n} t_{ij}x_j \qquad \text{for } 1 \leq i \leq n \tag{3.46}$$

For a projective geometry of dimension $n - 1$, what is the identity transformation? It must have the property that $T_P(Q) = Q$ for all points Q in P. Certainly the transformation cI, with $c \neq 0$, takes every point into itself. More generally we have

Theorem 3.30 If $T' = cT$, then the corresponding collineations T'_P and T_P are equal.

Proof This follows immediately from $cT(Q) = T'(Q)$.

But at this point it is not quite clear that the identity collineation corresponds to the set of linear transformations cI, with $c \neq 0$, since it might happen that the constant could vary from vector to vector. For instance, if $T(\alpha) = 2\alpha$ and $T(\beta) = 3\beta$, then T takes the corresponding points A and B into themselves. We now lead up to another theorem which can be used to show the converse of Theorem 3.30.

In the n-dimensional case, we see from (3.46) that to get the fixed points of a collineation we use the equations

$$0 = \sum_{j=1}^{n} t_{ij}x_j - ex_i \qquad \text{for } 1 \leq i \leq n \tag{3.47}$$

Thus the procedure for finding the fixed points in a projective geometry is just the same as that for finding the characteristic vectors as developed in Sec. 3.7. We first find the numbers e such that

$$M(T) - eI$$

is singular. One way to do this is to find the minimum polynomial of T. (In Chap. 4 we shall find another method.) Once the characteristic roots are found one may proceed to find the characteristic vectors which are the fixed points of our projective geometry. It should be noted that while we have at most n characteristic roots, we can have more than n fixed points. For instance, if

$$T(\alpha_1) = 2\alpha_1 \qquad T(\alpha_2) = 2\alpha_2 \qquad T(\alpha_3) = 3\alpha_3$$

we see that all points on the line $a_1\alpha_1 + a_2\alpha_2$ are fixed. On the other hand, if

$$T(x_1,x_2,x_3) = (x_2, \; -x_1 + 2x_2, \; 3x_3)$$

we see that the only fixed points are (1,1,0), corresponding to the characteristic root 1, and (0,0,1), corresponding to the characteristic root 3.

Another important result in projective geometry is described in the following theorem:

Theorem 3.31 Let A_i, where $1 \leq i \leq n + 1$, be $n + 1$ distinct points in an $(n - 1)$-dimensional projective geometry and α_i the corresponding vectors. Suppose that no subsequence of n different α_i is a linearly dependent sequence. Let E_i be the points whose n-tuples ϵ_i, for $1 \leq i \leq n$, are the rows of the identity matrix and E_{n+1} the point whose n-tuple ϵ_{n+1} is the sum of the rows of the identity matrix. Let T and T' be two linear transformations of V^n into itself such that

$$T(\epsilon_i) = d_i\alpha_i \qquad T'(\epsilon_i) = d'_i\alpha_i \tag{3.48}$$

for some nonzero numbers d_i, d'_i, with $1 \leq i \leq n + 1$. Then $T' = cT$ for some nonzero scalar c. Furthermore, for any choice of nonzero numbers d_i, d'_i there is a linear transformation T satisfying (3.48).

Proof Suppose there are linear transformations T and T' satisfying (3.48). Then by virtue of the definition of α_{n+1} and the linearity of T and T', we have

$$T(\epsilon_{n+1}) = d_1\alpha_1 + d_2\alpha_2 + \cdots + d_n\alpha_n = d_{n+1}\alpha_{n+1} \tag{3.49}$$

$$T'(\epsilon_{n+1}) = d'_1\alpha_1 + d'_2\alpha_2 + \cdots + d'_n\alpha_n = d'_{n+1}\alpha_{n+1} \tag{3.50}$$

Since α_{n+1} is a unique linear combination of α_i, for $1 \leq i \leq n$, we have $d_i/d_{n+1} = d'_i/d'_{n+1}$. Hence $d'_i/d_i = d'_{n+1}/d_{n+1}$ for $1 \leq i \leq n$. Since T and T' are both determined by the first n equations of (3.48), it follows that $T' = cT$. This proves the first part of the theorem.

Now we want to show that there is a linear transformation satisfying conditions (3.48). For this we choose $d_{n+1} = 1$ and see that (3.49) determines the rest of the d's. None of these can be zero since by hypothesis no subsequence of n of the α_i is linearly dependent. Then the first n equations of (3.48) determine a unique linear transformation T. This completes the proof.

We have a number of immediate corollaries, the first of which implies the converse of Theorem 3.30.

Corollary 1 Using the notation of Theorems 3.30 and 3.31, there is a unique collineation T_P such that $T_P(E_i) = A_i$, for $1 \leq i \leq n + 1$.

This follows since any two scalar multiples of a linear transformation correspond to the same collineation. It implies the converse of Theorem 3.30 since if T_P and T'_P are equal then they must have the same effect on the E_i.

Corollary 2 If in Theorem 3.31, $A_i = E_i$, for $1 \leq i \leq n + 1$, then all linear transformations T are scalar multiples of the identity, and the set of such transformations is the identity collineation.

This follows since we know one transformation which has the desired properties is the identity transformation. Almost a corollary is the following

Theorem 3.32 If A_i and B_i are two sets of $n + 1$ points in a projective geometry of dimension $n - 1$, with the property that any subset of n vectors defining the points A_i is linearly independent and similarly for the B_i, then there is a unique collineation taking one set into the other.

Proof This consists in going from the set A_i to the set B_i by way of E_i of Theorem 3.31. Let T_P be the collineation taking the points E_i into the A_i and T'_P that taking E_i into the B_i. Then

$$T'_P T_P^{-1}(A_i) = B_i \qquad \text{for } 1 \leq i \leq n + 1$$

Then if S_P is a collineation taking A_i into B_i, it follows that $S_P T_P(E_i) = B_i$, for $1 \leq i \leq n + 1$. Theorem 3.31 shows that $S_P T_P = T'_P$, and hence the unique collineation taking A_i into B_i is $T'_P T_P^{-1}$. This completes the proof.

Theorem 3.31 is sometimes called the *fundamental theorem of projective geometry.* For a projective geometry of dimension 1, it affirms that a collineation on a line is determined by the images of three distinct points. For dimension 2, it affirms that if S_1 is a set of four points no three of which are collinear and S_2 another such set, there is a unique collineation taking one set into the other.

Notice one peculiar contrast between transformations of V^n and collineations in a projective geometry of dimension $n - 1$. In the former, a transformation is determined by the images of n vectors. But for the latter, a collineation is determined by the images of $n + 1$ points. Though the dimension of P is one less than that of V^n, we have more leeway as far as transformations are concerned.

EXERCISES

1. Find the fixed points of each of the following transformations in one-dimensional projective geometry:

(*a*) $ex'_1 = x_1 + x_2,\ ex'_2 = 2x_2$

(*b*) $T(\alpha_1) = \alpha_1 + 3\alpha_2,\ T(\alpha_2) = 2\alpha_2$

(*c*) $ex'_1 = 2x_1 + x_2,\ ex'_1 = -x_1 + 4x_2$

(*d*) $ex'_1 = 2x_1 + x_2,\ ex'_1 = -3x_1 - x_2$

2. Let T and T' be two collineations (3.40) and P, Q, R three distinct points. Prove directly that if the images of P, Q, and R under the transformation T are the same as under transformation T', then T and T' are the same transformation.

3. Show that there is no linear transformation T such that

$$T(1,0,0) = (1,1,0) \qquad T(0,1,0) = (1,0,1)$$
$$T(0,0,1) = (0,1,1) \qquad \text{and} \qquad T(1,1,1) = (1,1,1)$$

On the other hand, suppose that each of the triples above represents a point in two-dimensional projective geometry. Show that there is a collineation T_P which takes the points whose coordinates are on the left of the equations into points whose coordinates are on the right.

4. If in Exercise 3 the last equation is replaced by $T(1,1,0) = (1,1,1)$, will there be a collineation in the sense of that exercise?

5. A collineation T is called an *involution* if $T^2(P) = P$ for all points in the space. Prove that the projective collineation transformation (3.40) is an involution if and only if it is either the identity transformation or $a + d = 0$.

The exercises below refer to an involution defined in Exercise 5.

6. Prove that if T is an involution, not the identity, in a one-dimensional projective space, then it is determined by the images of two distinct points. Why is the restriction "not the identity" necessary?

7. Suppose the collineation (3.40) is an involution. Prove that one of the following must happen:

(*a*) The collineation is the identity.
(*b*) The collineation has two distinct fixed points.
(*c*) The collineation has no (real) fixed points.

8. Let P, Q and R, S be two distinct *pairs* of points; that is, if P is one of R and S, then Q is not the other. Prove that there is one and only one involution T such that $T(P) = R$ and $T(Q) = S$, where $n = 2$.

4

DETERMINANTS

4.1 INTRODUCTION

In most walks of life styles come and go. Mathematics is no exception to this. For example, there was an era during which much time and energy was devoted to a study of the properties of determinants, as is shown by a monumental four-volume work by Thomas Muir. Nowadays there is much less concern with determinants, partly because of a shift of interest and partly because we have found that many things for which they were used can now be done more expeditiously without them. This latter can well be said of much of the material in the previous chapters.

But determinants do have their importance in linear algebra for a number of reasons. They enable us to calculate a polynomial whose zeros are the characteristic roots. Though a determinant is defined in terms of a matrix, it is an invariant of a linear transformation, that is, independent of the basis chosen to determine the matrix. The properties of a determinant enable us to prove rather elegantly important results not obviously connected with linear algebra. In short, we have now reached the point in this text where it would be much more labor to get along without determinants than to develop and use them. Furthermore, one does not have to be old-fashioned to appreciate some of their rather pretty properties.

There are at least three different approaches to the study of determinants. One can begin with a definition, then find the properties of determinants, and finally proceed to see how they are used. One can start with the idea of area and volume to get at a useable definition (see Birkhoff and MacLane, *A Survey of Modern Algebra*, listed in the bibliography). Or, finally one may decide what properties of a determinant it would be desirable to have and then try to devise a definition which fits the specifications as nearly as possible (see MacDuffee, *Vectors and Matrices*, listed in the bibliography). The third approach seems most interesting since it is an exploratory one. Also there is gain in economy of treatment.

So now we shall explore the possibilities for a useful definition. Basically, a determinant is to be a number associated with a square matrix whose elements are numbers. For each square matrix there should be exactly one number. From this point of view, a determinant is a function of the matrix. In fact, it will be convenient to define it as a certain polynomial in the elements of the matrix.

There are, of course, innumerable possible choices. One could, for instance, define the determinant of a matrix to be the element in the upper left-hand corner. If this were taken as the definition, the determinant of the sum of two matrices would be the sum of the determinants of the separate matrices. It is suggested that you write some of the properties which you feel might be useful before looking at the following list, not arranged in any special order of preference:

Possible properties of a determinant:

(*a*) A determinant might be additive; that is, the sum of the determinants of two matrices might be the determinant of the sum.

(*b*) A determinant might be multiplicative; that is, the product of two determinants might be the determinant of the product.

(*c*) It might be that a matrix is singular if and only if its determinant is zero.

(*d*) A determinant might be an invariant of a transformation, that is, independent of the basis chosen for the matrix representing the transformation.

(*e*) The determinant of the identity matrix might be 1.

(*f*) The determinant of a diagonal matrix might be the product of the elements along the diagonal.

One could go on with this list; and perhaps you have found other desirable properties. (Your list might not have been as biased as this one.) Certainly our first suggestion for the determinant of a matrix—the element in the upper left-hand corner—would not seem very fruitful for it singles out one element as more important than any other. For additivity, using the trace (see Sec.

3.12) would seem more sensible since it is additive and is unchanged under a similarity transformation—that is, it depends on the transformation without regard to the basis used. But there is no point in calling something a determinant which we have already dealt with under a different name.

Property (*c*) would certainly seem desirable. This points to the multiplicative property since the product of any singular matrix with another is singular and the product of two matrices is singular if and only if at least one of them is. So properties (*b*) and (*c*) seem consistent with each other. They are, however, inconsistent with property (*a*), since the sum of two singular matrices may be nonsingular. Once one decides to include the multiplicative property, one sees that the identity matrix must have determinant 1 if the determinant of the matrix IA is to be equal to that of I times that of A, for every matrix A, unless every matrix has zero determinant. Finally property (*f*) is consistent with property (*c*) since a diagonal matrix will be singular if and only if the product of its diagonal elements is zero. So properties (*b*) to (*f*) seem consistent with each other.

4.2 PROPERTIES OF A DETERMINANT

On the basis of our preliminary discussion, we now explore what the consequences would be of certain fundamental agreements on desirable properties. First of all, a determinant should be a function of a square matrix, that is, a unique number associated with a square matrix of numbers. To focus attention on the nature of this function, write the matrix as (x_{ij}), where the x_{ij} are variables or indeterminates. If the determinant is a function of the elements of the matrix, it will certainly be a function of the matrix itself. The simplest kind of function we know is a polynomial. So we first agree on the following property:

Property 1 The determinant of the square matrix $(x_{ij}) = X$ is to be a polynomial in the x_{ij} of degree at least 1 in each x_{ij}. We write it as $|(x_{ij})|$, $|X|$, or $\det(X)$.

Notice what we are doing here. We started out to find a number associated with a matrix of numbers, and here we begin to think of a matrix of variables x_{ij}. We are making this seeming shift in emphasis because we seek a *formula* for the determinant into which we can substitute the numbers of any particular matrix as needed. For instance, for the matrix X of order 2,

$$\begin{bmatrix} x_{11} & x_{12} \\ x_{21} & x_{22} \end{bmatrix}$$

the determinant turns out to be $x_{11}x_{22} - x_{12}x_{21}$. The formula tells us that to get the determinant of any matrix of order 2, we find the product of the elements on the diagonal and subtract the product of the other two elements. So in general, once we have a formula for a determinant, we can substitute numbers for the variables to get it. Actually, what happens in practice, at least for determinants of order greater than 3, is that we use the definition to prove properties which make computation possible without substitution in the formula.

The second property has been discussed above:

Property 2 The determinant is to be multiplicative, that is,

$$|AB| = |A| \cdot |B| \qquad \text{for all square matrices } A \text{ and } B$$

It is wise not to accumulate too many "desired properties" since the more there are, the less the likelihood of being able to satisfy them all. So let us see what two consequences of these two properties would be.

Consequence 1 The identity matrix would have determinant 1.

Proof $|X|$ is of positive degree in every element of X. Then

$$|X| \cdot |I| = |XI| = |X| \qquad \text{and} \qquad |X| \neq 0$$

imply

$$|I| = 1$$

Consequence 2 If a matrix B were obtained from A by adding a multiple of a row to another row, then A and B would have the same determinant.

Proof Use the notation of Sec. 2.9 and write $I_{ij}(k)$ for the matrix obtained from the identity matrix by adding k times the ith row to the jth row. We know that the inverse of this matrix is $I_{ij}(-k)$. So from the multiplicative property,

$$1 = |I_{ij}(-k)|\,|I_{ij}(k)|$$

But if 1 is to be a polynomial in k, the polynomial must be of zero degree, that is, independent of k. Hence

$$|I_{ij}(k)| = |I_{ij}(0)| = |I| = 1$$

Now, let us return to our list of possible properties in Sec. 4.1. We have items (c) and (f) to consider. It turns out that we can get both from one further desired property. If property (f) were to be true the determinant of $I_i(c)$ would be c, where $I_i(c)$ denotes the matrix obtained from the identity matrix by multiplying the ith row by c. Then the multiplicative property of determinants implies

$$|I_i(c)A| = c|A|$$

This is equivalent to the following third desired property:

Property 3 The determinant of (x_{ij}) is to be a linear homogeneous function of the elements of each row; that is, if one multiplies each element of any row of a matrix by c, one multiplies the determinant by c.

Notice that this property includes Property 1. Hence we have really only two properties to reckon with.

We shall see later that these two desired properties are sufficient to determine what the determinant must be. But first let us consider further consequences.

Consequence 3 If I_{ij} denotes the matrix obtained from the identity matrix by interchanging the ith and jth rows (or columns), then

$$|I_{ij}| = -1$$

Proof To simplify notation, take $i = 1, j = 2$, and consider just the 2 by 2 matrix in the upper left-hand corner of the identity matrix. Then we know that $|I_1(-1)| = -1$ by desired Property 3. Hence we have the following sequence of additions of multiples of rows to other rows:

$$\begin{vmatrix} -1 & 0 \\ 0 & 1 \end{vmatrix} = \begin{vmatrix} -1 & 0 \\ 1 & 1 \end{vmatrix} = \begin{vmatrix} 0 & 1 \\ 1 & 1 \end{vmatrix} = \begin{vmatrix} 0 & 1 \\ 1 & 0 \end{vmatrix}$$

We obtain the second determinant from the first by adding -1 times the first row to the second; next we add the second row to the first; and, finally, we subtract the first row from the second. It should be noted that, strictly speaking, it is a matrix—not a determinant—which has rows and columns. A determinant is a number or a function, neither of which has rows or columns. But there is probably not much harm in a little looseness of language in this connection where there is scant possibility of confusion.

From Property 3 and Consequences 2 and 3, we have the following

Corollary No elementary matrix has determinant zero.

Consequence 4 The determinant of a diagonal matrix would be the product of the elements of the diagonal.

The proof of this consequence is left as an exercise.

Consequence 5 If a row of a matrix consists of zeros, then the determinant of the matrix would be 0.

Proof If we multiply the zero row by 0, we leave the matrix unchanged and hence its determinant. But Property 3 affirms that by multiplying the row by 0, we multiply the determinant by 0. These conditions are both satisfied if and only if the determinant is zero.

Consequence 6 The determinant of a matrix would be zero if and only if it is singular.

Proof From Theorem 2.18 we know that if a matrix A is nonsingular it is expressible as a product of elementary matrices. None of these have zero determinant (see the Corollary above), and hence the determinant of their product, the determinant of A, cannot be zero.

On the other hand, if A is singular, we know that there is a product of elementary matrices E_i such that

$$E_1E_2 \cdots E_rA = B$$

is in echelon form, and since A is singular at least one row of the echelon form, B, is zero. Hence $|B| = 0$, and, since none of the determinants of the E_i is zero, it follows that the determinant of A must be zero.

Thus we have shown that on the basis of desired Properties 2 and 3 we would have all the possible properties listed in Sec. 4.1 except (*a*) and (*d*). We shall show in the next section that these two properties force us into a definition of a determinant. Then if we can show that the determinant has indeed these two desired properties, the others follow immediately without further proof.

4.3 DEFINITION OF A DETERMINANT

Here we show what the definition of a determinant must be if it is to have Properties 2 and 3. Using Property 3, let us see what a term of the determinant of $X = (x_{ij})$ must look like. Since $|X|$ is linear in the elements of each row, every term must contain x_{1i_1}, where i_1 is one of the numbers from 1 to n inclusive; since $|X|$ is linear in the elements of the second row, each term must contain x_{2i_2}, for some i_2 in the same range; and so on. Thus $|X|$ must be a sum of terms of the following form:

$$cx_{1i_1}x_{2i_2} \cdots x_{ni_n}$$

where c is a number which varies from term to term and each of the second subscripts is chosen from the set $1, 2, \ldots, n$. The second subscripts need not be distinct. Let us look more closely at c, which we write as a function of the second subscripts:

$$c(i_1, i_2, \ldots, i_n) \tag{4.1}$$

Consequence 3 tells us that interchanging the first two rows of a matrix changes the sign of its determinant. This is equivalent to interchanging i_1 and i_2 in (4.1). Hence

$$c(i_1, i_2, i_3, \ldots, i_n) = -c(i_2, i_1, i_3, \ldots, i_n) \tag{4.2}$$

It follows that if $i_1 = i_2$, then c must be zero (assuming that $1 + 1 \neq 0$ in the field). Similarly if any two i's of the function c are interchanged, c is replaced by its negative, and if any two i's are equal, c must be zero. This implies that $c(i_1, i_2, \ldots, i_n) = 0$ unless $i_1, i_2, \ldots, i_n$ is a permutation of the numbers from 1 through n.

At this point we need some information about sets of permutations. In order to disturb the continuity of our discussion as little as possible, we state the two needed results and a definition, relegating the proofs to Appendix B.

Theorem 4.1 Any permutation of the integers from 1 to n inclusive may be accomplished by a finite sequence of transpositions, that is, interchanges of two numbers.

Theorem 4.2 Given a permutation of the integers from 1 to n inclusive and suppose it is accomplished in two different ways as a sequence of transpositions. Then the number of transpositions is even in both cases or odd in both cases.

Definition A permutation is called an *even permutation* if it can be expressed as a sequence of an even number of transpositions; otherwise it is called an *odd permutation.* The evenness or oddness of a permutation is called its *parity;* that is, we say that two permutations have the same parity if they are both even or both odd.

It is shown in Appendix B that a permutation and its inverse have the same parity.

Thus from the above we see that

$$c(1, 2, \ldots, n) = (-1)^h c(i_1, i_2, \ldots, i_n) \tag{4.3}$$

where h is even if the permutation $i_1, i_2, \ldots, i_n$ is even and odd if the permutation is odd.

Hence we have almost proved the following

Theorem 4.3 If the determinant of (x_{ij}) has Properties 2 and 3, then it must be of the following form:

$$|(x_{ij})| = \Sigma(-1)^h x_{1i_1} x_{2i_2} \cdots x_{ni_n} \tag{4.4}$$

where the sum is over all permutations $(i_1, i_2, \ldots, i_n)$ of the numbers from 1 to n inclusive and h is even or odd according as the permutation is even or odd.

Proof To complete the proof of the theorem, in view of (4.3) we need only show that $c(1, 2, \ldots, n) = 1$. To do this, let $X = I$; that is, choose $x_{ii} = 1$ for all i and $x_{ij} = 0$ if $i \neq j$. Then there will be only one nonzero term in the expansion (4.4) and

$$|I| = c(1, 2, \ldots, n) = 1$$

This completes the proof.

The sum (4.4) is called the *expansion* of the determinant of X. The order of the matrix, n, is also called the *order* of the determinant.

One can calculate the determinant from the definition (4.4), but this is not the simplest method for orders greater than 3. However, before proceeding further, we need to show that the determinant we have so defined does indeed have Properties 2 and 3. Property 3 is apparent from the form of (4.4). It is Property 2 which must be dealt with. So we prove

Theorem 4.4 If the determinant is defined as in (4.4), then

$$|XY| = |X| \cdot |Y|$$

Proof Let us look at a typical element of $XY = Z = (z_{rs})$:

$$z_{ik} = \sum_{j=1}^{n} x_{ij}y_{jk}$$

This can be thought of as a linear polynomial in x_{ij} with coefficients y_{jk}. Thus any linear polynomial in z_{ik}, in particular the determinant of Z, will be a linear polynomial in x_{ij} whose coefficients are linear polynomials in the y's. So now the c in (4.1) becomes a polynomial in y_{ik} as well as depending on $i_1, i_2, \ldots, i_n$. Interchanging two rows of X is equivalent to interchanging two rows of Z. Thus as before, if two of the i's are equal, then c will be zero. Hence,

$$(4.5) \qquad |XY| = c(1, 2, \ldots, n)\Sigma(-1)^h x_{1i_1}x_{2i_2} \cdots x_{ni_n}$$

where the sum is over all permutations $(i_1, i_2, \ldots, i_n)$ of $(1, 2, \ldots, n)$ with h determined by the parity of the permutation. To find $c(1, 2, \ldots, n)$, which is now a function of the y_{ij}, we consider (4.5) as an identity in x_{ij}, and since $c(1, 2, \ldots, n)$ is independent of x_{ij} we can compute it by setting $x_{ii} = 1$ and $x_{ij} = 0$ for $i \neq j$, as before. Then X becomes the identity matrix and (4.5) becomes

$$|IY| = |Y| = c(1, 2, \ldots, n) \tag{4.6}$$

Now, since (4.6) gives us the coefficient c in (4.5), and the summation part of (4.5) is just the determinant of X, we have

$$|XY| = |Y| \cdot |X|$$

Now that we have shown that the determinant as given in (4.4) has Properties 2 and 3, we know that the consequences listed follow. We collect these in one place:

1. The identity matrix has determinant 1.
2. Adding a multiple of one row of a matrix to another row does not change its determinant.
3. Interchanging two rows of a matrix changes the sign of its determinant.
4. The determinant of a diagonal matrix is the product of the elements of the diagonal.
5. If a row of a matrix consists of zeros, then the determinant of the matrix is zero.
6. The determinant of a matrix is zero if and only if the matrix is singular.
7. The determinant of the product of two matrices is equal to the product of their determinants.

There is one other property which should be mentioned in this section since it shows us that in consequences 2, 3, and 5 above the word *row* may be replaced by *column*.

Theorem 4.5 The determinant of a matrix is the same as that of its transpose; that is, $|X| = |X^T|$.

Proof Our proof consists in matching the terms in the expansion (4.4) of $|X|$ with those of $|X^T|$. Each term is of the form

$$(-1)^h x_{1i_1} x_{2i_2} \cdots x_{ni_n} \tag{4.7}$$

where the second subscripts form a permutation p of the numbers from 1 through n and h is in accord with the parity of p; that is, h is even or odd according as p is even or odd. The term (4.7) contains one element from each row and column of X. Since (4.7) also contains one element from each column and row of X^T, it is a term in the expansion of X^T as well. It remains to show that the sign is the same whether it is regarded as a term of $|X|$ or of $|X^T|$. To determine the sign for the latter, we should rearrange the x's so that the second subscripts are in natural order and then determine the parity of q, the resulting permutation in the first subscripts. Now, every time we interchange two x's, we not only effect a transposition of first subscripts but also of second subscripts. Thus the number of transpositions required to go from permutation p to the natural order is the same as the number required to go from the natural order to q. Since the parity of a permutation and its inverse (see Appendix B) are the same, this shows that the permutations p and q have the same parity. Thus we have shown that term by term the expansion of $|X|$ is the same as that of $|X^T|$.

We state a useful theorem about direct sums (see Sec. 2.10), leaving the proof as an exercise:

Theorem 4.6 If a square matrix $A = B \oplus C$, then $|A| = |B| \cdot |C|$.

We close this section by stating two important consequences of the results of this section, whose proofs we leave as exercises.

1. If M is a nonsingular matrix, then $|M^{-1}| = |M|^{-1}$.
2. If C is a nonsingular matrix, then $|CMC^{-1}| = |M|$.

Notice that the second consequence is the same as desired property (*d*) in Sec. 4.1. So now we may write of the "determinant of a *transformation*" since it is independent of the particular basis chosen. We have thus far found the following numbers associated with a transformation: the characteristic roots, the trace, and now the determinant. We shall find other invariants later on.

EXERCISES

1. Show that the determinant of the matrix $X = (x_{ij})$ of order 2 is $a_{11}a_{22} - a_{12}a_{21}$.

2. Find the determinant of the matrix $X = (x_{ij})$ of order 3.

3. Find the determinants of each of the following matrices:

$$\begin{bmatrix} 2 & 3 \\ -4 & 5 \end{bmatrix} \quad \begin{bmatrix} 1 & 2 \\ 2 & 4 \end{bmatrix} \quad \begin{bmatrix} 1 & 4 & 5 \\ 0 & -1 & 3 \\ 1 & 0 & 4 \end{bmatrix} \quad \begin{bmatrix} 1 & 2 & 3 \\ 4 & 5 & 6 \\ 7 & 8 & 9 \end{bmatrix}$$

4. One way of computing the determinant of the fourth matrix in Exercise 3 is the following, which you should trace through geometrically on the matrix itself:

$$1 \cdot 5 \cdot 9 + 2 \cdot 6 \cdot 7 + 3 \cdot 8 \cdot 4 - (7 \cdot 5 \cdot 3 + 8 \cdot 6 \cdot 1 + 9 \cdot 2 \cdot 4)$$

Show that this gives the correct determinant.

5. Using the matrix $A = (a_{ij})$ instead of the numerical matrix in Exercise 4, write out a formula for the determinant of A and show that it is correct.

6. Would the method of computation of the determinant in Exercise 5 work also for a determinant of order 4? Why or why not?

7. Find the determinant of

$$\begin{bmatrix} 1 & 2 & 0 & 0 \\ 2 & 3 & 0 & 0 \\ 0 & 0 & 4 & 5 \\ 0 & 0 & 6 & 1 \end{bmatrix}$$

8. Find the sign of the term $x_{12}x_{31}x_{43}x_{24}$ in the expansion of $|(x_{ij})|$ of order 4, first by permuting the x's so that the first subscripts are in natural order, and second by permuting the x's so that the second subscripts are in natural order.

9. Prove that if one row of a square matrix is a linear combination of the other rows, then its determinant is zero.

10. Is the converse of the statement of Exercise 9 true? If so, prove it; if not, give an example.

11. Prove consequences 1 and 2 at the close of this section.

12. Let A be a matrix of order n and consider $|A - xI|$. Show that the determinant is a polynomial in x of degree n.

13. Show that $|C^{-1}AC - xI| = |A - xI|$, for any nonsingular matrix C.

14. How many terms are there in the expansion of $|X|$, if X is of order n?

15. Prove that $|cA| = c^n|A|$, where n is the order of the square matrix A.

16. Prove Theorem 4.6. Suggestion: First show it for $C = I$ and then note that $A = (B \oplus I_1)(I_2 \oplus C)$, where C and I_1 are of the same order as well as B and I_2.

17. Let a square matrix B be partitioned as follows (cf. Sec. 2.10):

$$B = \begin{bmatrix} C & (0) \\ D & E \end{bmatrix}$$

where C and E are square submatrices and (0) is a zero matrix. Prove that $|B| = |C| \cdot |E|$.

18. Prove Consequence 4 of Sec. 4.2.

19. Show that the following is the equation of a line through the points (x_1,y_1) and (x_2,y_2):

$$\begin{vmatrix} x & y & 1 \\ x_1 & y_1 & 1 \\ x_2 & y_2 & 1 \end{vmatrix} = 0$$

20. Prove that the area of the parallelogram, three of whose vertices are (0,0), (x_1,y_1), and (x_2,y_2), is the absolute value of the determinant

$$\begin{vmatrix} x_1 & y_1 \\ x_2 & y_2 \end{vmatrix}$$

21. Generalize the result of Exercise 19 to three dimensions.

22. Let (x_1,y_1), (x_2,y_2), and (x_3,y_3) be pairs of coordinates of three noncollinear points. Show that if the following determinant is set equal to zero, it is the equation of a circle through the three points indicated:

$$\begin{vmatrix} x^2 + y^2 & x & y & 1 \\ x_1^2 + y_1^2 & x_1 & y_1 & 1 \\ x_2^2 + y_2^2 & x_2 & y_2 & 1 \\ x_3^2 + y_3^2 & x_3 & y_3 & 1 \end{vmatrix}$$

23. What happens to the equation of Exercise 22 if the three points are collinear?

24. Show that the following determinant is the polynomial of order 3 whose zeros are x_1, x_2, and x_3.

$$\begin{vmatrix} x^3 & x^2 & x & 1 \\ x_1^3 & x_1^2 & x_1 & 1 \\ x_2^3 & x_2^2 & x_2 & 1 \\ x_3^3 & x_3^2 & x_3 & 1 \end{vmatrix}$$

25. Find a determinant which, when set equal to zero, gives that polynomial $y = ax^2 + bx + c$ which is satisfied by the coordinates of three points (x_i,y_i) for $i = 1, 2, 3$. What restrictions must be made on the three points?

4.4 COMPUTATION OF A DETERMINANT

While the definition of a determinant is useful for proving certain of its properties, the computation merely from this definition is quite complex except for determinants of orders 2 and 3. In this section we shall develop a method of computation which is perhaps already familiar to you. First we need a

Definition Let X be the square matrix (x_{ij}) of order n. The coefficient of x_{ij} in the expansion

$$|X| = \Sigma(-1)^h x_{1i_1} x_{2i_2} \cdots x_{ni_n} \tag{4.8}$$

is called its *cofactor* and is written X_{ij}.

Since, no matter what i is, every term in (4.8) will contain exactly one element from the ith row, it immediately follows that

$$|X| = x_{i1}X_{i1} + x_{i2}X_{i2} + \cdots + x_{in}X_{in} \tag{4.9}$$

We call this the *expansion of X by the* ith *row*. We can similarly expand a determinant by any column. We shall find that, except for sign, the X_{ij} are determinants of *submatrices* of X, where by the term *submatrix of X* we mean any matrix obtained from X by deleting some of its rows and/or columns.

This may be fine, but all we have done is to reduce the problem of computing a determinant to one of finding the cofactors. To do the latter, let us first see what the simplest cofactor looks like, namely, the cofactor of x_{11}. Referring back to Equation (4.8), we see that the cofactor of x_{11}, that is, the coefficient of x_{11}, is

$$X_{11} = \Sigma(-1)^h x_{2i_2}x_{3i_3} \cdots x_{ni_n}$$

where, since for this case $i_1 = 1$, the remaining second subscripts must be a permutation of the integers from 2 to n inclusive. What about h? It is the number of transpositions in a sequence of transpositions taking

$$i_1, i_2, \ldots, i_n \tag{4.10}$$

into $1, 2, \ldots, n$. But since $i_1 = 1$, h for the n numbers (4.10) is the same as for the $n - 1$ numbers of (4.10), omitting the first. Thus

The coefficient of x_{11} in (4.8) is the determinant of the submatrix obtained from X by deleting the first row and first column.

From this we can move without too much difficulty to a determination of the cofactor of any element of a matrix. Suppose we want to find the cofactor of x_{34}. From X we form a new matrix X' as follows:

Interchange the third and second rows, then the second and first rows; then interchange the fourth and third columns, the third and second columns, and the second and first.

This will bring x_{34} into the upper left-hand corner of X' without changing its position relative to any other element of X, nor the submatrix obtained by deleting the row and column containing x_{34}. That is, the submatrix obtained by deleting the first row and column of X' is the same as that obtained by deleting the third row and fourth column of X; also their determinants will be equal.

Since each time we interchange two rows or columns we change the sign of the determinant, and since we made five such interchanges above, we see that the cofactor of x_{34} in X is $(-1)^5 = -1$ times the determinant of the matrix obtained from X by deleting the third row and the fourth column.

From this we can state the general result as a theorem, leaving the formal proof as an exercise.

Theorem 4.7 The cofactor of x_{ij} in the expansion of the determinant (x_{ij}) is $(-1)^{i-1+j-1} = (-1)^{i+j}$ multiplied by the determinant of the matrix obtained from $X = (x_{ij})$ by deleting the ith row and the jth column.

The theorem shows that we can get the sign of the cofactor by "counting off" from the upper left-hand corner. To illustrate this, let us expand the following determinant by the second row:

$$\begin{vmatrix} 1 & 4 & 5 \\ 2 & -1 & 5 \\ -3 & 4 & 2 \end{vmatrix} = (-2)\begin{vmatrix} 4 & 5 \\ 4 & 2 \end{vmatrix} + (-1)\begin{vmatrix} 1 & 5 \\ -3 & 2 \end{vmatrix} + (-5)\begin{vmatrix} 1 & 4 \\ -3 & 4 \end{vmatrix}$$
$$= (-2)(-12) + (-1)(17) + (-5)(16) = -73$$

(The formula for a determinant of order 2 is given in Exercise 1 of Sec. 4.3.)

Thus we have reduced the computation of the determinant of a 3 by 3 matrix to finding three determinants of 2 by 2 matrices. The computation of the determinant of a 4 by 4 matrix would require the computation of four determinants of 3 by 3 matrices, and so on.

A better way to compute a determinant is to reduce it first by means of its properties. In fact, we apply the elementary transformations on its rows. If in the above matrix we subtract twice the first row from the second and add three times the first row to the third, we get

$$\begin{vmatrix} 1 & 4 & 5 \\ 2 & -1 & 5 \\ -3 & 4 & 2 \end{vmatrix} = \begin{vmatrix} 1 & 4 & 5 \\ 0 & -9 & -5 \\ 0 & 16 & 17 \end{vmatrix}$$

We may expand this by the first column or reduce it further by adding twice the second row to the third to get

$$\begin{vmatrix} 1 & 4 & 5 \\ 0 & -9 & -5 \\ 0 & -2 & 7 \end{vmatrix}$$

and then see that the determinant is equal to $(-9)(7) - (-2)(-5) = -73$.

Now that we have set up a method for computing determinants, there are two extra "dividends" which we cash in on. They both hinge on the following result, which includes (4.9) as a special case:

Theorem 4.8 If $X = (x_{rs})$ is an n by n matrix, then

$$|X|\delta_{ij} = x_{i1}X_{j1} + x_{i2}X_{j2} + \cdots + x_{in}X_{jn} \tag{4.11}$$

where δ_{ij}, called the *Kronecker delta*, is the element in the ith row and jth column of the identity matrix (see Sec. 3.10).

Proof If $i = j$, this result is merely (4.9) again. Suppose $i \neq j$. Let X' be the matrix which is the same as X except for the jth row, and whose jth row is the same as the ith row of X. Then the jth and ith rows of X' are the same, and hence $|X'| = 0$. On the other hand, if we expand X' by its jth row, we see that the right-hand side of (4.11) is the determinant of X'. That is, the sum on the right-hand side of (4.11) is zero if $i \neq j$. This completes the proof.

We have almost immediately the following corollary:

Corollary Let B be the matrix whose element in the ith row and jth column is X_{ji}, the cofactor of x_{ji}. Then

$$XB = I|X|$$

This can be shown by writing out the two matrices.

Definition If $X = (x_{ij})$, then we call the matrix $(x_{ij})^T$ the *adjunct* of X and write it as adj(X). (Classically, the term was *adjoint*, but we wish to reserve this term for another concept.)

Thus the corollary can be written

$$X(\text{adj } X) = I|X| \tag{4.12}$$

We leave as an exercise the proof that $(\text{adj } X)X = I|X|$.

Notice that the adjunct exists even when the matrix is singular. However, if X is nonsingular, (4.12) gives us the following formula for the inverse of X:

$$X^{-1} = (\text{adj } X)|X|^{-1} \tag{4.13}$$

One can deduce immediately the following theorem, which is sometimes useful even though the actual solution of a set of linear equations is usually easier without it.

Theorem 4.9 (Cramer's Rule) If the matrix A is nonsingular, then the solution of

$$A\xi^T = \gamma^T$$

where $\xi = (x_1, x_2, \ldots, x_n)$ and $\gamma = (c_1, c_2, \ldots, c_n)$, is

$$x_i = (c_1A_{1i} + c_2A_{2i} + \cdots + c_nA_{ni})|A|^{-1} \qquad \text{for } 1 \leq i \leq n \tag{4.14}$$

Partial Proof If we multiply the given matrix equation on the left by A^{-1} and use (4.13), we have

$$\xi^T = |A|^{-1}(\text{adj } A)\gamma^T$$

If you write out the given matrices and perform the indicated multiplication, you should obtain (4.14). The details are left as an exercise.

Let us look a little more carefully at the expression

$$c_1A_{1i} + c_2A_{2i} + \cdots + c_nA_{ni} \tag{4.15}$$

Compare it with

$$x_{1i}X_{1i} + x_{2i}X_{2i} + \cdots + x_{ni}X_{ni} \tag{4.16}$$

This is the expansion of the determinant of (x_{ij}) by the ith column. Thus (4.15) can be obtained from (4.16) by replacing x_{1i} by c_1, x_{2i} by c_2, . . . , x_{ni} by c_n and X by A. This means that we can think of (4.15) as the expansion by the ith column of the determinant of the matrix obtained from A by replacing the ith column of A by γ^T. For example, if $n = 3$, $\xi = (x_1,x_2,x_3)$, and $|A| \neq 0$,

$$x_1 = \frac{\begin{vmatrix} c_1 & a_{12} & a_{13} \\ c_2 & a_{22} & a_{23} \\ c_3 & a_{32} & a_{33} \end{vmatrix}}{|A|}$$

There is also a relationship between the rank of a matrix and certain determinants of submatrices which is sometimes useful. We have

Theorem 4.10 If the rank of a matrix M is r, it contains a nonsingular (square) submatrix of order r, and all square submatrices of order greater than r are singular.

Proof Let B be a square submatrix of order $s > r$. Then B is, in turn, a submatrix of a submatrix C consisting of certain s rows of M. Now the rows of C are linearly dependent since $s > r$. Thus the rank of C is less than s, and any s columns of C are linearly dependent. Since B consists of certain s columns of C, its columns are linearly dependent, and hence B is singular. This shows the second part of the theorem.

Let A be a matrix whose r rows are r linearly independent rows of M. Now the column rank of A is r, and hence A has r linearly independent columns. If D is a matrix consisting of r linearly independent columns of A, it is nonsingular as well as being a submatrix of N. This completes the proof.

EXERCISES

1. Compute each of the following determinants:

$$\begin{vmatrix} 1 & 2 & 3 & 4 \\ 0 & 2 & 0 & 5 \\ 1 & 0 & 0 & 2 \\ 0 & 0 & 1 & 6 \end{vmatrix} \qquad \begin{vmatrix} 2 & 3 & 4 \\ 5 & 6 & 7 \\ 8 & 9 & 10 \end{vmatrix}$$

$$\begin{vmatrix} 1 & -3 & 8 \\ 4 & 0 & 3 \\ 0 & 2 & 5 \end{vmatrix} \qquad \begin{vmatrix} 1 & 0 & 0 & 5 \\ 0 & 2 & 0 & 3 \\ 1 & -2 & -5 & 6 \\ 2 & 7 & 3 & -2 \end{vmatrix} \qquad \begin{vmatrix} 0 & 1 & 2 & 3 \\ 1 & 2 & 3 & 0 \\ 2 & 3 & 0 & 1 \\ 3 & 0 & 1 & 2 \end{vmatrix}$$

2. Express in terms of determinants the condition that the three points (a_1,a_2), (b_1,b_2), (c_1,c_2) be collinear.

3. Write out explicitly the inverse of the nonsingular matrix $A = (a_{ij})$ of order 2.

4. Use Cramer's rule to write the solution of the equation of Theorem 4.9 for the case $n = 2$.

5. Evaluate the following determinant where $w^3 = 1$, for $w \neq 1$:

$$\begin{vmatrix} 1 & w & w^2 \\ w & w^2 & 1 \\ w^2 & 1 & w \end{vmatrix}$$

If the matrix is singular, find its rank.

6. Prove that $(\text{adj } X)X = I|X|$.

7. Let (a_1,a_2), (b_1,b_2), (c_1,c_2), (d_1,d_2) be four pairs of coordinates, denote by (ab) the determinant $\begin{vmatrix} a_1 & a_2 \\ b_1 & b_2 \end{vmatrix}$, and similarly for the other pairs. Note that $(ac) = -(ca)$ and prove

$$(ab)(cd) + (bc)(ad) + (ca)(bd) = 0$$

Hint: Evaluate the following determinant:

$$\begin{vmatrix} (ad) & a_1 & a_2 \\ (bd) & b_1 & b_2 \\ (cd) & c_1 & c_2 \end{vmatrix}$$

8. Let A, B, C, D stand for the pairs in Exercise 7. In projective geometry, the cross ratio of these, written $R(A,B,C,D)$, is defined as

$$\frac{(ca)(db)}{(cb)(da)}$$

Prove that $R(A,B,C,D) = R(A',B',C',D')$ where, for some linear transformation T, $T(a_1,a_2) = (a_1',a_2')$, and similarly for the other three pairs, T being the same for all. Prove also that $R(A,B,C,D)$ is independent of the homogeneous coordinates used for any of the points (see Sec. 1.11); that is, R is unchanged if (a_1,a_2) is replaced by (ra_1,ra_2), for $r \neq 0$, and similarly for the other points.

9. Prove, using the notation in Exercises 7 and 8, that $R(A,B,C,D) = R(B,A,D,C) = 1 - R(A,C,B,D) = 1/R(B,A,C,D)$.

10. Prove the Corollary of Theorem 4.8.

11. Find a relationship between the determinant of a square matrix and that of its adjunct.

12. Let r be the rank of the matrix A of order n. Find, in terms of r and n, the rank of the adjunct of A.

13. Find adj(adj A).

14. Complete the proof of Theorem 4.9.

15. Prove that if A is a matrix of order 2, then $A + \text{adj } A = \text{tr}(A)I$, where $\text{tr}(A)$ is the trace of A.

16. Complete the proof of Theorem 4.7.

17. In matrix $A = (a_{ij})$, $a_{jj} = j$, for $1 \leq j \leq n$, and $a_{ij} = 1$ if $i \neq j$. Prove that $|A| = (n - 1)!$

18. Prove that it is impossible for all six terms of the determinant of a matrix of order 3 with real elements to be positive.

19. Show that

$$\begin{vmatrix} r & x & x \\ x & s & x \\ x & x & t \end{vmatrix}$$

is equal to $f(x) - xf'(x)$, where $f(x) = (r - x)(s - x)(t - x)$ and $f'(x)$ is the derivative of $f(x)$ with respect to x.

20. Let c be a root of the equation $x^2 + x + 1 = 0$. Evaluate the following determinant:

$$\begin{vmatrix} 1 & c & c^2 & c^3 \\ c & c^2 & c^3 & 1 \\ c^2 & c^3 & 1 & c \\ c^3 & 1 & c & c^2 \end{vmatrix}$$

21. The following is called *Vandermonde's determinant:*

$$\begin{vmatrix} 1 & x_1 & x_1^2 & \cdots & x_1^{n-1} \\ 1 & x_2 & x_2^2 & \cdots & x_2^{n-1} \\ \cdot & \cdot & \cdot & \cdots & \cdot \\ 1 & x_n & x_n^2 & \cdots & x_n^{n-1} \end{vmatrix}$$

Prove that this determinant is zero if $x_i = x_j$ for any two i and j and hence that it is divisible by $x_i - x_j$, for $i \neq j$. Prove that the determinant is equal to ± 1 times the following product:

$$\prod_{1 \leq i < j \leq n} (x_i - x_j)$$

22. Given n points in the plane with coordinates (x_i, y_i) such that no two x_i are equal. Prove that there is a unique polynomial $f(x)$ of degree n, with leading coefficient 1, such that $f(x_i) = y_i$ for $1 \leq i \leq n$.

23. Let C_n be a *continuant matrix* defined by $c_{ii} = a_i$, $c_{i,i+1} = 1$, $c_{i+1,i} = -1$, and all other c_{ij} are zero. Show that $C_n = a_n C_{n-1} + C_{n-2}$.

4.5 THE CHARACTERISTIC EQUATION

At the beginning of this chapter we stated that one reason for introducing determinants is to find a better means for computing the characteristic roots and minimum polynomial of a transformation. Some of the exercises in the previous section pointed toward the connection. Here we explore it further.

We showed in Sec. 3.7 that the characteristic roots of a transformation T are the numbers e such that $T - eI$ is singular. Then, if M is the matrix of T with respect to a certain basis, by Sec. 3.10, $T - eI$ is singular if and only if $M - eI$ is singular, that is, if $|M - eI| = 0$.

This leads us to an important polynomial. Suppose we replace e in $M - eI$ by the variable x. Then the matrix $M - xI$ will look like this:

$$M - xI = \begin{bmatrix} m_{11} - x & m_{12} & m_{13} & \cdots & m_{1n} \\ m_{21} & m_{22} - x & m_{23} & \cdots & m_{2n} \\ m_{31} & m_{32} & m_{33} - x & \cdots & m_{3n} \\ \cdots & \cdots & \cdots & \cdots & \cdots \\ m_{n1} & m_{n2} & m_{n3} & \cdots & m_{nn} - x \end{bmatrix}$$

Its determinant is a polynomial $g(x)$ in x, with coefficients expressed in terms of the m_{ij}. The zeros of this polynomial are the values of x for which the determinant of $M - xI$ is zero, that is, the characteristic roots of M. In other words, the roots of $g(x) = |M - xI| = 0$ are the characteristic roots of the transformation or matrix.

What is the degree of $g(x)$? In the expansion of the determinant, the term which will give the maximum degree in x is the product of the diagonal elements; for this product the coefficient of x^n is $(-1)^n$. So $g(x)$ is of degree n, and its leading coefficient is $(-1)^n$. It is convenient to avoid such a leading coefficient, and we can do this by replacing $M - xI$ by $xI - M$. This has the effect of replacing $g(x)$ by $f(x) = (-1)^n g(x)$ and does not change the set of zeros of the polynomial. It does have the advantage that the leading coefficient of $f(x)$ is now 1. So we have

Theorem 4.11 The characteristic roots of a transformation T with matrix M are the roots of the equation

$$|xI - M| = 0 \tag{4.17}$$

We call Equation (4.17) the *characteristic equation of* T, and the polynomial $f(x) = |xI - M|$ the *characteristic polynomial of* T. We have the right to call it the characteristic polynomial of the transformation since, from a statement at the close of Sec. 4.3 and Exercise 13 of that section, a change of coordinates does not change the characteristic equation.

You will find in some texts that the characteristic polynomial is defined to be the determinant of $M - xI$. As noted above, this has at most the effect of multiplying the polynomial by -1 and does not change the characteristic roots.

Since we can compute the determinant in (4.17) we have now a means of computing the characteristic roots. This is where determinants are most useful to us in dealing with transformations. Since all the characteristic roots of the transformation are zeros of the minimum polynomial (see the definition in Sec. 3.7), it follows that the following theorem and its corollary are true:

Theorem 4.12 If T is a linear transformation of V^n into itself, the zeros of its minimum polynomial are the same as the zeros of its characteristic polynomial.

Corollary If the minimum polynomial of T has n distinct zeros, then the minimum and characteristic polynomials are the same.

There is an extension of this theorem which is easy to prove using definitions and results from Appendix C.

The proof of this theorem uses the idea of an extension of a field. If this would cause difficulty, it is suggested that you omit this theorem and its proof in favor of Theorem 4.19 at the end of this section which, on the one hand, has this theorem as a corollary and, on the other hand, does not make use of the idea of the extension of a field.

Theorem 4.13 If F is the field of coefficients of $f(x)$ and $m(x)$, the characteristic and minimum polynomials, respectively, of a linear transformation of V into itself, then any irreducible factor of $f(x)$ is also a factor of $m(x)$.

Proof Let $g(x)$ be any factor of $f(x)$ which is irreducible over F. We assume that $g(x)$ is not a factor of $m(x)$ and arrive at a contradiction. Now $g(x)$ being irreducible and not a factor of $m(x)$ implies that the two polynomials are relatively prime, and hence there are polynomials $h(x)$ and $k(x)$ in F such that

$$m(x)h(x) + g(x)k(x) = 1 \tag{4.18}$$

Let x_0 be a zero of $g(x)$ in some extension of F. Then $f(x_0) = 0$ and, by Theorem 4.12, $m(x_0) = 0$. If we replace x by x_0 in (4.18) we get the contradiction $0 = 1$. This completes the proof.

Recall that an important property of the minimum polynomial $m(x)$ of a transformation T is that $m(T) = 0$. If for $f(x)$, the characteristic polynomial of T, $f(T)$ were zero, we know from Theorem 3.12 that $m(x)$ would be a factor of $f(x)$. Conversely, if $m(x)$ were a factor of $f(x)$ it would follow that $f(T) = 0$. In fact, if all the zeros of $m(x)$ are simple zeros, that is, if the linear factors are distinct, then $m(x)$ is a factor of $f(x)$ and $f(T) = 0$. All of this seems to point toward the question: If $f(x)$ is the characteristic polynomial of a transformation T, is $f(T) = 0$?

We have shown, above, cases in which the answer to the question is in the affirmative. But it is not clear at this point why it might not be possible, for instance, for $(x - e)^2$ to be a factor of $m(x)$ and for $(x - e)$ to appear only to the first power in $f(x)$. We now state the theorem which shows that the answer to the question is always "yes," and, to demonstrate that we need to exercise care in the proof, we give two bogus proofs before embarking on a genuine one.

Theorem 4.14 (***Cayley-Hamilton Theorem***) A linear transformation T of a vector space V into itself satisfies its characteristic equation; that is,

$$f(T) = 0$$

where $f(x) = |xI - M|$ for M some matrix of T with respect to a basis of V.

Bogus Proof 1 $|xI - M| = 0$ is given. So replace x by M and get $|MI - M|$ which certainly is zero.

Bogus Proof 2 In Equation (4.12) replace X by $xI - M$ and see that

$$(xI - M)[\operatorname{adj}(xI - M)] = I|xI - M| = If(x) \tag{4.19}$$

Now replace x by M throughout, since this is an identity, and get

$$0 = If(M)$$

The trouble with the procedure in both cases is, of course, that if we replace x in xI by M, we have a matrix with n^2 rows and columns which is to be added to M which has only n rows and columns. This is quite different from our procedure of replacing x by a matrix in a polynomial. But we can salvage something from (4.19). First, look at $\operatorname{adj}(xI - M)$. Since its elements are cofactors of the elements of $xI - M$, that is, determinants of matrices of order $n - 1$, each element of $\operatorname{adj}(xI - M)$ will be a polynomial in x of degree $n - 1$ or less. We can therefore write $\operatorname{adj}(xI - M)$ as a polynomial in x with matrix coefficients, that is,

$$\operatorname{adj}(xI - M) = B_{n-1}x^{n-1} + B_{n-2}x^{n-2} + \cdots + B_1x + B_0$$

Let $f(x) = x^n + a_{n-1}x^{n-1} + \cdots + a_1x + a_0$. Then, since $xI - M$ and $\operatorname{adj}(xI - M)$ are commutative, Equation (4.19) can be written in the form

$$(4.20) \quad (B_{n-1}x^{n-1} + B_{n-2}x^{n-2} + \cdots + B_1x + B_0)(xI - M) = (x^n + a_{n-1}x^{n-1} + \cdots + a_1x + a_0)I$$

So, if we equate coefficients of the powers of x on both sides of (4.20), we obtain the following equations:

The coefficient of	*gives*
x^n	$B_{n-1} = I$
x^{n-1}	$-B_{n-1}M + B_{n-2} = a_{n-1}I$
x^{n-2}	$-B_{n-2}M + B_{n-3} = a_{n-2}I$
. . . .	. . . .
x	$-B_1M + B_0 = a_1I$
x^0	$-B_0M = a_0I$

Now if we multiply the above equations on the right from top to bottom by the following in order: $M^n, M^{n-1}, M^{n-2}, \ldots, M, I$, and then add, the sum on the right will, by a telescoping effect, be the zero matrix, and the sum on the left is $f(M)$. Thus we have proved the Cayley-Hamilton theorem.

Recall that in Sec. 3.7 we called the set of (distinct) characteristic roots the spectrum of a transformation. Here it is convenient to extend the idea of a set to allow it to have repeated elements; that is, we call

$$x_1, x_2, x_3, \ldots, x_n$$

the *complete spectrum* of a linear transformation whose characteristic polynomial is

$$f(x) = (x - x_1)(x - x_2)(x - x_3) \cdots (x - x_n)$$

where each number x_i in the complete spectrum is included as many times as the corresponding linear factor in the characteristic polynomial. Since the characteristic polynomial determines the complete spectrum and is determined by it, it is useful to explore this relationship. To see this, write $f(x)$ in the following form:

$$(4.21) \quad f(x) = x^n - b_1x^{n-1} + b_2x^{n-2} - \cdots + (-1)^r b_r x^{n-r} + \cdots + (-1)^{n-1}b_{n-1}x + (-1)^n b_n$$

First recall that if we write $f(x)$ as a product of linear factors

$$(4.22) \qquad f(x) = (x - x_1)(x - x_2) \cdots (x - x_n)$$

then the coefficient b_1 in (4.21) is the sum of the x_i, that is, the sum of the characteristic roots; for, to get the coefficient of x^{n-1} in $f(x)$, we choose $n - 1$ x's and one x_i in all possible ways and take their sum. Similarly, b_2 in (4.21) si the sum of all products of two characteristic roots, that is,

$$\sum_{1 \leq i < j \leq n} x_i x_j$$

Also b_3 is the sum of all possible products of three characteristic roots, and so on. Finally, b_n is the product of the characteristic roots. Of course in these sums, each root is repeated as many times as it occurs in the factorization (4.22).

The coefficients in (4.21) can also be expressed in terms of determinants of certain submatrices. To do this we need a new term: A *principal* minor matrix of a square matrix A is a submatrix obtained by deleting a set of rows of A and the corresponding set of columns, for instance, the first, third, and fifth rows and the first, third, and fifth columns. A principal minor matrix has the property that its principal diagonal coincides with part of the principal diagonal of the given matrix. The trace of a square matrix is the sum of its principal minors of order 1. We call the determinant of a principal minor matrix a *principal minor determinant.* Now, with this terminology, we can describe relationships among the coefficients of (4.21), the characteristic roots, and the sums of principal minor determinants in the following theorem. We have proved the first part of the theorem and leave the proof of the second part as an exercise.

Theorem 4.15 If the characteristic polynomial of a matrix M is written as in (4.21), then

1. The coefficient b_r is the sum of all the products of the characteristic roots taken r at a time.

2. The coefficient b_r is the sum of the principal minor determinants of M of order r. In particular, b_1 is the trace of M and b_n is its determinant.

Notice that though we describe the b_i in part 2 of the theorem in terms of the matrix M, they are invariants of the transformation which M represents, that is, are independent of the basis chosen.

The following is a useful result.

Theorem 4.16 If the nullity of $aI - A$ is s, then $(x - a)^s$ is a factor of the characteristic polynomial of A.

Proof We first prove this for the case $a = 0$ Write the characteristic polynomial of A in the form (4.21), where r is the rank of A. Since A is of rank r, we know from Theorem 4.10 that all square submatrices of order greater than r are singular. But b_{r+1} is the sum of determinants of submatrices of order $r + 1$, and hence $b_{r+1} = 0$. The same is true for all coefficients to the right of b_{r+1} in (4.21). Hence $x^{n-r} = x^s$ is a factor of $f(x)$, and the theorem is proved for this case.

For the general case, let $f(x)$ be the characteristic polynomial of $A - aI$, that is,

$$f(x) = |xI + aI - A| = |(x + a)I - A| = |yI - A| = g(y)$$

where $y = a + x$. Hence $f(y - a) = g(y)$, which is the characteristic polynomial of A. Thus, from the first case, if s is the nullity of $aI - A$, then x^s is a factor of $f(x)$: that is, $(y - a)^s$ is a factor of $g(y)$. This completes the proof.

It should be noted that the converse of this theorem is not true. That is, if $(x - a)^s$ is a factor of the characteristic polymomial and $(x - a)^{s+1}$ is not a factor, it does not follow that the nullity of $aI - A$ is s. The theorem tells us only that in this case the nullity of $aI - A$ is not greater than s. For instance, the characteristic polynomial of

$$A = \begin{bmatrix} 0 & 1 \\ -1 & 2 \end{bmatrix}$$

is $(x - 1)^2$, but the nullity of $1 \cdot I - A$ is only 1.

EXERCISES

1. Find the characteristic polynomial of the following matrix:

$$\begin{bmatrix} 1 & 3 & 4 \\ -2 & 0 & 3 \\ 1 & 2 & 0 \end{bmatrix}$$

2. Find the characteristic polynomial of a linear transformation T taking V^3 into itself, defined over a basis $\alpha_1, \alpha_2, \alpha_3$ as follows:

$$T(\alpha_1) = \alpha_2 \qquad T(\alpha_2) = \alpha_3 \qquad T(\alpha_3) = a\alpha_1 + b\alpha_2 + c\alpha_3$$

3. Prove that in Exercise 2 the characteristic and the minimum polynomials are the same.

4. Find the matrices B_i in the proof of the Cayley-Hamilton theorem (Theorem 4.14) for the following matrix:

$$A = \begin{bmatrix} 1 & 0 & 4 \\ -1 & 1 & 0 \\ 0 & 1 & 2 \end{bmatrix}$$

5. Prove that if $f(x) = |xI - M|$, where M is a square matrix of order n, then $f(0) = (-1)^n|M|$.

6. Show that the characteristic polynomial of $A \oplus B$ is the product of the characteristic polynomials of A and B.

7. Let $f(x)$, the characteristic polynomial of a transformation T, be written in the form (4.21). Prove that

$$b_2 = \sum_{1 \le i < j \le n} \begin{vmatrix} a_{ii} & a_{ij} \\ a_{ji} & a_{jj} \end{vmatrix}$$

What relationship has b_2 to the characteristic roots?

8. Prove that if $f(x)$ is written in form (4.21) and if $b_n \neq 0$, then b_{n-1}/b_n is the sum of the reciprocals of the characteristic roots of the transformation whose characteristic polynomial is $f(x)$.

9. Let R and S be two matrices of order n. Prove that if $I - RS$ is nonsingular, so is $I - SR$ and $(I - SR)^{-1} = I + S(I - RS)^{-1}R$.

10. Using Exercise 9 or by other means, prove that for any two matrices A and B of the same order, AB and BA have the same characteristic roots.

11. Prove that if a transformation T is nonsingular, its inverse is expressible as a polynomial in T of degree $n - 1$ or less.

12. Complete the proof of Theorem 4.15.

4.6 COMPUTATION OF THE MINIMUM POLYNOMIAL

We have seen in previous sections how to find the characteristic polynomial of a transformation. Sometimes this is the minimum polynomial as well. In Sec. 3.8 a method was shown for computing the minimum polynomial, but it left something to be desired not only from a computational point of view but conceptually as well. Here we develop a method which has some advantages from both points of view.

Recall Equation (4.19):

$$(xI - M)[\text{adj}(xI - M)] = If(x) \tag{4.23}$$

where $f(x)$ is the characteristic polynomial of M. We know that $m(x)$, the minimum polynomial, is a factor of $f(x)$. Also, since the elements of $\text{adj}(xI - M)$ are polynomials in x, we know that their greatest common divisor must divide $f(x)$. The following theorem links the two facts:

Theorem 4.17 If $g(x)$ is the g.c.d. (greatest common divisor) of the elements of $\text{adj}(xI - M)$ and $f(x)/g(x) = h(x)$, then $h(x)$ is the minimum polynomial of M.

Proof Let $B(x) = \text{adj}(xI - M)/g(x)$ and see that

$$(xI - M)B(x) = Ih(x) \tag{4.24}$$

As in the proof of the Cayley-Hamilton theorem (Theorem 4.14) in Sec. 4.5, one can prove that $h(M) = 0$, from (4.24). Hence $m(x)$, the minimum polynomial of M, is a factor of $h(x)$. Let $k(x)$ be their quotient, that is,

$$h(x) = m(x)k(x) \tag{4.25}$$

It remains to prove that $k(x)$ is of degree zero.

Now by the factor theorem we know that $m(x) - m(y)$ is divisible by $x - y$, and hence

$$m(x) - m(y) = (x - y)g(x,y) \tag{4.26}$$

for $g(x,y)$, a polynomial in x and y. Since M and xI are commutative, we can replace x by xI and y by M in (4.26) to get

$$m(xI) - m(M) = (xI - M)g(xI,M)$$

that is,

$$m(x)I = (xI - M)g(xI,M) \tag{4.27}$$

Since in (4.24) $B(x)$ and $(xI - M)$ are commutative, we multiply (4.27) on the left by $B(x)$, use (4.24), and get

$$B(x)m(x) = h(x)g(xI,M) \tag{4.28}$$

Using (4.25) in (4.28) and dividing both sides by $m(x)$, we get

$$B(x) = k(x)g(xI,M) \tag{4.29}$$

Now $g(xI,M)$ is a matrix whose elements are polynomials in x. This shows that every element of $B(x)$ is divisible by $k(x)$. But 1 is the g.c.d. of the elements of $B(x)$ by our choice above. Therefore $k(x)$ is a constant, and hence $h(x) = m(x)$. This completes the proof.

Actually you may wonder what progress we have made computationally, since getting the adjunct of $xI - M$ could itself involve much computation. But we can often use Theorem 4.17 and its proof to find the minimum polynomial without computing the adjunct. This is illustrated in Exercise 4 at the end of this section. The following theorem is very useful later on.

Theorem 4.18 Given the polynomial $f(x) = x^n + a_{n-1}x^{n-1} + \cdots + a_1x + a_0$, the following matrix not only has this as its characteristic polynomial but its minimum polynomial as well:

$$M = \begin{bmatrix} 0 & 0 & 0 & \cdots & 0 & -a_0 \\ 1 & 0 & 0 & \cdots & 0 & -a_1 \\ 0 & 1 & 0 & \cdots & 0 & -a_2 \\ \cdots & \cdots & \cdots & \cdots & \cdots & \cdots \\ 0 & 0 & 0 & \cdots & 0 & -a_{n-2} \\ 0 & 0 & 0 & \cdots & 1 & -a_{n-1} \end{bmatrix} \tag{4.30}$$

Proof We know that M is the matrix of a transformation T with respect to a certain basis of the vector space V. Call $\alpha_1, \alpha_2, \ldots, \alpha_n$ the basis, and, using the matrix or its transpose, we can read off the images of the basis as follows:

$$\begin{aligned} T(\alpha_1) &= 0 \cdot \alpha_1 + 1 \cdot \alpha_2 + 0 \cdot \alpha_3 + \cdots + 0 \cdot \alpha_{n-1} + 0 \cdot \alpha_n \\ T(\alpha_2) &= 0 \cdot \alpha_1 + 0 \cdot \alpha_2 + 1 \cdot \alpha_3 + \cdots + 0 \cdot \alpha_{n-1} + 0 \cdot \alpha_n \\ &\cdots\cdots\cdots \\ T(\alpha_{n-1}) &= 0 \cdot \alpha_1 + 0 \cdot \alpha_2 + 0 \cdot \alpha_3 + \cdots + 0 \cdot \alpha_{n-1} + 1 \cdot \alpha_n \\ T(\alpha_n) &= -a_0\alpha_1 - a_1\alpha_2 - a_2\alpha_3 - \cdots - a_{n-2}\alpha_{n-1} - a_{n-1}\alpha_n \end{aligned}$$

This can be written more compactly as

$$\begin{aligned} T(\alpha_i) &= \alpha_{i+1} \qquad \text{for } 1 \le i \le n-1 \\ T(\alpha_n) &= -a_0\alpha_1 - a_1\alpha_2 - a_2\alpha_3 - \cdots - a_{n-1}\alpha_n \end{aligned} \tag{4.31}$$

The first $n - 1$ equations in (4.31) can also be written $T^i(\alpha_1) = \alpha_{i+1}$, and from this we see that

$$\alpha_1, T(\alpha_1), T^2(\alpha_1), T^3(\alpha_1), \ldots, T^{n-1}(\alpha_1) \tag{4.32}$$

is a linearly independent set of vectors, since they are the α_i in order. Thus, if $g(x)$ is a polynomial such that $g(T)(\alpha_1) = \theta$, the degree of $g(x)$ must be more than $n - 1$.

On the other hand, the last equation in (4.31) can be written

$$T(\alpha_n) = -a_0\alpha_1 - a_1T(\alpha_1) - a_2T^2(\alpha_1) - \cdots - a_{n-1}T^{n-1}(\alpha_1) \tag{4.33}$$

But $T(\alpha_n) = T[T^{n-1}(\alpha_1)]$ shows that (4.33) is equivalent to

$$[T^n + a_{n-1}T^{n-1} + a_{n-2}T^{n-2} + \cdots + a_1T + a_0I](\alpha_1) = \theta \tag{4.34}$$

This is just $f(T)(\alpha_1) = \theta$, where $f(x)$ is the polynomial written in the hypothesis of the theorem. The following sequence of equations shows that $f(T)$ takes each element of the basis and hence every vector of V into the null vector:

$$\theta = T^i[f(T)](\alpha_1) = f(T)[T^i(\alpha_1)] = f(T)(\alpha_{i+1})$$

So now we have shown that $f(T) = 0$ for the polynomial given above. We know that no polynomial in T of degree less than n can be zero because of the linear independence of the vectors of (4.32). Hence $f(x)$, having leading coefficient 1, must be the minimum polynomial of the matrix.

It remains to show that $f(x)$ is also the characteristic polynomial. Suppose $h(x)$ is the characteristic polynomial of T. We know that it is of degree n and also that it is divisible by $f(x)$, the minimum polynomial. Since both have leading coefficient 1, they must be equal. So the theorem is proved.

We call the matrix M in (4.30) the *companion matrix* of the given polynomial. We shall show in Chap. 6 that, by a proper choice of basis, any matrix is similar to a direct sum of matrices like M in (4.30).

Prior to Theorem 4.13 we promised a proof not depending on the idea of an extension of a field. Here is a theorem which has Theorem 4.13 as an immediate corollary and which uses an equation in the proof of Theorem 4.17 in place of the idea of an extension of a field.

Theorem 4.19 If T is a transformation of a vector space of dimension n into itself and if $m(x)$ is its minimum polynomial, then the characteristic polynomial of T is a factor of $[m(x)]^n$.

Proof Here we use Equation (4.27) in the proof of Theorem 4.17; namely, $m(x)I = (xI - A)g(xI,A)$, where $g(xI,A)$ is a matrix whose elements are polynomials in x. Since $|m(x)I| = [m(x)]^n$ we have, taking the determinant of both sides of the previous equation,

$$[m(x)]^n = |xI - A| \cdot |g(xI,A)| = f(x)q(x) \tag{4.35}$$

where $f(x)$ is the characteristic polynomial and $q(x)$ is the polynomial $|g(xI,A)|$.

Theorem 4.13 follows since any irreducible factor of $f(x)$ must be a factor of $[m(x)]^n$ and hence of $m(x)$.

Let us see now how all these results apply to a particular transformation. Let a transformation T of V^3 into itself be defined in terms of the canonical basis as follows:

$$T(\epsilon_1) = \epsilon_2 - \epsilon_3 \qquad T(\epsilon_2) = -\epsilon_1 + 2\epsilon_2 - \epsilon_3 \qquad T(\epsilon_3) = \epsilon_3$$

Then the matrix of T with respect to this basis is

$$M = \begin{bmatrix} 0 & -1 & 0 \\ 1 & 2 & 0 \\ -1 & -1 & 1 \end{bmatrix}$$

To find the characteristic polynomial, form the matrix $xI - M$ and compute its determinant. If you do this you will find that $f(x) = (x - 1)^3$ is the characteristic polynomial of T; hence the only characteristic root is 1. This is perhaps the most important contribution which the determinant makes to the information about a transformation.

As we have done previously, we seek the characteristic vectors associated with the characteristic root 1, that is, the solutions of $MX^T = X^T$, where X is the vector (x_1,x_2,x_3). The last equation is equivalent to $(M - I)X^T = 0$, and if we write this out, it looks like

$$(4.36) \qquad (M - I)X^T = \begin{bmatrix} -1 & -1 & 0 \\ 1 & 1 & 0 \\ -1 & -1 & 0 \end{bmatrix} \begin{bmatrix} x_1 \\ x_2 \\ x_3 \end{bmatrix} = \begin{bmatrix} 0 \\ 0 \\ 0 \end{bmatrix}$$

Now $M - I$ is of rank 1, and hence its nullity is $3 - 1 = 2$. That is, the kernel of $T - I$ is of dimension 2. If we denote by K the kernel of $T - I$, we see that it consists of the vectors ω such that $(T - I)(\omega) = 0$, that is $T(\omega) = \omega$. So V has a two-dimensional subspace K consisting of vectors unchanged by T.

The above is the most useful information about T, but one can go further and find a basis of V which contains one of K and thereby simplify the matrix of T. The solutions of Equation (4.36) are $x_1 + x_2 = 0$, that is, the set of triples $(x_1,-x_1,x_3)$. If we choose as a basis of K the vectors $\alpha_1 = (1,-1,0) = \epsilon_1 - \epsilon_2$ and $\alpha_2 = (0,0,1) = \epsilon_3$, we see that any element of K is a linear combination of α_1 and α_2. If we then complete the basis of V by ϵ_1, for example, the transformation T can be defined by

$$T(\alpha_1) = \alpha_1 \qquad T(\alpha_2) = \alpha_2 \qquad T(\epsilon_1) = -\alpha_1 - \alpha_2 + \epsilon_1$$

and the matrix with respect to this basis is

$$N = \begin{bmatrix} 1 & 0 & -1 \\ 0 & 1 & -1 \\ 0 & 0 & 1 \end{bmatrix}$$

One can also find the minimum polynomial. For this one could use Theorem 4.17, but it is probably simpler to notice that since $(x - 1)^3$ is the characteristic polynomial, the minimum polynomial must be a factor of this. So we compute $N - I$ or $M - I$ and see that neither is the zero matrix; this shows that the minimum polynomial is not $x - 1$. But $(N - I)^2$ is the zero matrix, and hence $(x - 1)^2$ is the minimum polynomial.

EXERCISES

1. Find a matrix whose characteristic and minimum polynomials are both $f(x) = x^3 + x^2 - x - 1$. What is the complete spectrum? By use of the factors of $f(x)$ or by other means, show that $f(M) = 0$, where M is the matrix which you found.

2. Suppose the characteristic polynomial of a matrix is $f(x)$ in Exercise 1 but its minimum polynomial is not this polynomial. What must the minimum polynomial be?

3. Determine the minimum polynomial of the matrix

$$\begin{bmatrix} 0 & 1 & -1 \\ -2 & -3 & 2 \\ -1 & -1 & 0 \end{bmatrix}$$

4. For the companion matrix of Theorem 4.18 note that 1 is the determinant of the submatrix of order $n - 1$ in the lower left-hand corner of M. Using Theorem 4.17 or by other means, prove independently of the proof of Theorem 4.18 that the minimum polynomial and characteristic polynomial of M are the same.

5. For the transformation whose matrix you found in Exercise 1, find the two subspaces of V which are invariant under T.

6. For the transformation whose matrix is described in Exercise 2, find two invariant subspaces of V.

7. Let M_1 and M_2 be two companion matrices as in Theorem 4.18. What is the characteristic polynomial of their direct sum? Under what conditions will M_1 and $M_1 \oplus M_2$ have the same minimum polynomial?

8. In the proof of Theorem 4.18 just beyond Equation (4.34), we showed by a sequence of equations that $f(T)(\alpha_i) = \theta$ for all basis elements α_i. Why was this unnecessary?

9. Find the minimum and characteristic polynomials of the following matrix of order n:

$$\begin{bmatrix} a & 0 & 0 & \cdots & 0 & 0 \\ 1 & a & 0 & \cdots & 0 & 0 \\ 0 & 1 & a & \cdots & 0 & 0 \\ \cdot & \cdot & \cdot & \cdots & \cdot & \cdot \\ 0 & 0 & \cdots & \cdots & 1 & a \end{bmatrix}$$

10. Let $A(x)$ be a square matrix whose elements are differentiable functions of x. Denote the rows of $A(x)$ by $\alpha_1, \alpha_2, \ldots, \alpha_n$. Prove that the derivative of the determinant of $A(x)$ with respect to x is

$$\begin{vmatrix} \alpha_1' \\ \alpha_2 \\ \alpha_3 \\ \cdot \\ \cdot \\ \cdot \\ \alpha_n \end{vmatrix} + \begin{vmatrix} \alpha_1 \\ \alpha_2' \\ \alpha_3 \\ \cdot \\ \cdot \\ \cdot \\ \alpha_n \end{vmatrix} + \cdots + \begin{vmatrix} \alpha_1 \\ \alpha_2 \\ \alpha_3 \\ \cdot \\ \cdot \\ \cdot \\ \alpha_n' \end{vmatrix}$$

where α_i' denotes the row vector obtained from α_i by replacing each element by its derivative with respect to x.

11. Let $f(x) = |xI - A|$, for some matrix A with real elements. Prove that the derivative of $f(x)$ with respect to x is equal to the trace of the adjunct of $xI - A$.

5

INNER PRODUCT SPACES; BILINEAR, QUADRATIC, AND HERMITIAN FORMS

5.1 PRODUCTS OF VECTORS

So far in this book, except in certain examples from Euclidean geometry, we have made no mention of distance. Nor have we defined a product of vectors in any sense. In this chapter we shall define both products of vectors and a kind of distance function and explore the consequences of these ideas. Here the most accessible application will be to n-dimensional Euclidean geometry.

To point the way to a fruitful set of postulates, let us first consider the usual vector product in the field of real numbers, called an *inner product* or *dot product*. The latter name comes from the notation $\xi \cdot \eta$, for this kind of a product of the two vectors ξ and η. In V^2 the dot product is defined as follows:

$$(x_1,x_2) \cdot (y_1,y_2) = x_1y_1 + x_2y_2$$

This can be given a geometric meaning by considering the two number pairs to be coordinates of the points X and Y and by computing in two different ways the distance XY. Using the Pythagorean theorem we have

$$(5.1)\qquad \begin{aligned}(XY)^2 &= (y_1 - x_1)^2 + (y_2 - x_2)^2\\ &= y_1^2 + x_1^2 + y_2^2 + x_2^2 - 2(x_1y_1 + x_2y_2)\end{aligned}$$

On the other hand, if we use the law of cosines for the triangle OXY, where O is the origin, we have

$$(5.2)\qquad (XY)^2 = (OX)^2 + (OY)^2 - 2(OX)(OY)\cos\varphi$$

where φ is the angle from the vector OX to the vector OY. Hence, if we equate the right sides of (5.1) and (5.2), we have

$$(OX)^2 + (OY)^2 - 2(OX)(OY)\cos\varphi = y_1^2 + x_1^2 + y_2^2 + x_2^2 - 2(x_1y_1 + x_2y_2)$$

Since

$$(OX)^2 = x_1^2 + x_2^2 \qquad \text{and} \qquad (OY)^2 = y_1^2 + y_2^2$$

we have

$$(5.3)\quad x_1y_1 + x_2y_2 = (OX)(OY)\cos\varphi = \sqrt{x_1^2 + x_2^2}\sqrt{y_1^2 + y_2^2}\cos\varphi$$

Thus the dot product of the vectors associated with the points X and Y is the product of the lengths of the line segments OX and OY by the cosine of the angle between them.

More generally, in n dimensions, if X is the point associated with the n-tuple $(x_1, x_2, \ldots, x_n)$ and Y with $(y_1, y_2, \ldots, y_n)$, then by definition

$$(5.4)\quad (x_1, x_2, \ldots, x_n) \cdot (y_1, y_2, \ldots, y_n) = x_1y_1 + x_2y_2 + \cdots + x_ny_n$$

Such a product is called a *dot product* or *inner product.* If we define distance for n dimensions as we do for two, since the triangle OXY is two dimensional we have, analogous to (5.3), the following:

$$(5.5)\qquad (x_1, x_2, \ldots, x_n) \cdot (y_1, y_2, \ldots, y_n) = (OX)(OY)\cos\varphi$$

where φ is the angle between the two vectors X and Y.

To be more rigorous about n dimensions, we define the distance between any two points X and Y to be that given by the expression corresponding to (5.1), and we define the angle between the vectors to be that angle between 0 and $\pi/2$ determined by Equation (5.5). In particular, the inner product will be zero if and only if the angle φ is $\pi/2 + k\pi$ for some integer k. This is thus the condition that the two vectors be perpendicular (the more customary term is *orthogonal*). Formally we have the following:

Definition Two vectors are said to be *orthogonal* if their inner product is zero.

In the light of the above discussion, let us draw up a list of *some* of the properties of an inner product. These can serve as a generalization of the inner product. So let V and W be vector spaces over a field F, and let (ξ,η) denote a function defined for all ξ in V and η in W with the following properties:

1. (ξ,η) is in F.
2. (ξ,η) is a *bilinear function*, that is,
   (*a*) $(\xi_1 + \xi_2, \eta) = (\xi_1,\eta) + (\xi_2,\eta)$ and $(\xi, \eta_1 + \eta_2) = (\xi,\eta_1) + (\xi,\eta_2)$.
   (*b*) $(c\xi,\eta) = c(\xi,\eta)$ and $(\xi,c\eta) = c(\xi,\eta)$ for all c in F.

We call such a function a *vector product function*. It is a function of pairs of vectors of V and W into F, written $V \times W \to F$, which is bilinear. The property of being bilinear could be described loosely by saying that the function is linear in both parts of the pair. Notice that V and W may have different dimensions. However, much of our discussion centers around the case when V and W are the same space. When $V = W$ we refer to a *vector product function over* V.

There are two properties of the dot product as introduced above which we do not assume at this point. The first is that $(\xi,\xi) = 0$ does not imply that ξ is the null vector, as for inner products in the field of real numbers. Second, we do not assume the *commutative* or *symmetric* property:

$$(\xi,\eta) = (\eta,\xi)$$

One reason for not assuming symmetry is that the equality would force us to take V and W to be the same space. Another reason is that in the field of complex numbers this property does not hold. However, avoiding symmetry of the vector product forces us to postpone consideration of orthogonality since, under these circumstances, (ξ,η) could be zero without (η,ξ) being zero. Such a possibility would do violence to our intuitive idea that orthogonality should be a symmetric relation; that is, if one vector is orthogonal to a second one, then the second should be orthogonal to the first. In this connection it is interesting to note that if $V = W$ and in place of the symmetry property one assumes that $(\xi,\eta) = 0$ implies $(\eta,\xi) = 0$ for all ξ and η in V (along with bilinearity), then either the symmetric property holds for all pairs of vectors of the space or $(\xi,\eta) = -(\eta,\xi)$ for all ξ and η. A proof of this is given in Appendix D.

We shall assume these additional properties in due course, but at present there are advantages in proceeding without them.

5.2 BILINEAR AND QUADRATIC FORMS

It is apparent that there are many possible vector product functions of $V \times W$ into F. By means of the bilinear property we can characterize all such functions by describing them in terms of bases of the vector spaces. That is, suppose

$$\alpha_1, \alpha_2, \ldots, \alpha_n \qquad \text{and} \qquad \beta_1, \beta_2, \ldots, \beta_s$$

are bases of the spaces V and W, respectively, and let

$$\xi = \sum_{i=1}^{n} x_i\alpha_i \qquad \text{and} \qquad \eta = \sum_{j=1}^{s} y_j\beta_j$$

be two arbitrary vectors in V and W where x_i and y_i are in F. Then, from the bilinear property of the vector product,

$$(5.6) \qquad (\xi,\eta) = \Sigma x_i y_j(\alpha_i,\beta_j)$$

where the sum is over $1 \leq i \leq n$ and $1 \leq j \leq s$.

If $(\alpha_i,\beta_j) = a_{ij}$, Equation (5.6) becomes

$$(5.7) \qquad (\xi,\eta) = \Sigma a_{ij}x_i y_j$$

where the sum is over i and j in the ranges indicated above. This can also be written in matrix form as

$$(5.8) \qquad (\xi,\eta) = XAY^T$$

where $A = (a_{ij})$, $X = (x_1, x_2, \ldots, x_n)$, and $Y = (y_1, y_2, \ldots, y_s)$.

The sum on the right of (5.7) is called a *bilinear form* since it is linear in the components of X and also in the components of Y. Since there is no reason why (α_i,β_j) should be equal to (α_j,β_i), it is not true that $a_{ij} = a_{ji}$ for the general bilinear form. Notice that the relationship (5.7) between the bilinear form and the vector product goes in both directions. That is, given two bases, not only does the vector product of pairs of elements of the bases determine a bilinear form but if we instead start with a bilinear form, the vector products are determined.

An important special case of the vector product function occurs when the spaces V and W are the same, that is, when the function takes $V \times V$ into the field F. Then it is natural to choose the same basis for V in both parts of the Cartesian product $(V \times V)$. We also take the vector product function in this case to be symmetric; that is, $(\xi,\eta) = (\eta,\xi)$ for all ξ and η in V. Since at first blush this may seem rather arbitrary, consider (ξ,ξ) first for dimension 2. Then, from (5.7),

$$(5.9) \qquad (\xi,\xi) = a_{11}x_1{}^2 + a_{12}x_1x_2 + a_{21}x_2x_1 + a_{22}x_2{}^2$$

Since, for any ξ, the components x_1 and x_2 are considered to be commutative,

$$(5.9a) \qquad (\xi,\xi) = a_{11}x_1{}^2 + \tfrac{1}{2}(a_{12} + a_{21})x_1x_2 + \tfrac{1}{2}(a_{21} + a_{12})x_2x_1 + a_{22}x_2{}^2$$

The symmetric bilinear form in (5.9a) leads to the same (ξ,ξ) as (5.9).

More generally if we define the function $(\xi,\eta)'$ by

$$(\xi,\eta)' = \tfrac{1}{2}[(\xi,\eta) + (\eta,\xi)]$$

it is apparent that it is symmetric. It is not hard to show that it is a vector product function (left as an exercise) and that $(\xi,\xi)' = (\xi,\xi)$.

In general

$$(5.10) \qquad (\xi,\xi) = \sum_{i,j=1}^{n} a_{ij}x_ix_j \qquad \text{where } a_{ij} = a_{ji}$$

We call this a *quadratic form*. (It is called *quadratic* since it is of the second degree, and a *form* since every term is of the same degree.)

An important advantage in considering the bilinear form to be symmetric, as noted previously, is that orthogonality can be introduced without difficulty since now $(\xi,\eta) = 0$ if and only if $(\eta,\xi) = 0$. Note that we have tacitly assumed $1 + 1 \neq 0$. Formally we have the following:

Definition If a vector product function of $V \times V$ into F is symmetric and if $1 + 1 \neq 0$ in F, then we call two vectors ξ and η *orthogonal* if $(\xi,\eta) = 0$. If $(\xi,\xi) = 0$ with $\xi \neq \theta$, we call ξ *isotropic*. (Some authors include the null vector among the isotropic ones.)

As noted above, the matrix associated with a vector product function depends on the basis used. Later in this section we shall explore the effect which a change of basis has on the matrix of the bilinear or quadratic form. But the important properties of a bilinear form are those which are independent of the particular basis chosen. One such property is that of singularity defined as follows:

Definition A vector product function of $V \times W$ into F is said to be *singular* over $V \times W$ if there is a nonzero vector α such that

$$(\alpha,\xi) = 0$$

or all vectors ξ in W. Otherwise the function is called *nonsingular*.

Another way to describe a singular vector product function is to affirm the existence of a nonzero vector α which is orthogonal to all vectors of V including itself. The connection between the singularity of a vector product and its matrix with respect to a particular basis is given by the following:

Theorem 5.1 Let A be the matrix of a vector product function with respect to some bases of V and W, where V and W have the same dimension n. Then A is singular if and only if the vector product function is singular over $V \times W$.

Proof Write

$$\xi = \sum_{i=1}^{n} x_i\alpha_i \qquad \text{and} \qquad \eta = \sum_{j=1}^{n} y_j\beta_j$$

where the α_i and the β_j form bases of V and W, respectively. The vector product (ξ,η) will be singular over $V \times W$ if and only if for some $\xi \neq \theta$ we have $(\xi,\beta_j) = 0$ for $1 \leq j \leq n$; that is,

$$\sum_{i=1}^{n} x_i(\alpha_i,\beta_j) = 0$$

for some set of x_i not all zero. If we write $(\alpha_i,\beta_j) = a_{ij}$ this condition becomes

$$\sum_{i=1}^{n} a_{ij}x_i = 0 \qquad \text{for } 1 \leq j \leq n$$

This is a set of n homogeneous linear equations in n unknowns. It has a nontrivial solution if and only if the matrix $A = (a_{ij})$ is singular. This completes the proof.

Notice that the theorem affirms that the singularity of the matrix of a vector product function is independent of the bases chosen. Since a matrix is singular if and only if its transpose is singular, we have the following

Corollary For a vector product function, there is a nonzero vector α in V such that $(\alpha,\eta) = 0$ for all η in W, if and only if there is a nonzero vector β in W such that $(\xi,\beta) = 0$ for all ξ in V. We are assuming that V and W have the same dimension.

It is enlightening to verify this theorem by the use of matrices. Referring back to Equation (5.8), suppose that a change of basis replaces X by XR and Y by YS, where R and S are nonsingular matrices. Then XAY^T becomes

$$XRAS^TY^T$$

Hence A is replaced by RAS^T. But since R and S are nonsingular, A is singular if and only if RAS^T is singular. This also shows, then, that the nonsingularity of the matrix of the vector product is independent of the bases chosen.

There are a number of reasons why we shall largely confine our discussion to vector product functions which are nonsingular over their vector spaces. On the one hand, we shall see that certain desirable theorems are just not true when the vector product is singular. On the other hand, when such a function is singular we can, in a certain sense, replace the vector spaces by subspaces over which the vector product is nonsingular. To see why this is so, consider a vector product function which is singular over $V \times W$, let α_1 be a vector of V such that $(\alpha_1,\eta) = 0$ for all η in W, and let β_1 be a vector of W

such that $(\xi,\beta_1) = 0$ for all ξ in V. Then we can take $\alpha_1, \alpha_2, \ldots, \alpha_n$ to be a basis of V and $\beta_1, \beta_2, \ldots, \beta_n$ to be a basis of W. It follows that

$$(x_1\alpha_1 + x_2\alpha_2 + \cdots + x_n\alpha_n, y_1\beta_1 + y_2\beta_2 + \cdots + y_n\beta_n) = (x_2\alpha_2 + \cdots + x_n\alpha_n, y_2\beta_2 + \cdots + y_n\beta_n)$$

for all vectors of V and W. Then if V_1 is the space spanned by $\alpha_2, \ldots, \alpha_n$ and W_1 that spanned by $\beta_2, \ldots, \beta_n$, it follows that every vector product over $V \times W$ is equal to the vector product of corresponding vectors in V_1 and W_1. We can continue this process until we arrive at subspaces V_0 and W_0 over which the vector product function is nonsingular.

It should be clear that dealing with a vector product function will be facilitated if we can choose a basis so that the matrix will be in some sense simple. And certainly it is natural to consider a diagonal matrix simpler than a nondiagonal one. So suppose $V = W$, and the bilinear form is symmetric. We would hope to be able to choose a basis of V so that the matrix of the vector product is diagonal. If the vector product is nonsingular, then none of the diagonal elements of the diagonal matrix can be zero. So we make the following definition:

Definition A basis $\alpha_1, \alpha_2, \ldots, \alpha_n$ of a vector space V with a symmetric vector product function over V is called an *orthogonal basis* if

$$(5.11) \qquad \begin{array}{ll} (\alpha_i,\alpha_j) = 0 & \text{for } i \neq j \\ (\alpha_i,\alpha_i) \neq 0 & \text{for } 1 \leq i \leq n \end{array}$$

The following theorem states that such a basis always exists for nonsingular vector product functions.

Theorem 5.2 If a symmetric vector product function over V is nonsingular and if $1 + 1 \neq 0$ in F, then V has an orthogonal basis.

Proof Let α be a nonzero vector. Then the nonsingularity of the vector product function implies that there is a vector β such that $(\alpha,\beta) \neq 0$. Then at least one of the following is different from zero:

$$(\alpha,\alpha) \qquad (\beta,\beta) \qquad (\alpha + \beta, \alpha + \beta) = (\alpha,\alpha) + (\beta,\alpha) + (\alpha,\beta) + (\beta,\beta)$$

since if the first two are zero, the third is equal to $(\alpha,\beta) + (\beta,\alpha) = 2(\alpha,\beta) \neq 0$ unless $1 + 1 = 0$ in F. So now there is a vector α_1 such that $(\alpha_1,\alpha_1) \neq 0$.

Now we proceed to a proof by induction on n. If $n = 1$, and $V = \text{sp}(\alpha)$, then nonsingularity implies that $(\alpha,\alpha) \neq 0$ and the theorem holds. Assume the theorem true for all vector spaces of dimension less than n. Starting with α_1, such that $(\alpha_1,\alpha_1) \neq 0$, we complete the basis of V to

$$\alpha_1, \beta_2, \beta_3, \ldots, \beta_n$$

We then seek a basis $\{\alpha_i\}$ in which α_1 is orthogonal to $\alpha_2, \alpha_3, \ldots, \alpha_n$. To this end let $\alpha_i = c_i\alpha_1 + \beta_i$, with c_i to be determined. Then

$$(\alpha_1,\alpha_i) = c_i(\alpha_1,\alpha_1) + (\alpha_1,\beta_i)$$

Since $(\alpha_1,\alpha_1) \neq 0$ we can choose c_i so that $(\alpha_1,\alpha_i) = 0$ for $2 \leq i \leq n$. (In some cases c_i may be zero.) The α_i so defined form a linearly independent set, and we define $V_1 = \text{sp}(\alpha_2, \alpha_3, \ldots, \alpha_n)$. The vector product function is nonsingular over V_1 because if V_1 contained a vector γ orthogonal to all elements of V_1 it would be orthogonal to α_1 as well and therefore to all vectors of V, which is impossible. Thus, by the induction hypothesis, V_1 has an orthogonal basis. This, together with α_1, constitutes an orthogonal basis of V, and the proof is complete.

We shall see that a completion theorem for orthogonal bases holds. That is, if a subspace of V has an orthogonal basis $\mathcal{B}$, then V has an orthogonal basis containing $\mathcal{B}$. But it is convenient to introduce on the way the concept of annihilator, which is of interest in its own right. Suppose we have a symmetric vector product function over V and let W be a subspace of V. The set of all vectors of V orthogonal to all vectors of W forms a vector space (the proof is left as an exercise). We call this vector space the *annihilator* of W and write it $W^\perp$.

For example, let $n = 2$ and let V have a basis α_1, α_2 such that $(\alpha_i,\alpha_j) = 1$ for $i = 1, 2$ and $j = 1, 2$. Then let $W = V$ and see that $W^\perp = \text{sp}(\alpha_1 - \alpha_2)$. In this case $\dim W + \dim W^\perp = 3 > 2 = \dim V$. This vector product is singular over V. When the vector product is nonsingular, $\dim W + \dim W^\perp$ cannot be greater than $\dim V$. Specifically we can prove

Theorem 5.3 If a symmetric vector product is nonsingular over V and W is any subspace of V, then

$$\dim W + \dim W^\perp = \dim V \tag{5.12}$$

Proof Let $\alpha_1, \alpha_2, \ldots, \alpha_r$ be a basis of W and complete it to a basis $\{\alpha_i\}$ of V. Now $\xi = \sum_{i=1}^{n} x_i\alpha_i$ is an element of $W^\perp$ if and only if $(\alpha_j,\xi) = 0$ for $1 \leq j \leq r$, that is,

$$\begin{aligned} (\alpha_1,\alpha_1)x_1 + (\alpha_1,\alpha_2)x_2 + \cdots + (\alpha_1,\alpha_n)x_n &= 0 \\ (\alpha_2,\alpha_1)x_1 + (\alpha_2,\alpha_2)x_2 + \cdots + (\alpha_2,\alpha_n)x_n &= 0 \\ \cdots\cdots\cdots\cdots\cdots\cdots\cdots\cdots & \\ (\alpha_r,\alpha_1)x_1 + (\alpha_r,\alpha_2)x_2 + \cdots + (\alpha_r,\alpha_n)x_n &= 0 \end{aligned} \tag{5.13}$$

The set of coefficients of the r equations (5.13) constitutes the first r rows of the matrix A of the vector product with respect to the given basis. Since A

is nonsingular, its first r rows are linearly independent. Hence, by Theorem 2.13, the set of solutions of (5.13) is of dimension $n - r$. This completes the proof.

Now if the vector product function over V is nonsingular over the subspace W as well as V, then $W \cap W^\perp = \theta$ since any vector common to $W^\perp$ and W will be an element of W orthogonal to every element of W. (The converse of this statement is also true and is left as an exercise.) Thus we have

Theorem 5.4 If a symmetric vector product function is nonsingular over V and W, a subspace of V, then

$$V = W \oplus W^\perp$$

Definition Under the conditions of Theorem 5.4, $W^\perp$ is called the *orthogonal complement* of W in V. Then we write $V = W \perp W^\perp$. (One should note that the definition implies its uniqueness.)

We have immediately the completion theorem for orthogonal bases.

Theorem 5.5 If the vector product function is symmetric and nonsingular and $\alpha_1, \alpha_2, \ldots, \alpha_r$ is an orthogonal set of vectors of V, there is an orthogonal basis of V which contains α_i for $1 \leq i \leq r$.

We leave the proof of this as an exercise. We state and prove the following:

Theorem 5.6 Under the conditions of Theorem 5.4, $(W^\perp)^\perp = W$.

Proof All elements of W are orthogonal to all elements of $W^\perp$, and hence W is contained in $(W^\perp)^\perp$. But W and $(W^\perp)^\perp$ have the same dimension, $n - \dim W^\perp$, where $n = \dim V$.

The proof of the following is left as an exercise:

If a set of $r \leq n$ vectors α_i satisfy condition (5.11), *they must be linearly independent.*

Finally we should notice that without the condition that the vector product is nonsingular over V_1, the completion theorem (Theorem 5.5) is false. To give an example, suppose α_1, α_2 is a basis of V, and

$$(\alpha_1,\alpha_1) = 1 \qquad \text{and} \qquad (\alpha_1,\alpha_2) = (\alpha_2,\alpha_2) = (\alpha_2,\alpha_1) = 0$$

Then if $W = \operatorname{sp}(\alpha_1)$, α_1 cannot be part of an orthogonal basis of V, for if $\beta = r\alpha_1 + s\alpha_2$, we have $(\alpha_1,\beta) = r$, which implies $(\alpha_1,\beta) = 0$ if and only if

$\beta = s\alpha_2$. But in this case, $(\beta,\beta) = 0$. So there is no orthogonal basis of V containing α_1. In fact, there is no orthogonal basis of V.

EXERCISES

1. Let $\alpha_1 = (1,0)$ and $\alpha_2 = (1,1)$ be a basis of V and let $\beta_1 = (0,1)$ and $\beta_2 = (2,1)$ be another basis. Define (α_i,β_j) to be the inner product of the vectors α_i and β_j as in the beginning of Sec. 5.1, for example, $(\alpha_1,\beta_2) = 2$. Find the matrix of the vector product with respect to these bases.

2. In Exercise 1 find the matrix of the vector product if α_i and β_j are interchanged.

3. Let

$$R = \begin{bmatrix} 1 & -1 \\ 1 & 0 \end{bmatrix} \qquad S = \begin{bmatrix} 0 & -1 \\ 1 & 0 \end{bmatrix} \qquad \begin{bmatrix} \alpha_1 \\ \alpha_2 \end{bmatrix} R = \begin{bmatrix} \alpha_1' \\ \alpha_2' \end{bmatrix} \qquad \begin{bmatrix} \beta_1 \\ \beta_2 \end{bmatrix} S = \begin{bmatrix} \beta_1' \\ \beta_2' \end{bmatrix}$$

Find the matrix of the vector product of Exercise 1 with respect to the bases α_1', α_2' and β_1', β_2'.

4. In Exercise 3, interchange R and S. Are the results in Exercise 3 still valid?

5. Find an orthogonal basis of V^3 which contains $\alpha_1 = (1,1,0)$.

6. In Exercise 1, replace the basis vectors by

$$\alpha_1 = (0,1) \qquad \alpha_2 = (1,1) \qquad \beta_1 = (1,0) \qquad \beta_2 = (1,2)$$

Compute the matrix of the inner product and compare it with that for Exercise 1.

7. Let α_1, α_2 be one basis of V and β_1, β_2 be another basis of V. Let A be the matrix of order 2 whose ijth element is (α_i,β_j). Take $A = \begin{bmatrix} 2 & 1 \\ 2 & 5 \end{bmatrix}$. Find $(x_1\alpha_1 + x_2\alpha_2,\ y_1\beta_1 + y_2\beta_2)$. Is there a vector $a_1\beta_1 + a_2\beta_2$ with not both a_1 and a_2 zero such that $(x_1\alpha_1 + x_2\alpha_2,\ a_1\beta_1 + a_2\beta_2) = 0$ for all x_1 and x_2?

8. Repeat Exercise 7 for $A = \begin{bmatrix} 1 & 2 \\ 2 & 4 \end{bmatrix}$

9. If in Exercise 7, $A = \begin{bmatrix} 1 & 2 \\ 2 & 7 \end{bmatrix}$ and $\alpha_i = \beta_i$, for $i = 1, 2$, is there a nonzero vector $x_1\alpha_1 + x_2\alpha_2$ whose vector product with itself is 0?

10. Repeat Exercise 9 for $A = \begin{bmatrix} 1 & 2 \\ 2 & 3 \end{bmatrix}$

11. Would your answers in Exercises 9 and 10 be the same if the condition $\alpha_i = \beta_i$ for $i = 1, 2$ were omitted?

12. Let α_1, α_2 and β_1, β_2 be two bases of V and let the vector product be determined by the matrix $A = \begin{bmatrix} a_{11} & a_{12} \\ a_{21} & a_{22} \end{bmatrix}$. Prove that there is a nonzero vector $b\alpha_1 + c\alpha_2$ such that

$$(b\alpha_1 + c\alpha_2, y_1\beta_1 + y_2\beta_2) = 0$$

for all y_1, y_2 if and only if the determinant of A is zero.

13. Let (ξ,η) be a vector product function of $V \times V$ into F in which $1 + 1 \neq 0$. Prove that $(\xi,\eta)' = \frac{1}{2}[(\xi,\eta) + (\eta,\xi)]$ is also a vector product function.

14. Prove that for a symmetric vector product function

$$(\xi,\eta) = \tfrac{1}{4}(\xi + \eta, \xi + \eta) - \tfrac{1}{4}(\xi - \eta, \xi - \eta)$$

15. Let S and T be two subspaces of a vector space V. Prove

$$(S + T)^{\perp} = S^{\perp} \cap T^{\perp} \qquad \text{and} \qquad (S \cap T)^{\perp} = S^{\perp} + T^{\perp}$$

Note that the conditions of Theorem 5.4 are assumed.

16. Prove the statement after Theorem 5.6.

17. Let V be a vector space with a symmetric vector product function and W a subspace of V. Show that $W^{\perp}$, the set of all vectors of V orthogonal to W, forms a vector space.

18. Let V be a vector space with a symmetric vector product function and W a subspace of V. Prove that if $W \oplus W^{\perp} = V$, then the vector product function is nonsingular over W.

19. Prove Theorem 5.5.

20. Is the following theorem true? If V is a vector space with a vector product function which is singular over V, then there is a basis for which the matrix of the vector product is a diagonal matrix. Establish your conclusion by proof or counterexample.

21. Suppose in Theorem 5.1 we assume that the dimension of W is greater than that of V; can the vector product function be singular? Can it be nonsingular? Answer the same questions for the case in which the dimension of W is less than that of V.

22. Is the Corollary of Theorem 5.1 true when the dimensions of V and W are not the same? Give reasons for your answer.

5.3 THE ADJOINT AND FUNCTIONALS

We have seen what the effect of a change of basis is on the matrix of a symmetric vector product. What about the effect of a transformation on a vector product itself? By way of introduction, let us look at an example from three-dimensional analytic geometry. We know that the equation $ax + by + cz = 0$ represents the equation of a plane through the origin since it may be thought of as the set of vectors (x,y,z) orthogonal to the vector (a,b,c). The equation may be written in the form of an inner product as follows:

$$(x,y,z) \cdot (a,b,c) = XA^T \tag{5.14}$$

where $X = (x,y,z)$ and $A = (a,b,c)$. The question we ask is: If a linear transformation S replaces X by $X' = (x',y',z')$, what effect does this have on the coefficients of the equation $ax + by + cz = 0$? More precisely, for what matrix R is

$$(XS)A^T = X(AR)^T$$

This is easily answered by the following manipulation:

$$(XS)A^T = X(SA^T) = X(AS^T)^T$$

Hence replacing X by XS has the same effect on (5.14) as replacing A by AS^T.

Now, in a general context, we have a symmetric vector product function over $V \times W$ and a linear transformation S. The question is now

(5.15) Is there a linear transformation S^* such that $(S(\xi),\eta) = (\xi,S^*(\eta))$ for all ξ and η in V and W, respectively?

In the example above, V is the set of triples of coordinates of points in three dimensions and W is the space of coefficients of equations of planes through the origin. Thus V and W are different spaces but have the same dimension.

The answer to question (5.15) is not always in the affirmative, as is shown by the following example. Let $V = W$ and let α_1, α_2 be a basis of V, where the vector product is defined as follows:

$$(\alpha_1,\alpha_1) = 1 \qquad (\alpha_1,\alpha_2) = (\alpha_2,\alpha_1) = (\alpha_2,\alpha_2) = 0$$

and let the transformation S be defined by $S\alpha_1 = \alpha_2$, $S\alpha_2 = \alpha_1$. Then suppose there were a transformation S^* such that

$$(S\alpha_1,\alpha_j) = (\alpha_1,S^*\alpha_j) \qquad \text{for } j = 1, 2$$

Since $S\alpha_1 = \alpha_2$, the vector product function is zero for both values of j. This means that $S^*\alpha_j$ is a scalar multiple of α_2 for $j = 1, 2$. Then

$$(S\alpha_2,\alpha_1) = (\alpha_1,\alpha_1) = 1$$

on the one hand, and

$$(S\alpha_2,\alpha_1) = (\alpha_2,S^*,\alpha_1) = (\alpha_2,k\alpha_2) = 0$$

on the other. This is a contradiction, showing that S^* does not exist. We shall see that if the vector product function is nonsingular, the answer to question (5.15) is in the affirmative.

Before dealing with the determination of the transformation S^*, we need the idea of a linear functional, which is a special case of a linear transformation of a vector space. We have the

Definition Let V be a vector space over a field F. A linear transformation taking each vector of V into an element of F is called a *linear functional.* In other words, a linear functional is a linear mapping of V into F.

One should be careful to note the importance of the adjective *linear* in the definition. Otherwise one might be tempted to think of a determinant as a linear functional since the set of matrices of order n defines an n^2-dimensional vector space (see Sec. 2.5) and the determinant gives a mapping into the field. But the mapping is not linear since

$$|A| + |B| \neq |A + B|$$

We shall give three examples, the last of which is of fundamental importance.

Example 1 Let V be the vector space of n-tuples $(x_1, x_2, \ldots, x_n)$. Define $f(x_1, x_2, \ldots, x_n)$ to be $x_1 + x_2 + \cdots + x_n$. This is a function, or mapping into F, and it is not hard to see that it is linear.

Example 2 Let V be the vector space of matrices M of order n. Then let $f(M) = \text{tr}(M)$ (see Sec. 3.12). This function is also linear, into F, and therefore a linear functional.

Example 3 Let (ξ,η) be a vector product function of $V \times V$ into F. (This is, in itself, a kind of functional on pairs of vectors instead of vectors.) Let α be a fixed vector of V and define $f(\xi) = (\xi,\alpha)$. Then $f(\xi)$ is a linear functional since the vector product function (ξ,α) is linear in ξ. A special case is the ordinary inner product over the field of reals.

A natural question to ask in this connection would be: Does Example 3 include all linear functionals? That is, given a linear functional of V into F and a vector product function of $V \times V$ into F, is there a vector α such that

$$f(\xi) = (\xi,\alpha)$$

for all ξ in V?

The answer is "not necessarily" if the product function is singular. For example, let β_1, β_2 be a basis of V and let $f(\beta_1) = 1$, $f(\beta_2) = 2$. Further, let $(\beta_1 - \beta_2, \xi) = 0$ for all ξ in V. Suppose that under these conditions there were a vector α such that

$$(\beta_1,\alpha) = f(\beta_1) \qquad \text{and} \qquad (\beta_2,\alpha) = f(\beta_2)$$

Then
$$(\beta_1 - \beta_2, \alpha) = f(\beta_1) - f(\beta_2) = 1 - 2 = -1$$

which is a contradiction. An α as described above can exist even if the vector product function is singular, as you are asked to show in Exercise 3 at the end of this section. However, when such an α exists for a singular vector product, it is not unique, as you are asked to show in Exercise 4.

But if we impose the condition that the vector product function is nonsingular, the answer is "yes" as is affirmed in the following:

Theorem 5.7 Let f be a linear functional of V into F and let (ξ,η) be a symmetric vector product function which is nonsingular over V; then there is a unique vector α such that

$$f(\xi) = (\xi,\alpha) \qquad \text{for all } \xi \text{ in } V$$

Proof Let $\alpha_1, \alpha_2, \ldots, \alpha_n$ be an orthogonal basis of V. In view of the linearity of the functional and the bilinearity of the vector product, we need merely choose α so that $f(\alpha_i) = (\alpha_i,\alpha)$ for $1 \leq i \leq n$. Let $f(\alpha_i) = a_i$ and $\alpha = \sum_{j=1}^{n} x_j\alpha_j$ with the x's to be determined. We then want

$$a_i = \left(\alpha_i, \sum_{j=1}^{n} x_j\alpha_j\right) = x_i(\alpha_i,\alpha_i) \qquad \text{for } 1 \leq i \leq n$$

Thus we merely need to choose $x_i = a_i/(\alpha_i,\alpha_i)$, which is possible since $(\alpha_i,\alpha_i) \neq 0$. This choice is unique. One can check this by noting that

$$(\xi,\alpha) = (\xi,\alpha') \qquad \text{for all } \xi$$

would imply that $(\xi, \alpha - \alpha') = 0$ for all ξ. Then the nonsingularity implies that $\alpha - \alpha' = 0$.

Now we are in a position to answer in the affirmative the question (5.15), that is, to prove the following:

Theorem 5.8 Let (ξ,η) be a symmetric vector product function nonsingular over V and T a linear transformation of V into itself. There is a unique linear transformation T^* of V into itself such that

$$(T(\xi),\eta) = (\xi,T^*(\eta)) \tag{5.16}$$

for all ξ and η in V.

Proof We note that, for each β, $(T(\xi),\beta) = f_\beta(\xi)$ is a linear functional since T is linear and the vector product is bilinear. Then by Theorem 5.7 there exists a unique vector $\alpha = \varphi(\beta)$, dependent on β, such that $f_\beta(\xi) = (\xi,\alpha)$ for all ξ.

Now φ is linear since

$$\begin{aligned}(\xi, \varphi(\beta_1 + \beta_2)) &= (T(\xi), \beta_1 + \beta_2) = (T(\xi),\beta_1) + (T(\xi),\beta_2)\\ &= (\xi,\varphi(\beta_1)) + (\xi,\varphi(\beta_2))\\ &= (\xi, \varphi(\beta_1) + \varphi(\beta_2))\end{aligned}$$

The uniqueness of φ shows that $\varphi(\beta_1 + \beta_2) = \varphi(\beta_1) + \varphi(\beta_2)$. Also

$$(T(\xi),c\beta_1) = c(T(\xi),\beta_1)$$

shows

$$\varphi(c\beta_1) = c\varphi(\beta_1)$$

Hence we may set $\varphi(\beta) = T^*(\beta)$, which is thus uniquely determined.

Corollary Under the conditions of the theorem, $(T^*)^* = T$, for

$$(\xi,(T^*)^*(\eta)) = (T^*(\xi),\eta) = (\eta,T^*(\xi)) = (T(\eta),\xi) = (\xi,T(\eta))$$

Since this holds for all ξ and η, $(T^*)^* = T$.

We call T^* defined by (5.16) the *adjoint* of T.

Finally let us see what is the relationship between the matrices of T and T^*. Suppose the bases are chosen and as in (5.8),

$$(\xi,\eta) = XAY^T$$

If M is the matrix corresponding to T for the given bases, then the matrix for $T(\xi)$ will be MX^T. Thus

$$(T(\xi),\eta) = XM^TAY^T \tag{5.17}$$

Similarly if N is the matrix for T^*, we have

$$(\xi,T^*(\eta)) = XANY^T \tag{5.18}$$

Hence if the two are equal for all X and Y, we have

$$M^TA = AN \qquad \text{or} \qquad A^{-1}M^TA = N \tag{5.19}$$

Notice how the nonsingularity of A enters in. In particular if $A = I$, that is, if $(\alpha_i,\beta_j) = \delta_{ij}$, then $M^T = N$. This would occur if the same orthonormal basis is used for both parts of the vector product. It immediately follows from (5.19) that a transformation and its adjoint have the same characteristic polynomials, since a transformation and its transpose have the same characteristic polynomial (see Exercise 12 below).

Notice that the matrix associated with T^* is quite different from the adjunct of the matrix associated with T as defined in Sec. 4.4.

We list the following properties of the adjoint function and leave the proofs as exercises:

$$(R + S)^* = R^* + S^* \qquad (cR)^* = cR^* \qquad (RS)^* = S^*R^* \tag{5.20}$$

for linear transformations R and S and a vector product function which is nonsingular over V.

A transformation of special interest, which we deal with more completely later in this chapter, is a reflection or symmetry which can be defined in two ways. We introduce this transformation here since its adjoint has interesting properties (see Exercises 5 to 7, 10, and 13 below).

Definition Given a nonsingular symmetric vector product function of $V \times V$ into F and α a nonisotropic vector. A transformation T_α of V into itself is called a *reflection* or *symmetry* if

$$T_\alpha(\alpha) = -\alpha \qquad \text{and} \qquad T_\alpha(\beta) = \beta$$

when
$$(\alpha,\beta) = 0$$

Whenever we deal with a symmetry we assume that the vector product function is symmetric.

It is left as an exercise to show that such a symmetry can be written explicitly in the following form:

$$T_\alpha(\xi) = \xi - \frac{2(\alpha,\xi)}{(\alpha,\alpha)}\,\alpha \tag{5.21}$$

It is also left as an exercise to show

$$T^2 = I \qquad \text{and} \qquad (\xi,\eta) = (T_\alpha(\xi),T_\alpha(\eta)) \qquad \text{for all } \xi \text{ and } \eta \text{ in } V \tag{5.22}$$

Geometrically the imagery of a reflection is this. The set of vectors orthogonal to α form an $(n - 1)$-dimensional space W which we call a *hyperplane.* Every vector in W is left invariant by the symmetry, and every vector orthogonal to W changes direction. So the reflection is in the hyperplane W. In the two-dimensional case W is a line, and in three dimensions it is a plane.

Notice that the hyperplane is the perpendicular bisector of the segment from α to $-\alpha$, the image of α.

EXERCISES

1. Let α_1, α_2 be a basis of V^2 and let $A = ((\alpha_i,\alpha_j))$ be the matrix

$$\begin{bmatrix} 1 & 2 \\ 2 & 0 \end{bmatrix}$$

Suppose f is a linear functional defined by $f(\alpha_1) = 1, f(\alpha_2) = -1$. Find numbers c and d such that for $\alpha = c\alpha_1 + d\alpha_2$, $f(\alpha_i) = (\alpha_i,\alpha)$ for $i = 1, 2$.

2. In Exercise 1, suppose that the matrix is changed to

$$\begin{bmatrix} 1 & 2 \\ 2 & 4 \end{bmatrix}$$

and the functional unaltered. Then can c and d be found? Give reasons for your conclusion.

3. Let a symmetric vector product function over a vector space V have the matrix given in Exercise 2. For what linear functionals f does there exist a vector α such that $f(\xi) = (\xi,\alpha)$ for all ξ in V? Show that when the functional has the property that such an α exists, then α is not uniquely determined by f.

4. Suppose a symmetric vector product function over a vector space V of dimension n is singular and that the linear functional f has the property that there is a vector α such that $f(\xi) = (\xi,\alpha)$ for all α in V. Show that such an α is not uniquely determined by f.

5. In V^3 let $\alpha = (1,1,0)$, and, letting $\xi = (x_1,x_2,x_3)$, find the matrix of the symmetry (5.21).

6. In V^3 let $\alpha = (1,1,0)$ and find a basis $\mathcal{B}$ of the orthogonal complement of α. Then, using the definition of a symmetry, find the matrix of the symmetry T_α with respect to the basis of V consisting of α and $\mathcal{B}$.

7. Prove that if T is a symmetry, then $T = T^*$.

8. In V^2 let $T(\alpha_1) = \alpha_1 + \alpha_2$ and $T(\alpha_2) = \alpha_1 - \alpha_2$. Find the matrix of T and of T^* when the vector product $(\alpha_i,\alpha_j) = \delta_{ij}$, the Kronecker delta.

9. Let α_1, α_2 be a basis of V^2 and let $A = ((\alpha_i,\alpha_j))$ be

$$\begin{bmatrix} 1 & 2 \\ 2 & 0 \end{bmatrix}$$

Let T be the transformation defined by $T(\alpha_1) = \alpha_1 + 2\alpha_2$, $T(\alpha_2) = \alpha_1 - \alpha_2$. Find the matrix of the adjoint of T with respect to the given basis.

10. Using the notation of Exercise 9 but letting T be defined by

$$T(\alpha_1) = \frac{\alpha_1 - \alpha_2}{\sqrt{2}} \qquad T(\alpha_2) = \frac{-(\alpha_1 + \alpha_2)}{\sqrt{2}}$$

prove that T is a symmetry.

11. Let (ξ,η) be a nonsingular symmetric vector product of $V \times V$ into F. Prove that there are bases $\alpha_1, \alpha_2, \ldots, \alpha_n$ and $\beta_1, \beta_2, \ldots, \beta_n$ of V such that

$$(\alpha_i,\beta_j) = \delta_{ij} \qquad \text{the Kronecker delta}$$

In other words, for these bases the matrix of the bilinear form is the identity matrix.

12. Prove that the characteristic polynomials of a transformation T and its adjoint T^* are the same.

13. Prove that the transformation exhibited in (5.21) is a reflection and that Equations (5.22) hold.

14. Prove that Equations (5.20) hold.

15. For a transformation T prove that, if $TT^* = I$ and if the determinant of T is 1, the adjoint and the adjunct of T are the same. Is the converse of this statement true?

16. Let a vector product function for a basis α_1, α_2 of V^2 be defined by the matrix

$$\begin{bmatrix} 1 & 2 \\ 2 & 4 \end{bmatrix}$$

and let T be the identity transformation. Find all linear transformations S such that $(T(\xi),\eta), = (\xi,S(\eta))$, for all ξ and η in V.

17. Given a symmetric vector product function of $V \times V$ into F and let R be the set of vectors α of V such that $(\alpha,\xi) = 0$ for all ξ in V. Prove that R is a vector space. (It is called the *radical* of V.) Prove that there is a subspace W such that $V = R \oplus W$, where the vector product function is nonsingular over W. Prove that the dimension of W is equal to the rank of the matrix associated with the vector product function.

5.4 INNER PRODUCTS OVER THE FIELD OF REALS

In this section we specialize the field over which a vector product function is defined to be that of real numbers, and impose another condition which brings us back to the idea of inner product with which we began this chapter. Notice first that if we merely take the field to be that of the real numbers, we still have the possibility of isotropic vectors, that is, vectors α such that $(\alpha,\alpha) = 0$. Indeed there might be vectors α such that $(\alpha,\alpha) < 0$. So, to revert to the idea of an inner product, we not only assume that the vector product function is symmetric but also that

$$(5.23) \qquad (\xi,\xi) \geq 0 \qquad \text{with equality if and only if } \xi = 0$$

If such a vector product function is over the field of real numbers, we call it an *inner product.* An inner product is a *nonsingular* vector product function perforce because any nonnull vector has a vector to which it is not orthogonal, namely, itself. If (ξ,η) is an inner product we write it more briefly as $\xi\eta$, omitting the dot.

For an inner product we can further specialize the orthogonal basis as follows. Let $\{\alpha_i\}$ constitute an orthogonal basis (Theorem 5.2). Then in view of (5.23) we have $(\alpha_i,\alpha_i) > 0$ for $1 \leq i \leq n$. If we replace each α_i by $\alpha_i/\sqrt{(\alpha_i,\alpha_i)}$, we have a basis such that

$$(\alpha_i,\alpha_j) = \delta_{ij} \qquad \text{the Kronecker delta} \tag{5.24}$$

Such a basis is called an *orthonormal basis.* What is the vector product function with respect to this basis? We have

$$\left(\sum_{i=1}^{n} x_i\alpha_i, \sum_{j=1}^{n} y_j\alpha_j\right) = \sum_{i=1}^{n} x_i y_i \tag{5.25}$$

This is the familiar inner product of vectors. If then we identify the vectors ξ and η with $(x_1, x_2, \ldots, x_n)$ and $(y_1, y_2, \ldots, y_n)$, respectively, we have

$$(\xi,\eta) = XY^T \tag{5.26}$$

where X and Y stand for the n-tuples.

As noted at the beginning of this chapter, we may identify n-tuples of real numbers with points in an n-dimensional Euclidean geometry. Choosing an orthonormal basis is equivalent to choosing "rectangular axes," that is, basis vectors of length 1 which are orthogonal in pairs. Having in mind the Pythagorean theorem for two or three dimensions, we could call $\sqrt{\xi^2}$ the *length* of the vector. This notation, however, is awkward since there is a temptation to equate $\sqrt{\xi^2}$ with ξ. The customary symbol for this length is

$$\|\xi\| = \sqrt{(\xi,\xi)}$$

when the vector product function is the inner product. This "length" is called the *norm* of ξ. Further, if A and B are two points designated by the n-tuples α and β, we call $\|\alpha - \beta\|$ the *distance* between A and B. This distance is zero if and only if $A = B$.

In Euclidean geometry there is an important inequality, known as the *triangle inequality,* which affirms

$$\begin{gathered}\text{If } A, B, C \text{ are three points then}\\ AB + BC \geq AC\end{gathered} \tag{5.27}$$

where AB denotes the length of the segment AB, and similarly for BC and AC; the equality holds if and only if the three points are collinear and B is between A and C.

Let us prove in vector notation what seems like a special case of (5.27). After this is done, we shall return to (5.27) to demonstrate it. Take B to be the origin and let $X = A$ and $Y = C$ in Fig. 5.1, where X and Y are the points

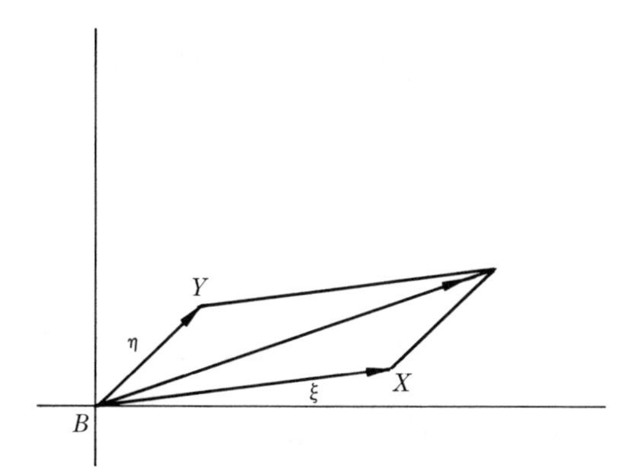

Figure 5.1

corresponding to the vectors ξ and η. Then (5.27) becomes

$$\|\xi\| + \|\eta\| \geq \|\xi - \eta\| \tag{5.28}$$

We now prove

Theorem 5.9 If ξ and η are any two nonzero vectors of V with an inner product (over the field of reals), then (5.28) above and

$$\|\xi\| \cdot \|\eta\| \geq |(\xi\eta)| \qquad \text{(Cauchy-Schwartz inequality)} \tag{5.29}$$

are both true for all ξ and η, where $|(\xi\eta)|$ is the absolute value of the inner product (ξ,η). Furthermore the equality holds if and only if ξ and η are linearly dependent.

Proof First we show that (5.28) and (5.29) are equivalent. Since the norm is a nonnegative real number, (5.28) is equivalent to what we get by squaring both sides; that is,

$$\begin{aligned} \|\xi\|^2 + 2\|\xi\| \cdot \|\eta\| + \|\eta\|^2 \geq \|\xi - \eta\|^2 &= (\xi - \eta, \xi - \eta) \\ &= (\xi,\xi) - (\eta,\xi) - (\xi,\eta) + (\eta,\eta) \\ &= \|\xi\|^2 - 2(\xi,\eta) + \|\eta\|^2 \\ &= \|\xi\|^2 - 2\xi\eta + \|\eta\|^2 \end{aligned} \tag{5.30}$$

which implies (5.29) if $\xi\eta \leq 0$. If $\xi\eta > 0$, we can replace η by $-\eta$ in (5.30) to get (5.29). Conversely, (5.28) is implied by (5.29). Hence the equivalence of (5.28) and (5.29) is established.

Now we prove (5.29). Since we are dealing with an inner product, it is true that $(\xi + t\eta)^2 > 0$ unless $\xi + t\eta = 0$. Thus

$$\begin{aligned} &0 \leq (\xi + t\eta)^2 = (\xi + t\eta, \xi + t\eta) = \xi^2 + 2t\xi\eta + t^2\eta^2 \\ &0 \leq \|\xi\|^2 + 2t\xi\eta + \|\eta\|^2t^2 \end{aligned} \tag{5.31}$$

Now the quadratic expression in t is nonnegative for all t if and only if its discriminant is not positive. Thus (5.31) holds for all t if and only if

$$\|\eta\|^2 \cdot \|\xi\|^2 \geq (\xi\eta)^2 \tag{5.32}$$

which is equivalent to (5.29).

It remains to deal with the equality. If ξ and η are linearly dependent, write $\xi = c\eta$, and (5.32) becomes

$$\|\eta\|^2 c^2 \|\eta\|^2 = c^2\|\eta\|^4$$

which is the equality desired. If, on the other hand, the equality in (5.32) holds, then for some $t = c$ the equality in (5.31) holds, that is, $0 = (\xi + c\eta)^2$ which implies that $\xi + c\eta = 0$. This completes the proof.

Now let us see why (5.29) implies the triangle inequality in the general case. Letting α, β, and γ be the vectors corresponding to the points A, B, and C, we see that the inequality (5.27) becomes

$$\|\alpha - \beta\| + \|\beta - \gamma\| = \|\alpha - \beta\| + \|\gamma - \beta\| \geq \|\alpha - \gamma\|$$

This is just (5.28) with $\xi = \alpha - \beta$, $\eta = \gamma - \beta$. Furthermore, the equality holds if and only if $\alpha - \beta = c(\gamma - \beta)$ for some c. Now $c = 1$ would imply $\alpha = \gamma$, which is impossible if A and C are different points. So $c \neq 1$ and

$$\beta = (1 - c)^{-1}\alpha - c(1 - c)^{-1}\gamma$$

Since 1 is the sum of the coefficients of α and γ, this means (see Exercise 13 of Sec. 1.6) that B is between A and C on the line AC.

Corollary The triangle inequality holds.

Note 1 Recall (Sec. 5.1) that in two dimensions the law of cosines shows us that $(\xi\eta) = \|\xi\| \cdot \|\eta\| \cos \varphi$, where φ is the angle between the vectors ξ and η. Inequality (5.29) merely affirms that the absolute value of the cosine is less than 1 unless the angle is a multiple of π. In the latter case the two vectors are linearly dependent.

Note 2 It is also worthy of notice that in proving Theorem 5.9 we used the fact that the field is real and the properties of an inner product, but not form (5.4) of the inner product.

5.5 ORTHOGONAL TRANSFORMATIONS

In Euclidean geometry we study properties of figures (sets of points) which are unchanged by rigid motions, that is, transformations which leave distance unchanged. In the present notation, Euclidean geometry is concerned with transformations T such that

$$\|T(\xi) - T(\eta)\| = \|\xi - \eta\|$$

or, more generally,

$$(T(\xi))(T(\eta)) = \xi\eta$$

We call a linear transformation with this property an *orthogonal transformation.* Notice that while a translation leaves distances unaltered, it is not linear since it does not take the null vector into itself and hence is not orthogonal. We shall deal with translations later in this chapter.

(At the end of the chapter we shall be considering such transformations for the vector product function. Then we shall use the term *isometry.*)

What are examples of orthogonal transformations in two dimensions? To be linear a transformation must leave the origin unchanged. To be orthogonal, it must leave unchanged the lengths of all vectors. Actually, to test for orthogonality it is only necessary to look at norms, as is asserted by the following:

Theorem 5.10 A linear transformation T is orthogonal if and only if $\|T(\xi)\| = \|\xi\|$ for all vectors ξ.

Proof Using the symmetric vector product notation, we have

$$(\xi + \eta, \xi + \eta) = (T(\xi + \eta), T(\xi + \eta))$$

which implies

$$(\xi,\xi) + 2(\xi,\eta) + (\eta,\eta) = (T(\xi),T(\xi)) + 2(T(\xi),T(\eta)) + (T(\eta),T(\eta))$$

so that

$$(\xi,\xi) = (T(\xi),T(\xi)) \qquad \text{and} \qquad (\eta,\eta) = (T(\eta),T(\eta))$$

imply that $(\xi,\eta) = (T(\xi),T(\eta))$.

Thus a rotation about the origin is an orthogonal transformation. A reflection in a line through the origin is also such a transformation. You will be asked to show in an exercise that these are the only orthogonal transformations in the plane. Equation (5.5) shows that if lengths and inner products are unchanged, then so is the angle between any two vectors. In particular, orthogonality of two vectors is preserved.

We leave as an exercise the proof of the following theorem and its corollary.

Theorem 5.11 If a linear transformation is orthogonal, it takes every orthonormal basis of V into an orthonormal basis. Furthermore a linear transformation is orthogonal if it takes some orthonormal basis into an orthonormal basis.

Corollary The set of orthogonal transformations of a vector space of n dimensions forms a multiplicative group.

Some consequences of this corollary are worth considering. Adopting the terminology of Euclidean geometry, we call two sets of points $\mathcal{P}_1$ and $\mathcal{P}_2$ *congruent* if one can be taken into the other by an orthogonal transformation. The fact that congruence is an equivalence relationship is logically equivalent to the corollary, as is shown by the following scheme:

Congruence	*Group property of orthogonality*
1. $\mathcal{P}_1$ is congruent to $\mathcal{P}_1$.	1. I, the identity, is an orthogonal transformation.
2. $\mathcal{P}_1$ congruent to $\mathcal{P}_2$ implies $\mathcal{P}_2$ congruent to $\mathcal{P}_1$.	2. If S is an orthogonal transformation, S^{-1} exists and is orthogonal.
3. $\mathcal{P}_1$ congruent to $\mathcal{P}_2$ and $\mathcal{P}_2$ congruent to $\mathcal{P}_3$ imply $\mathcal{P}_1$ congruent to $\mathcal{P}_3$.	3. If S_1 and S_2 are orthogonal transformations, so is S_1S_2.

What does the matrix of an orthogonal transformation look like? The answer depends of course on the basis used as well as on what we use as an inner product. Here it is best to let V be the space of n-tuples of real numbers and define the inner product as for vectors in the original sense; that is,

$$\alpha\beta = (a_1, a_2, \ldots, a_n)(b_1, b_2, \ldots, b_n) = a_1b_1 + a_2b_2 + \cdots + a_nb_n$$

Let M be the matrix of an orthogonal transformation S with respect to the canonical basis $\epsilon_1, \epsilon_2, \ldots, \epsilon_n$, the rows of the identity matrix. Then $M\epsilon_i^T = C_i$, the ith column of M. So $(S(\epsilon_i))(S(\epsilon_j)) = \delta_{ij}$ is equivalent to the matrix product

$$(M\epsilon_i^T)^T(M\epsilon_j^T) = C_i^TC_j = \delta_{ij}$$

The columns of M form an orthogonal basis of V; that is,

$$M^TM = I \tag{5.33}$$

Since (5.33) shows that M^T is the inverse of M, it follows that MM^T is also equal to I; the rows of M also form an orthogonal basis of V. For instance, let $n = 2$ and $M = (m_{ij})$; then M orthogonal means

$$\begin{bmatrix} m_{11} & m_{21} \\ m_{12} & m_{22} \end{bmatrix}\begin{bmatrix} m_{11} & m_{12} \\ m_{21} & m_{22} \end{bmatrix} = I$$

that is,

$$m_{11}^2 + m_{21}^2 = 1 = m_{12}^2 + m_{22}^2 \qquad \text{and} \qquad 0 = m_{11}m_{12} + m_{21}m_{22}$$

We can also arrive at the same form for M if we use the results of Sec. 5.3 as follows:

$$(\xi,\eta) = (S(\xi),S(\eta)) \qquad \text{for all } \xi \text{ and } \eta$$

implies, by the definition of the adjoint S^*, that

$$(\xi,\eta) = (\xi,S^*S(\eta)) \qquad \text{for all } \xi \text{ and } \eta$$

Thus $S^*S = I$ since the inner product is nonsingular. This is the general condition for orthogonality. If we choose the basis to be canonical, then the matrix of S^* becomes the transpose of that of S [see (5.19) with $A = I$], and we have, as a special case, condition (5.33) above.

We close this section with the statement of the completion theorem for orthonormal bases, leaving the proof, which follows almost immediately from Theorem 5.5, as an exercise.

Theorem 5.12 If V is a vector space over the field of reals with an inner product and if $\alpha_1, \alpha_2, \ldots, \alpha_r$ (for $r \leq n$) is an orthonormal basis of a subspace of V, then there is an orthonormal basis of V containing the set of r α's.

Geometrically this means that if there is a set $\mathcal{S}$ of r vectors of V of length 1 and perpendicular (orthogonal) in pairs, there is a set of n vectors of V containing $\mathcal{S}$, each of which is of length 1 and which is perpendicular (orthogonal) to all the others.

EXERCISES

1. Let M be an orthogonal matrix of order 2 over the field of reals. Prove that M is of one of the following forms:

$$\begin{bmatrix} \cos\varphi & -\sin\varphi \\ \sin\varphi & \cos\varphi \end{bmatrix} \quad \text{or} \quad \begin{bmatrix} \cos\varphi & -\sin\varphi \\ -\sin\varphi & -\cos\varphi \end{bmatrix}$$

for some angle φ and show that its determinant is either 1 or -1.

2. Show that if the determinant of M in Exercise 1 is $+1$, then M is a rotation through an angle φ. If the determinent is -1, show that there is an orthogonal basis α_1, α_2 of V such that $M\alpha_1 = \alpha_1$, $M\alpha_2 = -\alpha_2$, that is, M is a reflection (see Sec. 5.3).

3. Let S be an orthogonal transformation of V^3 into itself over the field of real numbers. Prove that 1 or -1 is a characteristic root of S. What can you say about the other characteristic roots?

4. The first two rows of an orthogonal matrix are

$$(\tfrac{1}{3},\tfrac{2}{3},\tfrac{2}{3}) \qquad \text{and} \qquad (\tfrac{2}{3},\tfrac{1}{3},-\tfrac{2}{3})$$

Find all possible third rows.

5. Prove that, for an inner product over the field of reals, if $\alpha^2 = \beta^2$, then $\alpha + \beta$ and $\alpha - \beta$ are orthogonal to each other.

6. Find an orthonormal basis for the space spanned by

$$(1,1,0,0) \qquad (1,0,-1,2) \qquad (0,1,-1,1)$$

7. Let α_1, α_2 be a basis of V and define $A = ((\alpha_i,\alpha_j))$ with

$$A = \begin{bmatrix} 1 & 3 \\ 3 & 10 \end{bmatrix}$$

Prove that this vector product satisfies property (5.23), that is, it is an inner product. Find an orthonormal basis.

8. Do Exercise 7 for $A = \begin{bmatrix} 2 & 1 \\ 1 & 3 \end{bmatrix}$.

9. Using the orthogonal matrices M of Exercise 1 and the matrix A of Exercise 7, find the matrices M^*.

10. Prove Theorem 5.12.

11. Given a symmetric vector product function over the field of reals. Let $(\alpha,\alpha) = a > 0$ and $(\beta,\beta) = b < 0$. Find a vector $x\alpha + y\beta$, for x and y real numbers, such that $(x\alpha + y\beta,\ x\alpha + y\beta) = 0$.

12. Show that the determinant of any orthogonal transformation is 1 or -1 and that those of determinant 1 form a multiplicative group.

13. Prove Theorem 5.11 and its Corollary.

14. Prove that if the determinant of an orthogonal transformation of V^2 into itself is -1, then the square of the transformation is the identity transformation.

15. Let M be a rotation as defined in Exercise 2. Prove that, given any reflection N, there is a reflection N' such that $M = NN'$.

16. Let T be a linear transformation of V onto W, a subspace of V, with an inner product. Prove that the kernel of T^* is the orthogonal complement of W.

5.6 EUCLIDEAN TRANSFORMATIONS

As was noted in the beginning of the previous section, if we identify the points of n-dimensional geometry with n-tuples of real numbers, then Euclidean geometry consists in a study of properties of sets of points which are invariant under all transformations E of V^n into itself having the property that, for any two points A and B, $E(A) = A'$ and $E(B) = B'$ implies that the distance AB is equal to the distance $A'B'$. Such transformations are called *rigid motions* or *Euclidean transformations*. We shall use the latter term.

The linear Euclidean transformations are by definition the orthogonal transformations. However, not all Euclidean transformations are linear, for example, a translation does not leave the origin fixed—it leaves no point fixed. So now let E be a Euclidean transformation and let $E(\theta) = \alpha \neq \theta$. Then define a transformation S so that $S(\xi) = E(\xi) - \alpha$. (We sometimes abbreviate this to $S = E - \alpha$.) We first show two properties of S:

1. $S(\theta) = \theta$.
2. S is a Euclidean transformation.

(Later we shall show that S is linear also.)

The first is easy since $S(\theta) = E(\theta) - \alpha = \theta$. For the second, let $\xi = (x_1, x_2, \ldots, x_n)$ and $\eta = (y_1, y_2, \ldots, y_n)$ and

$$E(\xi) = (x'_1, x'_2, \ldots, x'_n) \qquad E(\eta) = (y'_1, y'_2, \ldots, y'_n)$$

Then E, a Euclidean transformation, implies

$$\sum_{i=1}^{n} (x_i - y_i)^2 = \sum_{i=1}^{n} (x'_i - y'_i)^2 \tag{5.34}$$

If $\alpha = (a_1, a_2, \ldots, a_n)$, $S(\xi)$ can be obtained from $E(\xi)$ by replacing each component x'_i by $x'_i - a_i$, and $S(\eta)$ is obtained from $E(\eta)$ by replacing each component y'_i by $y'_i - a_i$. This replacement does not change the right-hand side of (5.34). Hence properties 1 and 2 hold.

Next we show that S takes lines into lines; that is, if α, β, γ represent collinear points, then $S(\alpha)$, $S(\beta)$, $S(\gamma)$ do also. Referring back to Sec. 5.4, we see that the points corresponding to α, β, γ are collinear with β between α and γ if and only if

$$\|\alpha - \beta\| + \|\beta - \gamma\| = \|\alpha - \gamma\|$$

But S, a Euclidean transformation, implies

$$\|S(\alpha) - S(\beta)\| = \|\alpha - \beta\|$$

and similarly for the other pairs, β, γ and α, γ. Hence the points $S(\alpha)$, $S(\beta)$, $S(\gamma)$ are collinear with $S(\beta)$ between $S(\alpha)$ and $S(\gamma)$. In other words, S preserves collinearity and betweenness.

Now a translation takes lines into lines without being linear, but it does not leave the origin invariant. Here we show that any Euclidean transformation which takes the origin into itself must be linear, that is, must be an orthogonal transformation. We state this as a theorem, including the above results:

Theorem 5.13 If E is a Euclidean transformation of V^n over the field of reals into itself and if S is defined by

$$S(\xi) = E(\xi) - E(\theta)$$

then S is a linear Euclidean transformation.

Proof We have already shown that $S(\theta) = \theta$ and that S is Euclidean. Let ξ and η be any linearly independent pair of vectors of V, and write $S(\xi) = \xi'$ and $S(\eta) = \eta'$. We want to prove

$$S(x\xi) = xS(\xi) = x\xi' \tag{5.35}$$

and

$$S(\xi + \eta) = \xi' + \eta' \tag{5.36}$$

To prove (5.35), notice that θ, ξ, $x\xi$ represent three collinear points if $x \neq 0$, and hence their images

$$S(\theta) = \theta \qquad S(\xi) = \xi' \qquad S(x\xi)$$

must be collinear. Thus $S(x\xi) = x'\xi'$ for some nonzero x'. Now S, a Euclidean transformation, implies $(x\xi,x\xi) = (x'\xi',x'\xi')$, and hence $x^2 = x'^2$. But betweenness is also preserved, and hence if x were negative the point θ would be between $x\xi$ and ξ, which implies that x' is also negative. Hence x and x' are both positive or both negative, which shows that they are equal.

To prove (5.36) first recall from Sec. 5.4 and earlier in this book that if points A and B are represented by n-tuples α and β, respectively, then the points of the line AB are represented by n-tuples $a\alpha + b\beta$, where $a + b = 1$. We show first

(5.37) If $x + y \neq 0$ then $S(x\xi + y\eta) = x'\xi' + y'\eta'$, where $x' + y' = x + y$

To see this, notice that

$$S\left(\frac{x}{x+y}\xi + \frac{y}{x+y}\eta\right)$$

being the image of a point on the line through the points ξ and η, must be on the line through the points ξ' and η'. Hence

$$S\left(\frac{x}{x+y}\xi + \frac{y}{x+y}\eta\right) = s\xi' + t\eta' \qquad \text{where } s + t = 1$$

Then, from property (5.35), we have

$$S(x\xi + y\eta) = (x + y)s\xi' + (x + y)t\eta'$$

and (5.37) is proved by letting $x' = (x + y)s$ and $y' = (x + y)t$, and noting that

$$x' + y' = (x + y)(s + t) = x + y$$

Thus we have shown that $S(\xi + \eta) = x\xi' + y\eta'$ with $x + y = 2$. We complete the proof by computing $S(\xi + \eta)$ another way:

$$x\xi' + y\eta' = S(\xi + \eta) = S(\xi + 2(\tfrac{1}{2}\eta)) = x'\xi' + y'(\tfrac{1}{2}\eta)'$$

where $x' + y' = 3$. Now, from (5.35), we have $(\frac{1}{2}\eta)' = \frac{1}{2}\eta'$. Thus ξ' and η' linearly independent shows that $x = x'$ and $y = \frac{1}{2}y'$. Then

$$3 = x' + y' = x + 2y \qquad \text{and} \qquad x + y = 2$$

imply $x = y = 1$ and the proof is complete.

Theorem 5.13 tells us that every Euclidean transformation is of the form $E(\xi) = S(\xi) + \alpha$, where S is an orthogonal transformation and α is a fixed vector. Thus Euclidean transformations form a subset of the affine transformations described in Sec. 3.13. We shall apply this to conic sections, that is, to quadratic forms and functions, in Sec. 5.10.

Notice that $(I + \alpha)S(\xi) = S(\xi) + \alpha$ and that $I + \alpha$ is a translation. Hence we have the following:

Corollary Any Euclidean transformation $E = S + \alpha$ can be written in the form $TS = E$ for a properly chosen translation T.

Note It can also be shown that E can be written as a product ST' for some translation T'. This is left as an exercise.

5.7 INNER PRODUCTS IN THE FIELD OF COMPLEX NUMBERS

Before considering what we can do with orthogonal and Euclidean transformations, we shall find it economical of effort to deal with an extension of the idea of inner product from the field of real numbers to that of complex numbers. Then to get results for the field of reals we can for the most part specialize results in complex numbers.

This treatment requires a modification in the basic vector product function as we now show. Suppose the vector α is the pair $(1,i)$, where i is the square root of -1. Then if we defined the inner product of this vector with itself as we did previously, we would have

$$(\alpha,\alpha) = (1,i) \cdot (1,i) = 1 - 1 = 0$$

and α would be a vector of length zero. This is inconvenient. But if we define

$$(x_1,x_2) \cdot (y_1,y_2) = x_1\bar{y}_1 + x_2\bar{y}_2$$

where the bar denotes complex conjugate, we have

$$(1,i) \cdot (1,i) = 1 + 1 = 2 \neq 0$$

In fact, for such a definition,

$$(x_1,x_2)(x_1,x_2) = x_1\bar{x}_1 + x_2\bar{x}_2$$

which is positive and real unless $x_1 = x_2 = 0$, in which case the inner product is zero.

Thus, for n-tuples of complex numbers, we define

$$(5.38) \quad (x_1, x_2, \ldots, x_n) \cdot (y_1, y_2, \ldots, y_n) = x_1\bar{y}_1 + x_2\bar{y}_2 + \cdots + x_n\bar{y}_n$$

This definition forces certain properties upon us. First, if we interchange x and y in the left side of (5.38), the right side is replaced by its conjugate. Thus the inner product is not commutative but

$$(5.39) \qquad \xi \cdot \eta = \overline{(\eta \cdot \xi)}$$

where ξ and η stand for the n-tuples. Furthermore, though $(c\xi) \cdot \eta = c(\xi \cdot n)$ we have

$$\xi \cdot (c\eta) = \overline{((c\eta),\xi)} = \overline{(c\eta,\xi)} = \overline{c(\eta,\xi)} = \bar{c}(\xi\eta) = \bar{c}(\xi \cdot \eta)$$

So now, using the more general notation, we can formally define the vector product function over the field C of complex numbers as a mapping of $V \times V$ into C with the following properties:

1. (ξ,η) is in C for all ξ and η in V.
2. $(\xi,\eta) = \overline{(\eta,\xi)}$.
3. (ξ,η) is a linear function of ξ and a *conjugate-linear function* of η; that is,
   (*a*) $(\xi, \eta + \rho) = (\xi,\eta) + (\xi,\rho)$ and $(\xi + \rho, \eta) = (\xi,\eta) + (\rho,\eta)$.
   (*b*) $(c\xi,\eta) = c(\xi,\eta)$ and $(\xi,c\eta) = \bar{c}(\xi,\eta)$.

For the inner product we add the following condition:

4. $(\xi,\xi) \geq 0$ for all ξ in V, where the equality holds if and only if $\xi = \theta$.

If all four properties hold we call it the *inner product* and write (ξ,η) as $\xi\bar{\eta}$ and (ξ,ξ) as $\xi\bar{\xi}$. As in the real case, we use the norm notation $\|\xi\| = \sqrt{(\xi,\xi)}$. If X and Y stand for the n-tuples, we can express the inner product in matrix form as

$$\xi\bar{\eta} = X\bar{Y}^T$$

This representation also shows property 2. We use the word *orthogonal* also for the complex numbers and have the following definition:

Definition For an inner product two vectors are said to be *orthogonal* if $(\xi,\eta) = \xi\bar{\eta} = 0$.

Notice that $\xi\bar{\eta} = 0$ if and only if $\eta\bar{\xi} = 0$. Hence, though the inner product is not symmetric, orthogonality is.

We could continue here as we did for the vector product function and proceed for a while without assuming property 4. But since we have already specialized the field and since the applications which we shall consider are solely to the inner product function, we shall confine ourselves to that case. Notice that we have assumed from the beginning that only the vector space V is involved.

First we have the notion analogous to that of an orthonormal basis, which is defined as follows:

Definition If a vector space V of dimension n over the field of complex numbers has an inner product, a set of vectors $\alpha_1, \alpha_2, \ldots, \alpha_n$ of V is called a *unitary basis* if

$$\alpha_i\bar{\alpha}_j = \delta_{ij} \qquad \text{the Kronecker delta} \tag{5.40}$$

Remark It can be shown, as for an orthonormal basis, that if a set of n vectors satisfies (5.40), it must be a basis. This is left as an exercise.

If V is the space of n-tuples of complex numbers, the set of rows of the identity matrix constitute a unitary basis. But, of course, this is not the only unitary basis. Furthermore, it is simpler notationally and conceptually to think of V as consisting of abstract vectors with an inner product having properties 1 to 4 of the beginning of this section. We next prove a completion theorem for a unitary basis. There are a number of ways of doing this. One can show that the inner product is nonsingular in the sense of Sec. 5.1 and then use the proof of Theorem 5.5. (For this, one needs to make alterations to take into account conjugate bilinearity.) Or one can give a constructive proof. We shall choose the latter since it is not difficult and is very similar to that of Theorem 5.2. You might like to try it without reading the proof given.

Theorem 5.14 If V is an n-dimensional vector space over the complex numbers with an inner product and if

$$\alpha_1, \alpha_2, \ldots, \alpha_r \tag{5.41}$$

is a unitary basis of a subspace of V, then there is a unitary basis of V containing the set (5.41).

***Proof* (*Gram-Schmidt Process*)** By the completion theorem for vector spaces (Theorem 1.3) there is a basis of V consisting of the vectors of (5.41) and $\beta_{r+1}, \ldots, \beta_n$. We want to choose the x's in

$$\alpha_{r+1} = \beta_{r+1} + x_1\alpha_1 + \cdots + x_r\alpha_r$$

so that $\alpha_{r+1}\bar{\alpha}_i = 0$ for $1 \leq i \leq r$. To this end, note that

$$\alpha_{r+1}\bar{\alpha}_i = \beta_{r+1}\bar{\alpha}_i + x_i\alpha_i\bar{\alpha}_i = \beta_{r+1}\bar{\alpha}_i + x_i$$

Hence we can determine x_i so that $\alpha_{r+1}\bar{\alpha}_i = 0$. Then, since $\alpha_{r+1}\bar{\alpha}_{r+1} = a > 0$, we can replace α_{r+1} by $\alpha_{r+1}\sqrt{a}$ to make $\alpha_{r+1}\bar{\alpha}_{r+1} = 1$. Then $\alpha_1, \alpha_2, \ldots, \alpha_{r+1}$ is a unitary basis of a subspace of V. We can continue in this way to construct a unitary basis of V which contains the set (5.41). This completes the proof.

Notice that from this theorem we can start with any nonzero vector, normalize it (that is, make its length 1), and then proceed step by step to construct a unitary basis. In practice this would be a very tedious process although it is very simple to describe. Fortunately we are usually interested in the existence of a unitary basis rather than in what it is explicitly.

A *unitary transformation* over the complex field is defined to be one which leaves unchanged all inner products. Since this is analogous to the orthogonal case (see Sec. 5.5), we leave as an exercise the proof of

Theorem 5.15 The matrix M of a unitary transformation over V with respect to a unitary basis has the property

$$M^T\bar{M} = I$$

The rows of the matrix of a unitary transformation form a unitary basis of V.

Corollary The determinant of the matrix of a unitary transformation has absolute value 1. It is convenient to write $\bar{M}^T$ as $\tilde{M}$. For the real case, $M^T = \tilde{M}$.

The definition of functional developed in Sec. 5.3 carries over with little modification to the complex inner product, and, with only slight alterations here and there, one can prove Theorem 5.8, which defines the adjoint of a transformation T for this extension. In place of Equation (5.19) we have the following: If M is a matrix of a transformation with respect to some basis and A is the inner product matrix, then the matrix of the adjoint transformation is

$$\bar{A}^{-1}\tilde{M}\bar{A} = N \tag{5.42}$$

In particular, if the basis is unitary, that is, A is the matrix for a unitary basis, then $A = I$ and $\tilde{M} = N$.

One can also set up a geometry for the complex field analogous to the Euclidean geometry for the real field, in which the points are n-tuples of complex numbers and distance is defined by means of the complex inner product. The triangle inequality and the Cauchy-Schwartz inequality of Sec. 5.4 can be established just as in the real case. This latter is left as an exercise.

When dealing with orthogonal transformations, we wrote nothing about their characteristic roots because it is simpler to deal with the complex case and specialize. We now prove

Theorem 5.16 All the characteristic roots of a unitary matrix have absolute value 1.

Proof Let S be a unitary transformation and α a characteristic vector such that $S(\alpha) = a\alpha$. Then S unitary implies

$$\alpha\bar{\alpha} = (S(\alpha),S(\alpha)) = (a\alpha,a\alpha) = a\bar{a}(\alpha,\alpha) = a\bar{a}\alpha\bar{\alpha}$$

But $\alpha\bar{\alpha} \neq 0$ implies that $a\bar{a} = 1$.

Corollary The characteristic roots of a (real) orthogonal transformation have absolute value 1. (Notice that the roots need not be real.)

Theorem 5.17 If, for S a unitary transformation, $S(\alpha) = a\alpha$ and $S(\beta) = b\beta$ with $a \neq b$, then α and β are orthogonal.

Proof Now

$$(S(\alpha),S(\beta)) = (\alpha,\beta) = (a\alpha,b\beta) = a\bar{b}(\alpha,\beta)$$

implies $(\alpha,\beta) = 0$ or $a\bar{b} = 1$. But $a\bar{b} = 1$ is impossible since, from Theorem 5.16, $b\bar{b} = 1$ and we know $a \neq b$. Hence $(\alpha,\beta) = 0$.

EXERCISES

1. Let $(1/\sqrt{2},i/\sqrt{2})$ be the first row of a unitary basis of V^2. Find all possible second rows.

2. Let the following be the first two rows of a unitary matrix:

$$\left(\frac{1}{\sqrt{6}},\frac{i}{\sqrt{6}},\frac{2}{\sqrt{6}}\right) \qquad \left(\frac{i}{\sqrt{7}},\frac{1+2i}{\sqrt{7}},\frac{-1}{\sqrt{7}}\right)$$

Find a third row.

3. Consider the Euclidean transformation $E = S + \beta$, where

$$S = \begin{bmatrix} \frac{3}{5} & -\frac{4}{5} \\ \frac{4}{5} & \frac{3}{5} \end{bmatrix} \quad \text{and} \quad \beta = (1,1)$$

Express E as a product of an orthogonal transformation and a translation.

4. Consider the Euclidean transformation

$$x' = \frac{1}{\sqrt{2}}x + \frac{1}{\sqrt{2}}y + 1$$

$$y' = -\frac{1}{\sqrt{2}}x + \frac{1}{\sqrt{2}}y - 1$$

Express this transformation as a product ST, where S is a rotation and T a translation.

5. Denote the transformation in Exercise 4 by $E_1 = S_1 + \beta_1$, and let E be defined as in Exercise 3. Find E_1E and EE_1. If S_2 and β_2 are defined by $E_1E = S_2 + \beta_2$ for S_2 an orthogonal transformation, find β_2 and S_2 in terms of S_1, β_1, S, and β.

6. Prove (5.42).

7. In Exercise 3, for what vectors α are $E(\alpha) = \alpha$? (This vector α can be considered to be a fixed point of E.)

8. In Exercise 4, what are the fixed points, that is, the pairs (x,y), such that $x = x'$ and $y = y'$?

9. In Exercise 4 can the Euclidean transformation be expressed as a product TS, where T is a translation and S an orthogonal transformation? If so, show how.

10. Prove the truth of the Remark following (5.40).

11. Prove Theorem 5.15.

12. Prove that for any vectors α and β in V over the complex field

$$4(\alpha\bar{\alpha} + \beta\bar{\beta}) = \sum_{k=1}^{4} \|\alpha + i^k\beta\|^2$$

13. Prove the equalities (5.20) for the complex case.

14. Let $T = T^*$ and $S = S^*$ for two transformations over a field with a complex inner product. Prove

$$(TS) = (TS)^* \qquad \text{if and only if } TS = ST$$

15. Prove the analog of Theorem 5.6 for the complex case.

16. Prove that the inner product defined in this section is nonsingular over V.

17. Prove, for the inner product over the complex field, that $(S^*)^* = S$ for any linear transformation S.

5.8 QUADRATIC AND HERMITIAN FORMS

Now that we have orthogonal and unitary transformations, what can we do with them? The chief application that we shall make is to the reduction and simplification of quadratic forms which we mentioned in Sec. 5.2 and the analogous treatment for Hermitian forms.

You will recall that in plane (Euclidean) analytic geometry the "standard" form of the equation of an ellipse is

$$\frac{x^2}{a^2} + \frac{y^2}{b^2} = 1 \tag{5.43}$$

In this form we know that the axes of the ellipse are parallel to the coordinate axes. For certain values of r, s, t, and c the following is also the equation of an ellipse:

$$rx^2 + sxy + ty^2 = c \tag{5.44}$$

If $s \neq 0$ the axes are not parallel to the coordinate axes. However, by a rotation of axes, that is, an orthogonal transformation, one may replace Equation (5.44) by one of the form (5.43). Once it is in the form (5.43) we can immediately read off the lengths of the axes and get other information about the curve.

In general, referring to (5.10), we have the quadratic form

$$\xi\xi = (\xi,\xi) = \sum_{i,j=1}^{n} a_{ij}x_ix_j \qquad \text{for } a_{ij} = a_{ji}$$

Using (5.8), we see that we can also write the quadratic form as XAX^T, where X is the n-tuple of x's and A is the matrix of the form. Setting the form equal to a number gives us the equation of a conic section in n-dimensions.

The immediate generalization of the quadratic form for the complex field is the so-called *Hermitian form*, defined as follows:

$$\xi\bar{\xi} = (\xi,\xi) = \sum_{i,j=1}^{n} a_{ij}x_i\bar{x}_j \qquad \text{for } a_{ij} = \bar{a}_{ji} \tag{5.45}$$

where the bar denotes complex conjugate. In matrix form this can be written

$$XA\tilde{X} \qquad \text{where } A = \tilde{A}$$

Note A remarkable property of a Hermitian form is that if the x's are replaced by any set of complex numbers, the number (5.45) as well as the determinant of the matrix A are real. The proof of this is left as an exercise.

For the real case we would hope to find an orthogonal transformation which would take the quadratic form into a diagonal form. That is, we want an orthogonal transformation S so that for $X^T = SY^T$ the quadratic form in Y is diagonal. In matrix form this is

$$XAX^T = YS^TASY^T = YBY^T \qquad \text{where } B = S^TAS \text{ and } S^TS = I$$

We wish to choose S so that B is a diagonal matrix.

For the complex field and Hermitian forms, we would seek a unitary matrix S such that for $\tilde{X} = S\tilde{Y}$

$$(5.46) \qquad XA\tilde{X} = YB\tilde{Y} \qquad \text{where } B = \tilde{S}AS \text{ is diagonal and } \tilde{S}S = I$$

We can proceed more efficiently if we consider the complex case first, because, once that is settled, the real case follows with little labor. First let us explore what the properties of S would have to be to make B diagonal in (5.46). Then we shall be in a position to show its existence. So suppose there is a unitary matrix S such that

$$(5.47) \qquad \tilde{S}AS = B$$

is a diagonal matrix. Since S is unitary, $\tilde{S} = S^{-1}$ and (5.47) can be written

$$(5.48) \qquad AS = SB$$

Then if S_i^T are the columns of S and the diagonal elements of B are b_i for $1 \leq i \leq n$, (5.48) is equivalent to

$$(5.49) \qquad AS_i^T = S_i^Tb_i \qquad \text{for } 1 \leq i \leq n$$

This means that the b_i are characteristic roots of A and the S_i are corresponding characteristic vectors. This shows that if there is a unitary matrix S which diagonalizes A, its columns must be characteristic vectors. Conversely, if there is a unitary basis of A consisting of characteristic vectors, then we can use this basis to construct a unitary transformation taking A into diagonal form.

With this guidance, we prove a series of theorems leading to the result we need. First notice that if a matrix is Hermitian, that is, $A = \tilde{A}$, a unitary transformation takes it into a Hermitian matrix, for

$$\widetilde{(\tilde{S}AS)} = \tilde{S}\tilde{A}S = \tilde{S}AS \qquad \text{if } \tilde{A} = A$$

Second, if a diagonal matrix is Hermitian it must be real, since $a_{ij} = \bar{a}_{ji}$ implies $a_{ii} = \bar{a}_{ii}$; in fact, the diagonal elements of *any* Hermitian matrix are real. These considerations would seem to make reasonable the following

Theorem 5.18 The characteristic roots of a Hermitian matrix are real.

Proof Suppose $A = \tilde{A}$ and $AX^T = cX^T$ for X a matrix and c a complex number. Then $\bar{X}AX^T = c\bar{X}X^T$. However, $AX^T = cX^T$ implies $\bar{X}\tilde{A} = \bar{c}\bar{X}$, and, since $A = \tilde{A}$, we have $\bar{X}\tilde{A}X^T = \bar{X}AX^T = \bar{c}\bar{X}X^T$. Since $\bar{X}X^T$ is a positive real number, $c\bar{X}X^T = \bar{c}\bar{X}X^T$ implies $c = \bar{c}$; that is, c is real.

Corollary The characteristic roots of a real symmetric matrix are real.

Now we can prove the existence of the diagonalizing matrix S.

Theorem 5.19 Given a Hermitian matrix A, there is a unitary matrix S such that $\tilde{S}AS$ is a diagonal matrix. The columns of S are characteristic vectors of A. (In Sec. 5.12 a theorem is proved which has this as a special case.)

Proof Let c be a characteristic root of A and S_1 a characteristic vector associated with c. By multiplication of S_1, if necessary, by a scalar we may assume $S_1\tilde{S}_1 = 1$. Then by Theorem 5.14 there is a unitary basis S_i of V of which the first vector is S_1. Let R^T denote the matrix whose columns are the $S_i{}^T$ in order and note that

$$\bar{R}AR^T = (\bar{S}_iAS_j{}^T)$$

Now $\bar{S}_iAS_1{}^T = c\bar{S}_iS_1{}^T = 0$ for $2 \leq i \leq n$, since the S_i form a unitary basis. Also $AS_1{}^T = cS_1{}^T$ implies $\bar{S}_1\tilde{A} = \overline{cS_1}$. But since $A = \tilde{A}$ and $c = \bar{c}$, we have $\bar{S}_1A = c\bar{S}_1$, and hence $\bar{S}_1AS_i{}^T = c\bar{S}_1S_i{}^T = 0$ for $2 \leq i \leq n$. Hence

$$\bar{R}AR^T = c \oplus A_0$$

where A_0 is a matrix of order $n - 1$. But $\bar{R}AR^T$ Hermitian implies that A_0 is Hermitian. Then we can use the same process on A_0 instead of A. Notice that the argument is not affected if $c = 0$.

More precisely, we have in the above a basis for an induction argument on n, the order of A. If $n = 1$, the theorem is trivial. Assuming the theorem true for $n - 1$, we know of the existence of a unitary transformation R_0 of order $n - 1$ such that

$$\bar{R}_0A_0R_0{}^T = B_0$$

where B_0 is diagonal. Then the matrix $T = 1 \oplus R_0$ is unitary, and

$$\begin{aligned}\bar{T}\bar{R}AR^TT^T &= (1 \oplus \bar{R}_0)(c \oplus A_0)(1 \oplus R_0)^T \\ &= c \oplus B_0\end{aligned}$$

which is a diagonal matrix. Since T and R are unitary matrices, $S = TR$ is also unitary, and we have completed the proof.

Corollary Let A be a real symmetric matrix. There is a (real) orthogonal transformation S such that $\tilde{S}AS$ is a diagonal matrix.

We leave the proof of the corollary as an exercise.

Note It should be recorded that the elements of the diagonal matrices referred to in the theorem and corollary are the characteristic roots of the matrix, because the elements of a diagonal matrix are its characteristic roots and a similarity transformation does not change the characteristic roots of a matrix.

Before giving two examples of the use of the theorem, we find helpful the following piece of information.

Theorem 5.20 If X_1 and X_2 are characteristic vectors of a Hermitian matrix belonging to different characteristic roots, then they are orthogonal. (Compare Theorem 5.17. An extension occurs in Exercise 24 of Sec. 5.12.)

Proof Suppose $AX_i^T = c_iX_i^T$ for $i = 1, 2$ where $c_1 \neq c_2$. Then $\bar{X}_1AX_2^T = c_2\bar{X}_1X_2^T$. Also, taking the transpose conjugate of the first equation of this proof, we get $\bar{X}_1A = \bar{c}_1\bar{X}_1$ since A is Hermitian. Thus $\bar{X}_1AX_2^T = \bar{c}_1\bar{X}_1X_2^T$. Now the c_i are real, so that $\bar{c}_1 = c_1$. Hence

$$c_1\bar{X}_1X_2^T = c_2\bar{X}_1X_2^T$$

Thus $c_1 \neq c_2$ implies $\bar{X}_1X_2^T = 0$; that is, the characteristic vectors X_1 and x_2 are orthogonal. The theorem is proved.

Corollary If X_1 and X_2 are characteristic vectors belonging to distinct characteristic roots of a real symmetric matrix, then they are orthogonal.

To show how these results apply, we shall consider two examples.

Example 1 Describe the curve whose equation is $2x^2 + 4xy + 5y^2 = 3$. Here, as is not unusual, the amount of work necessary depends on the completeness of the information desired. We know from the looks of the equation that the curve is a conic section. We also know from the Corollary to Theorem 5.19 that there is an orthogonal transformation taking the given equation into one of the form

$$ax'^2 + by'^2 = 3$$

where a and b are the characteristic roots of the matrix

$$A = \begin{bmatrix} 2 & 2 \\ 2 & 5 \end{bmatrix}$$

of the quadratic form in x and y. One can compute these roots in two different ways. One is by evaluating the determinant

$$|A - cI| = \begin{vmatrix} 2-c & 2 \\ 2 & 5-c \end{vmatrix} = c^2 - 7c + 6$$

Or, one may note that since A is similar to the diagonal matrix $a \oplus b$, they have the same trace and determinant. That is,

$$2 + 5 = a + b \qquad \text{and} \qquad 2 \cdot 5 - 2^2 = ab$$

Both ways of course lead to the same equation, and we see that the characteristic roots are 1 and 6. Hence by a rotation of axes the original equation becomes

$$x'^2 + 6y'^2 = 3$$

This tells us that the given equation represents an ellipse with axes of lengths $\sqrt{3}$ and $\sqrt{\frac{1}{2}}$.

This is usually all we want to know. But, to complete the example, we continue further to find an orthogonal transformation S which accomplishes the diagonalization. We know from Theorem 5.19 that the columns of S will be characteristic vectors of the matrix A. So we want to find characteristic vectors associated with the characteristic roots 1 and 6.

For the characteristic root 1, we seek a vector (x_1,x_2) such that

$$\begin{bmatrix} 2-1 & 2 \\ 2 & 5-1 \end{bmatrix} \begin{bmatrix} x_1 \\ x_2 \end{bmatrix} = \begin{bmatrix} 1 & 2 \\ 2 & 4 \end{bmatrix} \begin{bmatrix} x_1 \\ x_2 \end{bmatrix} = \begin{bmatrix} 0 \\ 0 \end{bmatrix}$$

A solution is $(-2,1)$. To normalize this we divide by $\sqrt{5}$. So the first column of S can be taken to be $(-2/\sqrt{5}, 1/\sqrt{5})^T$. Similarly a characteristic vector associated with the root 6 is $(1/\sqrt{5}, 2/\sqrt{5})$. Notice that the two characteristic vectors are orthogonal in accordance with Theorem 5.20. So one matrix S is

$$\frac{1}{\sqrt{5}} \begin{bmatrix} -2 & 1 \\ 1 & 2 \end{bmatrix} = S$$

This should be checked by multiplication. Actually in this case we have very little choice for S. It can be shown without great difficulty that we can replace one or both columns by their negatives. This is all the leeway we have, if the resulting quadratic form is to be $x'^2 + 6y'^2$.

Example 2 Find a diagonal form into which the following Hermitian form can be taken by a unitary transformation:

$$f = \tfrac{3}{2}x_1\bar{x}_1 + \tfrac{3}{2}x_2\bar{x}_2 + 3x_3\bar{x}_3 + \tfrac{1}{2}x_1\bar{x}_2 + \tfrac{1}{2}x_2\bar{x}_1 + \frac{1+i}{\sqrt{2}}\,x_1\bar{x}_3 + \frac{1-i}{\sqrt{2}}\,x_3\bar{x}_1 + \frac{1+i}{\sqrt{2}}\,x_2\bar{x}_3 + \frac{1-i}{\sqrt{2}}\,\bar{x}_2x_3$$

To write the matrix of the form $A = (a_{ij})$, recall that a_{ij} is the coefficient of $x_i\bar{x}_j$. So the matrix is

$$\begin{bmatrix} \frac{3}{2} & \frac{1}{2} & \frac{1+i}{\sqrt{2}} \\ \frac{1}{2} & \frac{3}{2} & \frac{1+i}{\sqrt{2}} \\ \frac{1-i}{\sqrt{2}} & \frac{1-i}{\sqrt{2}} & 3 \end{bmatrix}$$

and the characteristic polynomial is $(c - 1)^2(c - 4)$. Hence f can be taken by a unitary transformation into the form

$$f' = y_1\bar{y}_1 + y_2\bar{y}_2 + 4y_3\bar{y}_3$$

This again is as far as we are required to go. But to illustrate Theorem 5.19 completely, let us sketch the computation of a unitary matrix which accomplishes this. First, to find the characteristic vectors associated with the characteristic root 1, we determine the vectors (x_1,x_2,x_3) such that

$$\begin{bmatrix} \frac{1}{2} & \frac{1}{2} & \frac{1+i}{\sqrt{2}} \\ \frac{1}{2} & \frac{1}{2} & \frac{1+i}{\sqrt{2}} \\ \frac{1-i}{\sqrt{2}} & \frac{1-i}{\sqrt{2}} & 2 \end{bmatrix} \begin{bmatrix} x_1 \\ x_2 \\ x_3 \end{bmatrix} = \begin{bmatrix} 0 \\ 0 \\ 0 \end{bmatrix}$$

The last row of the matrix on the left above can be obtained from the first row by multiplying by $\sqrt{2}\,(1 - i)$. Thus the matrix is of rank 1 and the set $\mathcal{S}$ of solutions is of dimension 2. In fact, (x_1,x_2,x_3) is a solution if and only if

$$x_1 + x_2 + \sqrt{2}(1 + i)x_3 = 0 \tag{5.50}$$

So we seek a unitary basis of $\mathcal{S}$. We know that we can construct such a basis starting with any vector of length 1. One solution of (5.50) is $(1,-1,0)$, and hence we can take

$$\alpha_1 = \frac{1}{\sqrt{2}}\,(1,-1,0)$$

Then we want an $\alpha_2 = (a_1,a_2,a_3)$ in $\mathcal{S}$ such that $\alpha_1\bar{\alpha}_2 = 0$. That is, we wish to solve

$$\begin{aligned} a_1 + a_2 + \sqrt{2}(1 + i)a_0 &= 0 \\ a_1 - a_2 \qquad\qquad &= 0 \end{aligned}$$

One solution is $(1, 1, (i - 1)/\sqrt{2})$. The length of this vector is $\sqrt{3}$. Hence we choose

$$\alpha_2 = \frac{1}{\sqrt{3}}\left(1, 1, \frac{i - 1}{\sqrt{2}}\right)$$

Then we may take $\alpha_1{}^T$ and $\alpha_2{}^T$ to be the first two columns of a unitary matrix S. The third column must be a characteristic vector of length 1 associated with the characteristic root 4. Such a vector is

$$\alpha_3 = \frac{1}{\sqrt{6}}(1, 1, \sqrt{2}(1 - i))$$

It may be verified that

$$S = (\alpha_1{}^T, \alpha_2{}^T, \alpha_3{}^T)$$

has the property that

$$\tilde{\bar{S}}AS = \begin{bmatrix} 1 & 0 & 0 \\ 0 & 1 & 0 \\ 0 & 0 & 4 \end{bmatrix}$$

It might be noticed that once we chose α_1 above, we had very little leeway in choosing the other two vectors.

We close this section by calling attention to the fact that Theorem 5.19 shows that for any Hermitian matrix there is a set of characteristic vectors which spans the space. Recall that this is a necessary and sufficient condition (see Theorems 3.19 and 3.23) for similarity to a diagonal matrix.

EXERCISES

1. Check the computation of the characteristic vector associated with the characteristic root 4 in Example 2.

2. Show that if complex numbers are substituted for the x_i in the Hermitian form (5.45), the resulting number is real. Prove that the determinant of a Hermitian matrix is real.

3. Prove the Corollary of Theorem 5.19.

4. In each case below write the diagonal form into which the given form can be taken by an orthogonal transformation. Find such an orthogonal transformation and indicate how many there are.

(*a*) $x^2 + 4xy + y^2$
(*b*) $5x^2 + 4xy + 2y^2$
(*c*) $x^2 + 4xy + 4y^2$
(*d*) $2xy + 4xz - 3z^2 - 4yz$

5. Find a unitary transformation which takes the following Hermitian form into a diagonal form:

$$x_1\bar{x}_2 + ix_1\bar{x}_3 + x_2\bar{x}_1 + ix_2\bar{x}_3 - ix_3\bar{x}_1 - ix_3\bar{x}_2$$

6. Consider $ax^2 + 2bxy + cy^2 = k \neq 0$, where a, b, c, k are real numbers. Prove:

(*a*) If $ac - b^2 > 0$, the equation represents an ellipse which is real if $ka > 0$ and imaginary if $ka < 0$.

(*b*) If $ac - b^2 < 0$, the equation represents a hyperbola.

(*c*) If $ac - b^2 = 0$, the equation represents two straight lines.

7. Let α be a characteristic vector of the quadratic form (*a*) in Exercise 4. Define a transformation T so that $T(\alpha) = -\alpha$, and $T(\beta) = \beta$ if $\alpha \perp \beta$; that is, $a_1b_1 + a_2b_2 = 0$ where a_i and b_i are the components of α and β, respectively. Show that T takes the form (*a*) into itself. What is a geometric interpretation of this?

8. Consider the general Hermitian form in two variables:

$$f = ax\bar{x} + bx\bar{y} + \bar{b}\bar{x}y + cy\bar{y}$$

The determinant of its matrix is the real number $ac - b\bar{b}$. As in Exercise 6 consider the three possibilities for the determinant and find the diagonal form into which a unitary transformation takes f.

9. Suppose M is a matrix with complex elements and that $M = M^T$. Does this imply that the characteristic roots of M are all real? Establish your answer by proof or counterexample.

10. Show that any matrix with complex elements can be expressed uniquely as a sum of a Hermitian matrix and a skew-Hermitian matrix [$A = -\tilde{A}$].

5.9 CANONICAL FORMS

Now that we have some examples, it is well to pinpoint the term *canonical form*. For instance, we showed that any quadratic form over the field of real numbers can be taken into a diagonal form by a properly chosen orthogonal transformation. The diagonal form is unique except for the order of the numbers on the principal diagonal since these numbers are the characteristic roots of the matrix of the form. We call this diagonal matrix the *canonical form* for quadratic forms under an orthogonal transformation. The term means "simplest" in some sense. A canonical form does not have to be unique, but it should have some intrinsic relation to the matrix or transformation, independent of the basis chosen. One can sometimes use a canonical form to establish a relationship between two matrices or transformations. For instance, for the quadratic form, we know, from the transitivity property of orthogonal transformations, that if two quadratic forms have the same canonical form, there is an orthogonal transformation which takes one into the other. Specifically, suppose R and R' are orthogonal transformations which take quadratic forms f and f', respectively, into a diagonal form q. Then $R'R^{-1}$ takes f' into f, and $R'R^{-1}$ is an orthogonal transformation. However, a simpler way to establish this relationship would be to compare the characteristic roots of f and f'. The chief use of a canonical form in general is that we can read off from it certain properties of the vector product or inner product function which are independent of the bases chosen. Without such a canonical form, it might be hard to find out from the matrix of the quadratic form the constants that are independent of the bases chosen.

The canonical form for a Hermitian form is also a diagonal form. We have seen (Sec. 3.11) that if T is a transformation of V into itself, we can choose the basis of V and a basis of its image space so that the matrix of T consists of zeros except for 1s partially or wholly down the principal diagonal, the number of 1s being the rank of T. In matrix form this means that if M is the matrix of T with respect to one pair of bases, one can find nonsingular transformations R and S such that

$$RTS = D$$

for $D = I \oplus Z$, where I is the identity matrix whose rank is that of T and where Z is the zero matrix. Then it follows that all transformations of V into itself having the same rank also have the same diagonal form. In fact, one can get the rank by finding the canonical form.

We shall discuss other examples of canonical forms in the rest of this chapter.

5.10 REAL QUADRATIC FUNCTIONS AND EUCLIDEAN TRANSFORMATIONS

In this section we consider simplification, by means of Euclidean transformations, of the general quadratic function

$$f = \sum_{i,j=1}^{n} a_{ij}x_ix_j + 2\sum_{i=1}^{n} a_{n+1,i}x_i + a_{n+1,n+1} \tag{5.51}$$

in which linear terms occur. In order to bring to bear our knowledge of quadratic forms, it is convenient to consider the companion form

$$g = \sum_{i,j=1}^{n+1} a_{ij}x_ix_j \tag{5.51a}$$

The function f may be obtained from g by replacing x_{n+1} by 1. The matrix of g can be written

$$G = \begin{bmatrix} A & \alpha^T \\ \alpha & a_{n+1,n+1} \end{bmatrix} \tag{5.52}$$

where A is the matrix of the quadratic part of f and

$$\alpha = (a_{n+1,1}, \ldots, a_{n+1,n})$$

We can carry this further to the transformations involved, as is shown in the following theorem:

Theorem 5.21 A Euclidean transformation $S + \beta$ on f is equivalent to the linear transformation

$$S_0 = \begin{bmatrix} S & \beta^T \\ 0 & 1 \end{bmatrix}$$

on g in (5.51a). That is, if g is transformed by S_0 and x_{n+1} replaced by 1, we get the quadratic function f', which is obtained from f by the Euclidean transformation $S + \beta$, and conversely. To simplify notation in this theorem and the proof, we are identifying a transformation with its matrix.

Proof First let $X = (x_1, x_2, \ldots, x_n)$ and $X_0 = (X,1)$. Then matrix multiplication shows that $X_0GX_0^T = f$. Now the Euclidean transformation $\xi = S(\eta) + \beta$ is equivalent to the matrix equation $X^T = SY^T + B^T$, where B is the row matrix corresponding to β. Again, multiplication of matrices shows that

$$X^T = SY^T + B^T \qquad \text{and} \qquad X_0^T = S_0Y_0^T$$

are equivalent if $Y_0 = (Y,1)$. Then $X_0GX_0^T$ yields

$$Y_0S_0^TGS_0Y_0^T$$

This completes the proof.

This theorem then reduces the problem of transforming f by a Euclidean transformation to transforming g by a special kind of linear transformation. Multiplication of matrices shows that S_0 replaces A by S^TAS and, since S is orthogonal, it does not change either the characteristic roots of A or its determinant. Also the determinants of S_0 and S are equal and hence either 1 or -1. Thus S_0 does not change the determinant of g.

Now we know from the Corollary of Theorem 5.19 that there is an orthogonal transformation S which takes A into a diagonal matrix. So the corresponding S_0 will take g into

$$g' = \sum_{i=1}^{n} c_i y_i^2 + 2 \sum_{i=1}^{n} c_{n+1,i} y_i y_{n+1} + b_{n+1,n+1} y_{n+1}^2$$

and f into the quadratic function f' obtained from g' by replacing y_{n+1} by 1, that is,

$$f' = \sum_{i=1}^{n} c_i y_i^2 + 2 \sum_{i=1}^{n} c_{n+1,i} y_i + b_{n+1,n+1}$$

We can simplify this further for those c_i which are not zero by replacing each such y_i by $z_i - c_{n+1,i}/c_i$, a translation. Hence f' may be further reduced to the form

$$f'' = \sum_{i=1}^{k} d_i z_i^2 + 2 \sum_{i=k+1}^{n} d_{n+1,i} z_i + e \tag{5.53}$$

where $d_1, d_2, \ldots, d_k$ are the nonzero characteristic roots of A, if any, and e is a number of F. Remember that the quadratic form g'' corresponding to f'' has the same determinant as g.

Suppose two or more characteristic roots of A are zero. Then two rows of the matrix of (5.53) will have zeros in all places but the last, and hence these two rows will be linearly dependent and the determinant of g will be zero. Hence for $|g| \neq 0$, we have two possible cases:

1. No characteristic roots of A are zero, in which case $k = n$ and the matrix of (5.53) takes the form $D \oplus e$, where $D = d_1 \oplus d_2 \oplus \cdots \oplus d_n$.
2. Exactly one characteristic root of A is zero, that is, $k = n - 1$, in which case the matrix of (5.53) takes the form

$$D_0 \oplus \begin{bmatrix} 0 & d \\ d & e \end{bmatrix}$$

where D_0 is the diagonal matrix whose diagonal elements are the nonzero characteristic roots of A and $d = d_{n+1,n}$.

In the latter case the determinant of g is $-|D_0|d^2$, which shows that $d \neq 0$; in fact, d is determined by $|g| = -|D_0|d^2$. Furthermore, we can eliminate e by replacing y_n by $z_n - e/2d$. Hence forms 1 and 2 with $e = 0$ are the canonical forms for quadratic functions of nonzero determinants under Euclidean transformations.

Though we leave as an exercise the specialization of these results to dimension 2, it seems desirable to illustrate case 2 above for a particular quadratic function in three variables. Let

$$f = x_1^2 + 4x_1x_2 + 2x_1x_3 + 9x_2^2 + 4x_2x_3 + x_3^2 + 2x_1 + 2x_2 - 2x_3 + 3$$

We seek, by a Euclidean transformation, to reduce this to form 1 or 2 above. Now the matrix of the function, that is, the matrix of the companion form g in (5.51a), is

$$G = \begin{bmatrix} 1 & 2 & 1 & 1 \\ 2 & 9 & 2 & 1 \\ 1 & 2 & 1 & -1 \\ 1 & 1 & -1 & 3 \end{bmatrix}$$

where the first three elements of the last row and column are the halves of the coefficients of x_1, x_2, x_3, respectively, and the element in the lower right-hand corner is the constant term 3. Let A be the matrix obtained from G by deleting the last row and column; that is, A is the matrix of the quadratic part of f. Calculation shows that the characteristic roots of A are 1, 10, 0. And hence, by (5.53), a properly chosen Euclidean transformation replaces G by

$$G' = \begin{bmatrix} 1 & 0 & 0 & 0 \\ 0 & 10 & 0 & 0 \\ 0 & 0 & 0 & d \\ 0 & 0 & d & e \end{bmatrix}$$

for some numbers d and e. The quadratic function for G' is

$$f' = y_1^2 + 10y_2^2 + 2dy_3 + e$$

Since the Euclidean transformation has determinant 1 or -1, the determinants of G and G' are the same. Calculation shows that the determinant of G is -20. Since the determinant of G' is $-10d^2$, it follows that $d = \sqrt{2}$. (We arbitrarily choose d to be positive.) We then can, as above, eliminate the constant term by the replacements

$$y_1 = z_1 \qquad y_2 = z_2 \qquad y_3 = z_3 - \frac{e}{2d} = z_3 - \frac{e}{2\sqrt{2}}$$

The final form is then

$$f'' = z_1^2 + 10z_2^2 + 2\sqrt{2}\, z_3$$

EXERCISES

1. Find the canonical form under a Euclidean transformation for each of the following quadratic functions (see Exercise 4 of Sec. 5.8).

(a) $x^2 + 4xy + y^2 + 2x + 2y + 1$
(b) $5x^2 + 4xy + 2y^2 + 2x - 4y - 2$
(c) $x^2 + 4xy + 4y^2 + 4x - 6y + 3$
(d) $2xy + 4xz - 3z^2 - 4yz - 2x + 2y + 1$

2. Find the canonical form under a Euclidean transformation for each of the following quadratic functions:

(*a*) $f = 2x_1^2 + 2x_2^2 + 5x_3^2 + 4x_2x_3 + 2x_1 + 4x_2 + 5$

(*b*) $f = 2x_1^2 + 3x_2^2 + 3x_3^2 + 12x_2x_3 + 4x_2 + 2x_3 + 1$

3. Let f be the quadratic function (5.51) for $n = 2$. Let d be the determinant of the quadratic part of f, that is, $d = a_{11}a_{22} - a_{12}^2$, and let D be the determinant of the associated quadratic form g. If $D \neq 0$ prove

(*a*) If $d > 0$, f represents a real or imaginary ellipse.

(*b*) If $d < 0$, f represents an hyperbola.

(*c*) If $d = 0$, f represents a parabola.

4. What happens in the various cases of Exercise 3 if $D = 0$? That is, what curves does f then represent?

5. In Exercise 3(*a*) how can one tell from D and the coefficients of f whether the ellipse is real or imaginary?

5.11 THE GENERAL LINEAR TRANSFORMATION ON REAL QUADRATIC FORMS; INDEX, THEOREM OF INERTIA

In the previous sections we have been concerned, in the setting of Euclidean geometry, with orthogonal transformations on quadratic forms over the field of reals. Such transformations are special cases of the general linear transformation that is associated with projective geometry. Using a less restricted linear transformation, quite reasonably, gives a simpler canonical form for the quadratic form.

In Equation (5.46) we noted that if in the quadratic form XAX^T the linear transformation S takes X^T into Y^T, that is, $X^T = SY^T$, the matrix of the quadratic form in Y is S^TAS. We know from the Corollary of Theorem 5.19 that there is such a transformation S, in fact an orthogonal one, which takes A into a diagonal matrix B. Assuming for convenience that A is nonsingular, we can write B as

$$B = b_1 \oplus b_2 \oplus \cdots \oplus b_k \oplus (-b_{k+1}) \oplus \cdots \oplus (-b_n)$$

where $b_i > 0$ for $1 \leq i \leq n$. (Of course k may be zero or n.) The quadratic form associated with B is

$$f' = \sum_{i=1}^{k} b_i y_i^2 - \sum_{i=k+1}^{n} b_i y_i^2$$

Then the transformation T such that

$$T(y_i) = \frac{z_i}{\sqrt{b_i}} \qquad \text{for } 1 \leq i \leq n$$

takes f' into a form f'' with matrix

$$C = I_k \oplus (-I_{n-k}) \tag{5.54}$$

where I_j stands for the identity matrix of order j. Thus, taking $R = ST$, we have proved the following:

Theorem 5.22 If f is a nonsingular quadratic form over the field of reals, there is a linear transformation R taking f into a form with matrix (5.54) above, that is, $R^T A R = C$.

(This theorem is extended to include singular matrices below. It can be proved directly without use of Theorem 5.19.)

When there is a nonsingular matrix R such that $R^T A R = B$ we call A and B *congruent*. It is not hard to show that congruence is an equivalence relationship; that is,

1. A is congruent to itself.
2. If A is congruent to B, then B is congruent to A.
3. If A is congruent to B and B is congruent to C, then A is congruent to C.

(Compare Sec. 5.5 after Theorem 5.11.) So (5.54) is *a* canonical form for symmetric real matrices under congruence. We do not have the right to call it *the* canonical form until we show that k depends only on the original matrix. To be sure, k is the number of positive characteristic roots of A. Thus if we take A into C by the product of an orthogonal matrix and a diagonal matrix, k is determined. But not all linear transformations taking A into a diagonal form of 1s and -1s can be expressed as a product of an orthogonal transformation and a diagonal one. (See Exercise 4 at the end of this section.) Hence we must show that for *any* nonsingular linear transformation taking A into form C in (5.54), the number k is the same. We prove the following:

Theorem 5.23 (Sylvester's Law of Inertia) If two nonsingular linear transformations S and R take a nonsingular matrix A into a diagonal matrix of form C in (5.54), then k is the same for both.

Proof We assume that some basis of V has been chosen and that, with respect to this basis, A is the matrix of the quadratic form for a certain inner product. The transformation S will take this basis into an orthogonal basis β_i, where

$$\beta_i^2 = 1 \qquad \text{for } 1 \le i \le k$$

and

$$\beta_j^2 = -1 \qquad \text{for } k+1 \le j \le n$$

Let the transformation R take the basis into an orthogonal basis γ_i, where

$$\gamma_i^2 = 1 \qquad \text{for } 1 \le i \le s$$

and

$$\gamma_j^2 = -1 \qquad \text{for } s+1 \le j \le n$$

We want to prove that $k = s$. To do so let

$$W_1 = \text{sp}\{\beta_1, \beta_2, \ldots, \beta_k\} \qquad W_2 = \text{sp}\{\gamma_{s+1}, \gamma_{s+2}, \ldots, \gamma_n\}$$

Then $\xi_1 \neq \theta$, an element of W_1, implies $\xi_1\bar{\xi}_1 > 0$, and $\xi_2 \neq \theta$, an element of W_2, implies $\xi_2\bar{\xi}_2 < 0$. Hence the spaces W_1 and W_2 have the zero vector as their intersection, and, by Theorem 1.10,

$$\dim(W_1 + W_2) = \dim W_1 + \dim W_2 = k + n - s$$

But $W_1 + W_2$ is a subspace of V, and hence $k + n - s \leq n$, which shows that $k \leq s$. Interchanging the roles of the bases β_i and γ_i would prove that $s \leq k$. Hence our proof is complete.

This permits us to make the following definitions:

Definitions If a linear transformation takes a nonsingular symmetric real matrix A into one of the form (5.54), we call k the *index* of A and $k - (n - k) = 2k - n$ the *signature* of A. (Some authors call $n - k$ the index.)

Having taken care of the case when A is nonsingular, we can easily remove that restriction. Theorem 5.19 holds for singular matrices A, and since the characteristic vectors of A are real, it informs us that there is an orthogonal transformation taking A into $A' \oplus Z$, where Z is a zero matrix and A' is nonsingular. Then in place of form (5.54) we can take A into the form

$$C = I_k \oplus (-I_{r-k}) \oplus Z \tag{5.55}$$

for some zero matrix Z. Then the signature of A is

$$k - (r - k) = 2k - r$$

where r is the rank of A.

If $k = n$ in (5.55), the form f is called *positive definite;* if $r = k$, it is called *positive semidefinite*. If $k = 0$, f is called *negative semidefinite*, and if $k = 0$ and $r = n$, f is called *negative definite*. A form which is none of these is called *indefinite*.

Of special interest are positive-definite forms. First notice that all the characteristic roots of a positive-definite form are positive. Furthermore, we have

Theorem 5.24 A quadratic form with matrix A is positive definite if and only if for all n-tuples X of real numbers

$$XAX^T \geq 0 \tag{5.56}$$

with equality if and only if $X = \theta$.

Proof We know from Theorem 5.22 and the discussion leading to (5.55) that there is a nonsingular transformation S such that $Y = XS^{-1}$ yields

$$XAX^T = (XS^{-1})(SAS^T)(XS^{-1})^T = (XS^{-1})C(XS^{-1})^T = YCY^T$$

where C is of the form (5.55). Then if the index k is less than n, we can choose $Y = \epsilon_n$, the last row of the identity matrix of order n, and see that, for $X = \epsilon_n S$, $0 \geq YAY^T = XAX^T$. If, on the other hand, $k = n$, then $C = I$ and

$$XAX^T = YCY^T = YY^T \geq 0$$

with equality only if $T = \theta$, that is, $X = YS = \theta$. This completes the proof.

Another way to look at this result is to note that if A is considered to be the matrix $((\alpha_i,\alpha_j))$ for a basis α_i, the vector product is an inner product if and only if A is positive definite.

One way to find whether a real symmetric matrix is positive definite is to compute its characteristic roots. But there is a less laborious way using the idea of a principal minor. We call the ith order *principal minor* A_i the matrix obtained from A by eliminating all but the first i rows and columns of A. That is,

$$A_i = \begin{bmatrix} a_{11} & a_{12} & \cdots & a_{1i} \\ \cdots & \cdots & \cdots & \cdots \\ a_{i1} & a_{i2} & \cdots & a_{ii} \end{bmatrix}$$

Thus $A_1 = a_{11}$ and $A_n = A$. We shall prove the following:

Theorem 5.25 A symmetric real matrix A is positive definite if and only if the determinants of all its principal minors are positive.

Proof First suppose the determinant of A_i is not positive for some i. Then A_i is not positive definite, for its determinant is the product of its characteristic roots. Thus there is a nonzero i-tuple X_i such that $X_iA_iX_i^T \leq 0$. Then, letting X be the n-tuple whose first i elements are X_i and whose last $n - i$ elements are zero, we see that $XAX^T \leq 0$ and $X \neq (0)$. This completes the proof in one direction.

Now suppose the determinants of all the principal minors are positive. We wish to show that this implies that A is positive definite. We shall find the following lemma useful here as well as elsewhere:

Lemma A real symmetric matrix A is positive definite if and only if there is a real positive-definite symmetric matrix F such that $F^2 = A$.

Proof If $F^2 = A$, then $XAX^T = (XF)(XF)^T$. This is the square of the length of the vector XF, which must be positive since $XF \neq \theta$.

Conversely, if A is positive definite, there is a real orthogonal matrix S such that $S^TAS = D$, where D is a diagonal matrix whose diagonal elements are all positive. Hence if R is gotten from D by replacing each diagonal element by its positive square root, R is real symmetric and positive definite and $R^2 = D$. Then

$$A = S^{-T}DS^{-1} = S^{-T}R^2S^{-1} = (S^{-T}RS^{-1})(S^{-T}RS^{-1})$$

which proves the lemma with $F = S^{-T}RS^{-1}$.

Now we return to the proof that if the determinants of the principal minors of A are positive, then A is positive definite. We prove this by induction on n. If $n = 1$, then A is the single element a_{11} and the theorem holds. Now assume the theorem holds for $n = 1, 2, \ldots, k$. Then by the induction hypothesis, the determinants of A_i positive for $1 \leq i \leq k$ implies that A_k is positive definite. So we need to show:

If A_k is positive definite and $\det(A_{k+1}) > 0$, *then A_{k+1} is positive definite.*

Now write

$$A_{k+1} = \begin{bmatrix} A_k & \alpha^T \\ \alpha & a \end{bmatrix}$$

where $a = a_{k+1,k+1}$ and $\alpha = (a_{k+1,1}, \ldots, a_{k+1,k})$. If we let

$$X = (x_1, x_2, \ldots, x_k)$$

we see that

$$(X,x)A_{k+1}(X,x)^T = XA_kX^T + \alpha X^Tx + X\alpha^Tx + x^2a \tag{5.57}$$

On the other hand, if we take

$$R = \begin{bmatrix} I & 0 \\ -\alpha A_k^{-1} & 1 \end{bmatrix}$$

where I is the identity matrix of order k, we see first that the determinant of R is 1. Also

$$RA_{k+1}R^T = A_k \oplus (a - \alpha A_k^{-1}\alpha^T) = B \tag{5.58}$$

Since the determinants of B and A_{k+1} are the same and the determinant of A_k is positive, we have

$$a > \alpha A_k^{-1}\alpha^T \tag{5.59}$$

Thus, using the lemma to let $A_k = F^2$ and substituting in (5.57), we have

$$\begin{aligned}(X,x)A_{k+1}(X,x)^T &> XF^2X^T + \alpha X^Tx + X\alpha^Tx + x^2\alpha F^{-2}\alpha^T \\ &= (XF + \alpha F^{-1}x)(XF + \alpha F^{-1}x)^T \geq 0\end{aligned} \tag{5.60}$$

Thus $(X,x)A_{k+1}(X,x)^T > 0$ for all $(X,x) \neq (0)$. This completes the proof of Theorem 5.25.

There is a generalization of Theorem 5.25 which we shall state and illustrate but not prove. The proof may be found in the author's *The Arithmetic Theory of Quadratic Forms*, pp. 7–10, listed in the bibliography.

Theorem 5.26 Let A be a real symmetric nonsingular matrix such that no two successive principal minors are singular and consider the sequence

$$1, |A_1|, |A_2|, \ldots, |A_n| \tag{5.61}$$

of determinants of principal minors. If $|A_i| \cdot |A_{i+1}| > 0$, we say we have a *permanence in sign* from A_i to A_{i+1}. If $|A_i| = 0$, it can be proved that the determinants of A_{i-1} and A_{i+1} have opposite signs, and we agree that in going from A_{i-1} to A_{i+1} we have just one permanence in sign. Then the index of A is the number of permanences in sign in the sequence (5.61).

Let us merely check this for a diagonal form D in which the first k diagonal elements are positive and the last $n - k$ negative. We then see that the portion of the sequence (5.61) up to $|A_k|$ gives k permanences. There are no others.

For comparison purposes it is worthwhile recalling three kinds of transformations with which we have concerned ourselves. Two matrices A and B are said to be similar (see Sec. 3.12) if there is a nonsingular matrix S such that $S^{-1}AS = B$. Two symmetric matrices A and B are congruent if there is a nonsingular matrix S such that $S^TAS = B$. When S is orthogonal, $S^T = S^{-1}$ and the matrices A and B are both congruent and similar. We have found canonical forms for congruence and the orthogonal transformation. It will be the chief task of the next chapter to find a canonical form for the similarity transformation.

We state without proof an interesting theorem.

Theorem 5.27 (**Polar Form**) Any matrix T of order n over the complex field can be expressed in the form $T = UN$, where N is a Hermitian positive-semidefinite matrix and U is a unitary matrix.

A proof may be found in Hoffman and Kunze, *Linear Algebra*, pp. 279, 280, listed in the bibliography.

All the results of this section can be paraphrased for Hermitian matrices over the complex field. In the beginning, one notices that since the characteristic roots of a Hermitian matrix are all real, there is a matrix S such that, for A Hermitian, $\tilde{S}AS$ takes the form (5.55). Positive-definite and semidefinite Hermitian forms or matrices are defined just as for real symmetric forms or matrices. Two matrices A and B are called congruent (or *Hermitian congruent*) if there is a matrix R such that $\tilde{R}AR = B$. The statement and proof of Sylvester's law of inertia (Theorem 5.23) goes through without alteration except that one deals with a unitary instead of an orthogonal basis. The analogs of Theorems 5.24 and 5.25 are stated in the exercises, and the alterations in proofs are left to the reader.

EXERCISES

1. Find the index of each of the following matrices:

$$M = \begin{bmatrix} 1 & i & 1+i \\ -i & 2 & 2i \\ 1-i & -2i & 3 \end{bmatrix} \qquad N = \begin{bmatrix} 1 & -i & i \\ i & 0 & 1-i \\ -i & 1+i & 2 \end{bmatrix}$$

2. For what values of k are the following matrices positive definite?

$$M = \begin{bmatrix} 3 & 1 & 0 \\ 1 & 3 & 0 \\ 0 & 0 & k \end{bmatrix} \qquad N = \begin{bmatrix} 3 & 1 & 1 \\ 1 & 3 & 2 \\ 1 & 2 & k \end{bmatrix}$$

3. For $k = 3$ in matrix M of Exercise 2, find a positive-definite symmetric matrix F such that $F^2 = M$.

4. Relative to the discussion before Theorem 5.23, let

$$A = \begin{bmatrix} 1 & -1 \\ -1 & 0 \end{bmatrix} \qquad B = \begin{bmatrix} 1 & 1 \\ 0 & 1 \end{bmatrix}$$

Show that $B^TAB = 1 \oplus (-1)$ and yet B cannot be expressed as a product of an orthogonal matrix and a diagonal matrix.

5. Find the index of the following matrices:

$$M = \begin{bmatrix} 5 & -3 & 1 \\ -3 & 4 & -2 \\ 1 & -2 & 7 \end{bmatrix} \qquad N = \begin{bmatrix} 1 & 2 & 3 \\ 2 & 1 & 1 \\ 3 & 1 & 4 \end{bmatrix}$$

6. Prove Theorem 5.26 for any diagonal matrix.

7. Let A be a positive-definite real matrix. Prove that for every positive integer n, there is a real positive-definite symmetric matrix B such that $B^n = A$. Is B unique?

8. Let A and B be symmetric real matrices of order n. Prove that AB is symmetric if and only if A and B are commutative, that is, if $AB = BA$.

9. Let A and B be two commutative real positive-definite matrices. Is AB necessarily positive definite? Establish your result by proof or counterexample.

10. Prove that a matrix A of order n is positive-definite symmetric if and only if $A = BB^T$ for some nonsingular matrix B.

11. Analogous to Theorem 5.24 is the following theorem for the field of complex numbers: A Hermitian form with matrix A is positive definite if and only if for all n-tuples X of complex numbers

$$XA\tilde{X} \geq 0$$

with equality if and only if $X = (0)$. What modifications need to be made in the proof?

12. What modifications need to be made in the proof of the lemma of this section to prove the following: A Hermitian matrix A is positive definite if and only if there is a Hermitian nonsingular matrix F such that $F^2 = A$.

13. Since the determinants of the principal minors of a Hermitian matrix are all real, the analog of Theorem 5.25 is: A Hermitian matrix A is positive definite if and only if the determinants of all its principal minors are positive. What modifications need to be made in the proof?

14. Let S be any square matrix over the complex field. Prove that $\tilde{S}S$ is positive-semidefinite Hermitian.

15. Prove that if $AA^T = 0$ for a real matrix A, then $A = 0$.

16. Prove that if R and S are real symmetric matrices and S is positive definite, then the roots of $|R - xS| = 0$ are all real.

17. Prove that if M is a real nonsingular skew-symmetric matrix, then M^TM is positive-definite symmetric and there is an orthogonal matrix P and a matrix C whose square is diagonal such that $M = P^TCP$.

5.12 NORMAL MATRICES OVER THE COMPLEX FIELD

In Secs. 5.13 and 5.14 we shall be working toward a canonical form for the unitary and orthogonal transformations. It is helpful first to define and discuss so-called *normal* matrices and transformations. In this section we define normal matrices and prove the principal theorem about them. Then we give the definition and theorem for normal transformations. Since the results for normal matrices may be derived from those for normal transformations, some readers may prefer to omit the former.

Definition A matrix M in the complex field is called *normal* if it is commutative with its conjugate transpose, that is, if $M\tilde{M} = \tilde{M}M$.

The immediate value of this concept can be seen if we note that the class of normal matrices includes the following: real symmetric, Hermitian, orthogonal, unitary matrices as well as skew symmetric: $A = -A^T$; skew Hermitian: $A = -\tilde{A}$; and others as well. Theorem 5.19 informed us that a Hermitian matrix H can be taken into a diagonal form by a unitary transformation; in other words, there is a unitary basis of the vector space which consists of characteristic vectors of H. We have the same result for real symmetric matrices, as a special case. The following theorem affirms that all normal matrices have this property as well.

Theorem 5.28 A matrix M in the complex field is normal if and only if there is a unitary matrix S taking M into diagonal form, that is, $\tilde{S}MS$ is a diagonal matrix.

Proof First, if for a unitary matrix S we have $\tilde{S}MS = D$, a diagonal matrix, then $D\tilde{D} = \tilde{D}D$ implies $(\tilde{S}MS)(\widetilde{\tilde{S}MS}) = SM\tilde{M}S = S\tilde{M}MS$. Since S is nonsingular, this implies $M\tilde{M} = \tilde{M}M$.

Second, suppose $M\tilde{M} = \tilde{M}M$. Our proof is by induction on n. If $n = 1$, every matrix is normal and diagonal. We assume the theorem true for $n - 1$. Let α be a characteristic vector of M, that is, $M\alpha^T = a\alpha^T$ with $\alpha \neq \theta$. We may, by multiplying by a number if necessary, assume $\alpha\bar{\alpha} = 1$. Then by Theorem 5.14 there are $n - 1$ vectors α_i, where $2 \leq i \leq n$, such that

$$\alpha, \alpha_2, \alpha_3, \ . \ . \ . \ , \alpha_n$$

is a unitary basis of V. Let S_1 be the n by $n - 1$ matrix whose columns are $\alpha_i{}^T$ and V_1 the space spanned by the columns of S_1. Then, taking $S = (\alpha^T, S_1)$, we have

$$\tilde{S}MS = (\bar{\alpha}^T, \bar{S}_1)^T M(\alpha^T, S_1) = \begin{bmatrix} \bar{\alpha}M\alpha^T & \bar{\alpha}MS_1 \\ \tilde{S}_1M\alpha^T & \tilde{S}_1MS_1 \end{bmatrix} = B \tag{5.62}$$

Now $M\alpha^T = a\alpha^T$, and hence $\tilde{S}_1M\alpha^T = a\tilde{S}_1\alpha^T = a(\alpha\bar{S}_1)^T = \theta$. (Note that this holds even if $a = 0$.) Thus B may be written

$$B = \begin{bmatrix} a\bar{\alpha}\alpha^T & \rho \\ \theta & M_0 \end{bmatrix}$$

where ρ is an $(n - 1)$-tuple of complex numbers. But $M\tilde{M} = \tilde{M}M$ implies $B\tilde{B} = \tilde{B}B$. The upper left-hand element in the product $B\tilde{B}$ is

$$a\bar{a}(\bar{\alpha}\alpha^T)^2 + \rho\tilde{\rho} \tag{5.63}$$

and that of $\tilde{B}B$ is

$$a\bar{a}(\bar{\alpha}\alpha^T)^2 \tag{5.64}$$

Equating (5.63) and (5.64) gives us $\rho\bar{\rho} = 0$; hence $\rho = 0$ and

$$B = (a\bar{\alpha}\alpha^T) \oplus M_0$$

Now from the induction hypothesis, since M_0 is commutative with its transpose conjugate, we know that there is a unitary matrix S_0 of order $n - 1$ such that $\tilde{S}_0 M_0 S_0 = D_0$, a diagonal matrix. Thus

$$S(1 \oplus S_0)$$

takes M into the diagonal matrix $a\alpha\bar{\alpha} \oplus D_0$, and our proof is complete.

There are some extra facts which we can deduce from the theorem and its proof. Notice, from (5.62), that M normal and $M\alpha^T = a\alpha^T$ imply $\rho = \bar{\alpha}MS_1 = \theta$ and $(\bar{\alpha}M)\beta^T = \theta$ for all β^T in V_1. Hence $\bar{\alpha}M = \bar{c}\bar{\alpha}$ and $\tilde{M}\alpha^T = c\alpha^T$ for some number c, and $\bar{\alpha}\tilde{M}\alpha^T = c\bar{\alpha}\alpha^T$. On the other hand, $M\alpha^T = a\alpha^T$ implies $\bar{\alpha}M\alpha^T = a\bar{\alpha}\alpha^T$ and $\bar{\alpha}\tilde{M}\alpha^T = \bar{a}\bar{\alpha}\alpha^T$. This shows that $c = \bar{a}$. Thus we have the following:

Corollary If M is normal and $M\alpha^T = a\alpha^T$, then $\tilde{M}\alpha^T = \bar{a}\alpha^T$; that is, $\bar{\alpha}M = a\bar{\alpha}$.

Note 1 The corollary implies that V_1 is an invariant subspace of M since if β is in V_1 and $M\beta^T = c\alpha^T + \gamma^T$, where γ is in V_1, we have, on the one hand, $\bar{\alpha}M\beta^T = c\bar{\alpha}\alpha^T$ and, on the other, $\bar{\alpha}M\beta^T = a\bar{\alpha}\beta^T = 0$ which imply that c is zero.

Note 2 Since Hermitian and orthogonal matrices are normal, Theorem 5.19 and its Corollary follow from Theorem 5.28.

Note 3 Theorem 5.28 does not hold for all fields, as is shown by the following example in the field of integers modulo 5. Let

$$M = \begin{bmatrix} 1 & 2 \\ 2 & 3 \end{bmatrix}$$

M is normal since it is symmetric. The characteristic equation of M is $(x - 2)^2$, whereas the rank of $M - 2I$ is 1. Hence the null space (kernel) of $M - 2I$ is of dimension 1, and 2 is the only characteristic root. Thus the characteristic vectors of M cannot span the two-dimensional space.

Now we consider the linear transformation T over the complex field with an inner product (ξ,η) satisfying properties 1 through 4 of Sec. 5.7, which imply also that the inner product is nonsingular over V. Recall that the adjoint matrix of T was defined in Sec. 5.3 by $(T(\xi),\eta) = (\xi,T^*(\eta))$. Then we have the following:

Definition A transformation T over the complex field with an inner product is called *normal* if it is commutative with its adjoint, that is, if $TT^* = T^*T$.

We state the theorem analogous to Theorem 5.28 in somewhat different form, for convenience, to include Note 1 after that theorem as well as the theorem itself.

Theorem 5.29 If an inner product function over V and the complex field is defined as in Sec. 5.7 and T is a linear transformation of V into itself, then

1. If T is a normal transformation and $T(\alpha) = a\alpha$, for $\alpha \neq \theta$, then, first, $T^*(\alpha) = \bar{a}\alpha$; and, second, $(\alpha,\beta) = 0$ implies $(\alpha,T(\beta)) = 0$.
2. T is normal if and only if there is a unitary basis of V consisting of characteristic vectors of T.

Proof Let $T(\alpha) = a\alpha$, for $\alpha \neq 0$, and call V_1 the orthogonal complement of $\text{sp}(\alpha)$, that is, $V = \text{sp}(\alpha) \perp V_1$.

To prove part 1, assume $TT^* = T^*T$ and let $T^*(\alpha) = c\alpha + \beta$ where β is in V_1. We want to show $c = \bar{a}$ and $\beta = \theta$. First $(\alpha,T^*(\alpha)) = (T(\alpha),\alpha) = a\alpha\bar{\alpha}$ and $(\alpha,T^*(\alpha)) = (\alpha, c\alpha + \beta) = \bar{c}\alpha\bar{\alpha}$, and, since $\alpha\bar{\alpha} \neq 0$, we have $\bar{c} = a$. Second,

$$a\bar{a}\alpha\bar{\alpha} = (T(\alpha),T(\alpha)) = (\alpha,T^*T(\alpha)) = (\alpha,TT^*(\alpha)) = (T^*(\alpha),T^*(\alpha))$$

since, by the Corollary of Theorem 5.8, $(T^*)^* = T$. But

$$(T^*(\alpha),T^*(\alpha)) = (\bar{a}\alpha + \beta, \bar{a}\alpha + \beta) = a\bar{a}\alpha\bar{\alpha} + \beta\bar{\beta}$$

Hence $\beta\bar{\beta} = 0$ and $\beta = \theta$.

Next let $(\alpha,\beta) = 0$. Then

$$(\alpha,T(\beta)) = (T^*(\alpha),\beta) = \bar{a}(\alpha,\beta) = 0$$

and hence $(\alpha,T(\beta)) = 0$; that is, $T(\beta)$ is orthogonal to α. This proves part 1.

Now we proceed to show that if T is normal, there is a unitary basis of characteristic vectors. Let T_1 be T restricted to V_1, that is, $T_1(\beta) = T(\beta)$ for all β in V_1. Then T_1^* is T^* restricted to V_1 and we have $T_1T_1^* = T_1^*T_1$. We use an induction argument on the dimension n of V. If $n = 1$, the theorem is trivial. Then by the induction hypothesis, since T_1 is normal, we may assume that V_1 has a unitary basis $\beta_2, \beta_3, \ldots, \beta_n$ consisting of characteristic vectors of T_1 which are therefore characteristic vectors of T. Then we may choose α so that $\alpha\bar{\alpha} = 1$ and see that

$$\alpha, \beta_2, \beta_3, \ldots, \beta_n$$

is a unitary basis of V consisting of characteristic vectors of T.

Finally, to prove the converse, let $T(\alpha_i) = a_i\alpha_i$, where the α_i form a unitary basis of V, and write

$$T^*(\alpha_j) = \sum_{i=1}^{n} c_{ji}\alpha_i$$

Then
$$(T(\alpha_i),\alpha_j) = a_i(\alpha_i,\alpha_j) = a_i\delta_{ij}$$

and
$$(T(\alpha_i),\alpha_j) = (\alpha_i,T^*(\alpha_j)) = \left(\alpha_i, \sum_{k=1}^{n} c_{jk}\alpha_k\right) = \bar{c}_{ji}(\alpha_i,\alpha_i) = \bar{c}_{ji}$$

Hence $\bar{c}_{ji} = a_i\delta_{ij}$; that is, $\bar{c}_{ji} = 0$ if $j \neq i$ and $\bar{c}_{jj} = a_j$. Thus

$$T^*(\alpha_j) = \bar{a}_j\alpha_j$$

which is part of statement 1 of this theorem. Then

$$TT^*(\alpha_i) = T(\bar{a}_i)\alpha_i = \bar{a}_i a_i \alpha_i$$

and
$$T^*T(\alpha_i) = T^*(a_i\alpha_i) = \bar{a}_i a_i \alpha_i \qquad 1 \leq i \leq n$$

This shows that $T^*T = TT^*$ and completes the proof.

Theorem 5.28 can be deduced from this by choosing the matrix of T to be with respect to the canonical basis. The details are left as an exercise.

EXERCISES

1. Show that the matrix $A = \begin{bmatrix} 2 & i \\ 1 & 2 \end{bmatrix}$ is normal and find its characteristic roots. Find the characteristic vectors and a unitary matrix which takes A into diagonal form.

2. For what numbers b is the matrix $\begin{bmatrix} i & b \\ 1 & 2 \end{bmatrix}$ normal?

3. Prove that the matrix $A = \begin{bmatrix} a & b \\ c & d \end{bmatrix}$ is normal if and only if $b\bar{b} = c\bar{c}$ and $b(\bar{a} - \bar{d}) = \bar{c}(a - d)$.

4. In the two-dimensional vector space over the complex numbers, let the inner products be $\alpha_1\bar{\alpha}_1 = 1$, $\alpha_1\bar{\alpha}_2 = i$, $\alpha_2\bar{\alpha}_2 = 3$, and a transformation T be defined by $T(\alpha_1) = 2\alpha_1 + i\alpha_2$, $T(\alpha_2) = \alpha_1 + 2\alpha_2$. Find T^*. Is T a normal transformation? [Note that two possible methods of solution are the following: first, find the matrix of T^* using Equation (5.42) and then see whether $TT^* = T^*T$; second, use part 2 of Theorem 5.29.]

5. Repeat Exercise 4 for the following inner product and transformation:

$$\alpha_1\bar{\alpha}_1 = 1 \qquad \alpha_1\bar{\alpha}_2 = i \qquad \alpha_2\bar{\alpha}_2 = 0$$

and
$$T(\alpha_1) = 2\alpha_1 \qquad T(\alpha_2) = -i\alpha_1 + \alpha_2$$

6. For the transformation T in Exercise 4 is $T\tilde{T} = \tilde{T}T$?

7. For the transformation T in Exercise 5 is $T\tilde{T} = \tilde{T}T$?

8. Why is the normality (the property of being normal) of a transformation T independent of the basis chosen?

9. Show that M is a normal matrix if it satisfies any one of the following conditions:
 (*a*) M real and $M = -M^T$ (M skew symmetric)
 (*b*) M complex and $M = -\tilde{M}$ (M skew Hermitian)

10. Find a normal real matrix which is none of the following: symmetric real, Hermitian, orthogonal, unitary, or (*a*) or (*b*) of Exercise 9.

11. Show how Theorem 5.28 may be deduced from Theorem 5.29 by choosing the Hermitian form and the matrix of T to be with respect to the canonical basis.

12. Show that if M is a real skew-symmetric matrix, then iM with $i^2 = -1$ is Hermitian.

13. Using Exercise 12 and Theorem 5.18 (or by other means), prove that the characteristic roots of a real skew-symmetric matrix are all pure imaginary.

14. Using the proof of Theorem 5.18 (or by other means), show that if M is a skew-Hermitian matrix, then its characteristic roots are all pure imaginary.

15. Prove that if a Hermitian matrix is both positive definite and unitary, then it is the identity.

16. Prove that if $M^n = Z$, the zero matrix, for some positive integer n, and if M is a normal matrix, then $M = Z$.

17. Prove that if M is a normal matrix and $M^2\alpha = \theta$, then $M\alpha = \theta$.

18. Show that if M is a real normal matrix of order 2, then it is symmetric or skew symmetric.

19. Show that if M is a real normal matrix of order 2 with no zero element, then M is symmetric or a scalar multiple of an orthogonal matrix.

20. Prove that if M and R are normal matrices such that $MR = RM$, then MR is a normal matrix.

21. Let T be a normal transformation and W a subspace of V. If $M(W) \subset W$, prove that $M^*(W) \subset W$.

22. Prove that if M is a normal matrix, there is a polynomial $f(x)$ with complex coefficients such that $M^* = f(M)$. Hint: First consider the case in which M is diagonal and use Vandermonde's determinant (see Exercise 21, Sec. 4.4).

23. Let M be a normal matrix and P any matrix which commutes with M. Prove that P commutes with $\tilde{M}$.

24. Let T be a normal transformation such that $T(\alpha) = a\alpha$ and $T(\beta) = b\beta$. Prove that either α and β are orthogonal or $a\bar{b}$ is real. Compare your results to Theorems 5.17 and 5.20.

5.13 DECOMPOSITION OF ORTHOGONAL AND EUCLIDEAN TRANSFORMATIONS

We know from Theorem 5.19 that there is a unitary transformation taking a Hermitian form into a diagonal form whose diagonal elements are the characteristic roots of the matrix of the form. (This is the canonical form for a Hermitian matrix with respect to unitary transformations.) The same is true, by Theorem 5.28, of a unitary matrix. In fact, as a special case of the latter, if P is an orthogonal matrix, there is a unitary matrix S such that $\tilde{S}PS$ is a diagonal matrix whose diagonal elements are the characteristic roots of P. So from this point of view both unitary and orthogonal matrices have diagonal canonical forms.

However, for a (real) orthogonal matrix, if we are to remain in the field of real numbers, we should restrict the unitary matrix to be real, that is, orthogonal. So the first question we might ask is: Is there an *orthogonal* transformation R such that R^TPR is diagonal, for P an orthogonal matrix? The answer is "not always," for if P and R are real, the diagonal matrix R^TPR must be real. Its diagonal elements are the characteristic roots of P. So if P has imaginary characteristic roots (and it certainly can have), there is no orthogonal matrix which takes it into diagonal form. Thus if we are to confine ourselves to orthogonal matrices R, we have to settle for something less than a diagonal canonical form.

To describe what the situation is, let $n = 3$. Then since all the characteristic roots of P have absolute value 1 (see the Corollary of Theorem 5.16), the only possibilities are $1, -1, \cos\varphi + i\sin\varphi$, for some (real) angle φ. Suppose P has an imaginary characteristic root $\cos\varphi + i\sin\varphi$. Then, since the characteristic equation of P has real coefficients, $\cos\varphi - i\sin\varphi$ is also a characteristic root, and thus the characteristic roots of P will be $\cos\varphi + i\sin\varphi$, $\cos\varphi - i\sin\varphi$, ± 1. The following theorem affirms that for this case there is an orthogonal transformation R such that

$$R^TPR = R_0 \oplus e$$

where $$R_0 = \begin{bmatrix} \cos\varphi & \sin\varphi \\ -\sin\varphi & \cos\varphi \end{bmatrix} \quad \text{and} \quad e = 1 \text{ or } -1$$

This means that, for a proper choice of orthogonal basis, if $e = 1$, R is a rotation in a plane: $R_0 \oplus 1$; while if $e = -1$, R is a rotation $R_0 \oplus 1$, followed by a reflection: $I \oplus (-1)$.

Now we state and prove the general theorem.

Theorem 5.30 If P is a (real) orthogonal matrix, then

$$V = V_1 \oplus V_{-1} \oplus V_0 \tag{5.65}$$

where V_1 is the set of all vectors α such that $P\alpha = \alpha$, V_{-1} is the set of all vectors β such that $P\beta = -\beta$, and V_0 is the subspace spanned by the imaginary characteristic vectors of P. Also

$$P = P_1 \oplus P_{-1} \oplus Q \tag{5.66}$$

where P_1 is P restricted to V_1, and similarly for P_{-1} and Q. Furthermore, the dimension of V_0 is even, $2k$, and

$$V_0 = W_1 \oplus W_2 \oplus \cdots \oplus W_k \tag{5.67}$$

$$Q_0 = Q_1 \oplus Q_2 \oplus \cdots \oplus Q_k \tag{5.68}$$

where the W_i are two-dimensional subspaces invariant under P and Q_i is Q, that is, P, restricted to W_i. Each Q_i is orthogonal and all the subspaces in (5.65) and (5.67) are invariant under P.

Proof Since P is a normal matrix, we know by Theorem 5.29 that there is a set of characteristic vectors of P which span the space. This gives us the splittings (5.65) and (5.66) since the subspaces are invariant under P. Note that since Q is real with only imaginary characteristic roots, the dimension of V_0 is even, $2k$.

It remains to show (5.67) and (5.68). Now, as we have developed it, the vectors which span V_0 are imaginary. We wish to replace these vectors by real ones which, though not characteristic vectors, span the same space. So suppose $Q\alpha = a\alpha$, where a and α are imaginary. Then $Q\bar{\alpha} = \bar{a}\bar{\alpha}$, since Q is real, and $\bar{\alpha}$ is also a characteristic vector; α and $\bar{\alpha}$ are linearly independent since $a \neq \bar{a}$. So if $W = \text{sp}(\alpha,\bar{\alpha})$, then also $W = \text{sp}(\beta,\gamma)$ where

$$\beta = \alpha + \bar{\alpha} \qquad \text{and} \qquad \gamma = i(\alpha - \bar{\alpha})$$

are real vectors and $i^2 = -1$. Furthermore, any real vector in W is a linear combination of β and γ with real coefficients, since $a\alpha + b\bar{\alpha}$ real implies $b = \bar{a}$, and

$$a\alpha + \bar{a}\bar{\alpha} = \frac{1}{2}(a + \bar{a})(\alpha + \bar{\alpha}) - \frac{i}{2}(a - \bar{a})i(\alpha - \bar{\alpha})$$

Thus if we make this replacement for all conjugate pairs of characteristic roots of Q, we have the splitting (5.67). Furthermore, $W_j = \text{sp}(\alpha_j,\bar{\alpha}_j)$ is invariant under Q. Also, since P is orthogonal, each restriction Q_j is as well. This completes the proof.

We have the

Corollary For the choice of basis in the theorem, the matrix of P is

$$I_r \oplus (-I_s) \oplus R_1 \oplus R_2 \oplus \cdots \oplus R_k$$

where I_r is the identity matrix over V_1, I_s is the identity over V_{-1}, and

$$R_j = \begin{bmatrix} \cos \varphi_j & -\sin \varphi_j \\ \sin \varphi_j & \cos \varphi_j \end{bmatrix} \tag{5.69}$$

and the characteristic roots of Q_j are $\cos \varphi_j + i \sin \varphi_j$.

Proof Since the characteristic roots of Q_j are imaginary numbers a and $\bar{a}$ whose absolute value is 1, the determinant of R_j is 1. You were asked to show in Exercise 1 of Sec. 5.5 that the matrix of such a Q_j is of the form (5.69). This completes the proof of the corollary.

Notice that V_1 of Theorem 5.30 is the set of "fixed points" of the transformation, that is, the set of vectors α such that $P(\alpha) = \alpha$. An orthogonal transformation must have the origin as a fixed point, but it may have no others.

Now recall the definition of a symmetry or reflection given in Sec. 5.3. The matrix

$$R = \begin{bmatrix} \cos \varphi & -\sin \varphi \\ \sin \varphi & \cos \varphi \end{bmatrix}$$

can be written as the product ST, where

$$S = \begin{bmatrix} 1 & 0 \\ 0 & -1 \end{bmatrix} \qquad \text{and} \qquad T = \begin{bmatrix} \cos \varphi & -\sin \varphi \\ -\sin \varphi & -\cos \varphi \end{bmatrix}$$

The matrix S is a symmetry, as is apparent from its form. The matrix T is also, since its determinant is -1, showing that the characteristic roots are 1 and -1; hence $T(\alpha) = \alpha$ and $T(\beta) = -\beta$ for some orthogonal vectors α, β. Thus each Q_j is a product of two symmetries. We have almost proved the following theorem:

Theorem 5.31 Every real orthogonal transformation P of V^n into itself can be expressed as a product of $n - u$ symmetries, where u is the nullity of $P - I$.

Proof Let $\beta_1, \beta_2, \ldots, \beta_s$ be a basis of V_{-1} and define a symmetry T_i by $T_i(\beta_i) = -\beta_i$ and $T_i(\alpha) = \alpha$ for all vectors α in the orthogonal complement of β_i. Similarly, if we define Q_i' over V by

$$Q_i'(\beta_i) = Q_i(\beta_i) \qquad \text{for } \beta_i \text{ in } W_i$$

and $Q_i(\alpha) = \alpha$ for all α in the orthogonal complement of β_i, we have

$$P = T_1T_2 \cdot \cdot \cdot T_sQ'_1Q'_2 \cdot \cdot \cdot Q'_k$$

and each Q'_i is a product of two symmetries. This completes the proof.

It is not hard to extend these results to any Euclidean transformations. Recall from Sec. 5.6 that any Euclidean transformation can be represented as a product of a translation and an orthogonal transformation. Hence to show that any Euclidean transformation is a product of symmetries we need merely show that a translation has this property. It is easy to see that in two dimensions any translation is equivalent to a succession of two reflections in two parallel lines. In n dimensions, hyperplanes (see Sec. 5.3) correspond to lines in two dimensions. So we shall prove

Theorem 5.32 Any translation can be represented as a product of two symmetries in parallel hyperplanes.

Proof Consider the translation $T(\xi) = \xi + \alpha$ for all ξ in V, where α is a fixed nonzero vector. The vector α indicates the "direction" of the translation, and hence we would expect the hyperplanes to be orthogonal to α. So we define a symmetry S by

$$S(\alpha) = -\alpha \qquad \text{and} \qquad S(\beta) = \beta \qquad \text{for all } \beta \text{ orthogonal to } \alpha$$

We designate by H the hyperplane consisting of all vectors orthogonal to α. H is thus an $(n - 1)$-dimensional space which includes the origin.

Now we want to find a symmetry S' such that $S'S = T$, that is, $S'S(\xi) = \xi + \alpha$ for all ξ. Any vector ξ in V can be written $\xi = c\alpha + \beta$, where c is some number and β is in H. We have, on the one hand,

$$S'S(\xi) = S'S(c\alpha + \beta) = S'(-c\alpha + \beta)$$

and on the other, $S'S(\xi) = \xi + \alpha = c\alpha + \beta + \alpha$. So we want to choose S' to be a symmetry satisfying the condition

$$S'(-c\alpha + \beta) = (c + 1)\alpha + \beta$$

for all c and all β in H. Since S' is to be a symmetry, it must have a hyperplane of invariant vectors. A vector $-c\alpha + \beta$ will be invariant if and only if

$$-c\alpha + \beta = (c + 1)\alpha + \beta$$

that is, $c = -\frac{1}{2}$. The vectors $\frac{1}{2}\alpha + \beta$ for β in H are in a hyperplane H' parallel to H and containing the point corresponding to the vector $\frac{1}{2}\alpha$. So we have not only proved the theorem but have information about the hyperplanes, as stated in the following

Corollary 1 The two hyperplanes of the theorem are orthogonal to α, and the distance between them is half the norm of α.

Using Theorem 5.31 we have

Corollary 2 Every Euclidean transformation in n dimensions can be expressed as a product of $n + 2$ or fewer symmetries. (Actually, by Theorem 5.31 we can decrease n by u, where u is the nullity of $E - I$, for the Euclidean transformation E.)

It is true that $n + 2$ in Corollary 2 may be reduced to $n + 1$. A proof of this is indicated in Exercises 19 and 20 below. That the number cannot in general be reduced further is shown by an example which you are asked to supply in Exercise 10 below; namely, that in two-dimensional space there is a Euclidean transformation which cannot be expressed as a product of less than three symmetries.

To clarify the above theory let us use it to investigate the Euclidean transformations for dimension 2. We showed in Sec. 5.6 that any Euclidean transformation E can be represented in the form TR, where T is a translation and R is an orthogonal transformation. Immediately before Theorem 5.30 we showed that R is either a reflection (when its determinant is -1) or a rotation (when its determinant is 1). We also showed above that in the latter case the rotation R can be expressed as a product of two reflections. So now we divide our discussion into two parts according to the determinant of E.

First, if -1 is the determinant of E, we know from Corollary 2 of Theorem 5.32 that E can be expressed as a product of four or fewer reflections. But since the determinant of E is -1 it must be the product of an odd number of reflections. Hence it is either a reflection or a product of three reflections. In the former case it has a line of fixed points; in the latter case it has no such line. To show how this may be determined in a special case, consider the following:

Example 1 Describe the Euclidean transformation E defined by

$$y_1 = -x_1 + 1 \qquad y_2 = x_2 + 1$$

Notice that -1 is the determinant of E. To find the fixed points, we should set $x_i = y_i$ for $i = 1, 2$. But this yields $x_2 - x_2 = 1$, which is a contradiction. Thus the transformation has no fixed points, and hence, as we showed above, it cannot be a reflection. So E must be a product of three reflections. It is enlightening to find such a set of three reflections. Let R be the reflection $x_1 = -z_1$, $x_2 = z_2$, that is, a reflection in the line $x_1 = 0$. Then E is TR, where T is the translation $I + \alpha$ with $\alpha = (1,1)$. From Theorem 5.32, the translation T is a product of two reflections in lines orthogonal to the vector

(1,1). One such line is the set of points $(a,-a)$, that is, the line $x_1 + x_2 = 0$. So we choose S_1 to be the reflection in this line and seek a reflection S_2 so that $T = S_1S_2$. Since $S_1^2 = I$, the equations $T = S_1S_2$ and $S_1T = S_2$ are equivalent. So our procedure is to calculate S_1T and show that it is a reflection. Using the canonical basis ϵ_1, ϵ_2, we see that S_1 is described by the following equations:

$$S_1(\epsilon_1,-\epsilon_2) = (\epsilon_1,-\epsilon_2) \qquad S_1(\epsilon_1,\epsilon_2) = -(\epsilon_1,\epsilon_2)$$

Hence

$$S_1[(\epsilon_1,-\epsilon_2) + (\epsilon_1,\epsilon_2)] = (\epsilon_1,-\epsilon_2) + (-\epsilon_1,-\epsilon_2)$$

implies the first equation below, and the second is obtained by subtraction:

$$S_1(\epsilon_1) = -\epsilon_2 \qquad S_1(\epsilon_2) = -\epsilon_1$$

Now we calculate S_1T as follows:

$$\begin{aligned} S_1T(x_1,x_2) &= S_1T(x_1\epsilon_1 + x_2\epsilon_2) = S_1(x_1\epsilon_1 + x_2\epsilon_2) + S_1(1,1) \\ &= (-x_1\epsilon_2 - x_2\epsilon_1) + (-1,-1) = (-x_2 - 1,\ -x_1 - 1) \end{aligned}$$

This transformation has as its fixed points all those (x_1,x_2) for which $x_1 + x_2 = -1$; that is, the transformation S_1T has a whole line of fixed points. Hence $S_1T = S_2$ is a reflection in the line $x_1 + x_2 = -1$. Thus we can write E as

$$E = S_1S_2R$$

where R is the reflection in the line $x_1 = 0$, S_1 is the reflection in the line $x_1 + x_2 = 0$, and S_2 is the reflection in the line $x_1 + x_2 = -1$.

Of course this is not the only set of three reflections which yields the transformation E. We could have chosen another reflection in place of R, or, having chosen R as above, we could have taken S_1 to be a reflection in any line parallel to $x_1 + x_2 = 0$.

Second, suppose 1 is the determinant of E. Corollary 2 of Theorem 5.32 assures us that E is either the identity transformation, the product of two reflections, or the product of four reflections. In fact, the third possibility can be excluded. We show this in a particular example and leave the general result as an exercise.

Example 2 Describe the Euclidean transformation E defined by

$$y_1 = \frac{3}{5}x_1 + \frac{4}{5}x_2 + 2 \qquad y_2 = \frac{-4}{5}x_1 + \frac{3}{5}x_2 + 2$$

Since the determinant of E is 1, we know from the above discussion that this must be a rotation, or a product of four reflections. It is not, however, a rotation about the origin since the origin is not left fixed by the transformation. So, in the hope that we can show it to be a single rotation, we seek a fixed point (or fixed points) of the transformation. This is done by setting $x_1 = y_1$ and $x_2 = y_2$ and solving for (x_1,x_2). Calculation shows that $(3,-1)$ is the only fixed point. So this must be the center of the rotation. To verify this, we can transform (x_1,x_2) and (y_1,y_2) by the translation which takes $(3,-1)$ into $(0,0)$, namely,

$$x_1 = x_1' + 3 \qquad y_1 = y_1' + 3 \qquad x_2 = x_2' - 1 \qquad y_2 = y_2' - 1$$

If we substitute these in the equations of the given transformation, we get

$$y_1' = \frac{3}{5}x_1' + \frac{4}{5}x_2' \qquad y_2' = \frac{-4}{5}x_1' + \frac{3}{5}x_2'$$

which is a rotation about the "new origin," that is, about the point $(3,-1)$. Thus E is a rotation about $(3,-1)$. Notice that the rotation is the same as the orthogonal part of the original transformation. You are asked to generalize this in an exercise below.

EXERCISES

1. Let P be the orthogonal matrix

$$\frac{1}{3}\begin{bmatrix} 1 & 2 & 2 \\ 2 & 1 & -2 \\ 2 & -2 & 1 \end{bmatrix}$$

Find k, r, and s as defined in the Corollary of Theorem 5.30.

2. Let S be the rotation R of (5.69), with $\varphi_j = \pi/4$. Express S as a product of two reflections.

3. Let S be the rotation R_j of (5.69) and let S_1 be any symmetry (or reflection) in V^2. Prove that S_1S and SS_1 are symmetries.

4. Generalize Example 2 worked out in the text previous to these exercises. That is, let $\eta = TR(\xi)$ be a Euclidean transformation with R an orthogonal transformation and T the translation $I + \alpha$ for some vector α. Prove that if γ is a fixed point of TR, then $\eta = \eta' + \gamma$ and $\xi = \xi' + \gamma$ imply that $\eta' = R(\xi')$.

5. Let β be the vector (1,1) and express the translation $I + \beta$ in two dimensions as a product of two reflections.

6. Choosing S and β as in Exercises 2 and 5, express $S + \beta$ as a product of three or fewer reflections.

7. The transformation $S + \beta$ of Exercise 6 is equivalent to a rotation about some point P. Find the coordinates of the point P.

8. Prove that if S is an orthogonal transformation of a vector space V of dimension n over the field of real numbers and if 1 is not a characteristic root of S, then the Euclidean transformation $S + \beta$ has a fixed point.

9. Given the rotation and a reflection, R and R_1, respectively,

$$R = \begin{bmatrix} \cos\varphi & \sin\varphi \\ -\sin\varphi & \cos\varphi \end{bmatrix} \qquad R_1 = \begin{bmatrix} \cos\omega & \sin\omega \\ \sin\omega & -\cos\omega \end{bmatrix}$$

then the transformation RR_1 is a reflection in what line?

10. Give an example of a Euclidean transformation in two dimensions which is a product of three reflections but not fewer.

11. Let (1,2) and (3,−1) be two points in the plane. Find the reflection which takes one into the other.

12. In Exercise 2, let S be the rotation R, with $\varphi = \pi/6$. Express S as a product of two reflections or symmetries.

13. Let $\mathcal{S}$ denote the set of Euclidean transformations of the form $S + \beta$ where $\det(S) = 1$. Prove that $\mathcal{S}$ is a group. (This set is sometimes called the *set of rotations*.)

14. Let S_1 and S_2 be two symmetries in n dimensions, with $n > 2$, and suppose $S_i(\alpha_i) = -\alpha_i$, for $i = 1, 2$. Prove that if α_1 and α_2 are linearly dependent, S_1S_2 is the identity, while if α_1 and α_2 are linearly independent, then S_1S_2 is a rotation in the two-dimensional space spanned by α_1 and α_2. Also show that S_1S_2 leaves invariant all vectors orthogonal to α_1 and α_2.

15. Let S_1, S_2, S_3 be three symmetries which leave invariant the vectors of an $(n-2)$-dimensional subspace W of V. Prove that there is a symmetry S_4 leaving each vector of W invariant and such that $S_1S_2 = S_3S_4$.

16. Let α and α' denote two points in the plane. What is the algebraic form of the reflection which takes α into α'?

17. Generalize the result of Exercise 16 to three dimensions.

18. If in the plane R is a rotation and T a translation, is RT always a rotation, always a translation, or sometimes one and sometimes the other? If RT is a rotation, must TR be also?

19. Let R be an orthogonal transformation of V^n into itself and S be any symmetry of V^n into itself. Complete the proof of the following theorem: There is an orthogonal transformation R' which is the product of $n-1$ or fewer symmetries such that $R = R'S$. Partial proof: Suppose n is even. Let R be the product of r symmetries. We know (why?) that we can assume that $r \leq n$. Then RS is a product of $r+1$ symmetries. If $r < n$, take $R' = RS$, and $S^2 = I$ implies $R = R'S$. If $r = n$, then RS has determinant -1 (why?). Since RS is a product of n or fewer symmetries and n is even, RS is a product of $n-1$ or fewer symmetries. (Notice that this result may be described loosely by saying that, in expressing an orthogonal transformation as a product of symmetries, one may be chosen arbitrarily.)

20. Prove that any Euclidean transformation $E = RT$ of V^n into itself, where R is an orthogonal transformation and T a translation, can be expressed as a product of $n + 1$ or fewer symmetries. Hint: Take $T = S_1S_2$, where S_1 and S_2 are symmetries, and use the results of Exercise 19.

5.14 QUADRATIC FORMS AND ISOMETRIES OVER ARBITRARY FIELDS

In this section we deal with a generalization of the orthogonal transformation without restriction as to field. We acquire some information about the characteristic polynomial of this generalization and make some remarks about its decomposition. This also serves as a review of the chapter from a slightly more general point of view.

So we start with a vector product for a vector space V and an arbitrary field. We assume throughout the section that this is nonsingular and symmetric over V, but we do not assume that $(\alpha,\alpha) = 0$ implies $\alpha = \theta$. Given a basis $\alpha_1, \alpha_2, \ldots, \alpha_n$, the vector product defines and is defined by a symmetric bilinear form (see Sec. 5.2)

$$(\xi,\eta) = \sum_{i,j=1}^{n} a_{ij}x_iy_j \qquad \text{where } a_{ij} = (\alpha_i,\alpha_j) = a_{ji}$$

A transformation T of V into itself is called an *isometry* (see Sec. 5.5) if

$$(T(\xi),T(\eta)) = (\xi,\eta) \qquad \text{for all } \xi, \eta \text{ in } V$$

It should be noticed that if F is taken to be the field of real numbers and if the vector product is the inner product, T is an orthogonal transformation.

All this may be expressed in terms of matrices for the given basis. The quadratic form may be written XAY^T, and if $X' = XM$ and $Y' = YM$, where M is the matrix of the transformation T, then

$$X'BY'^T = XMBM^TY^T$$

where $A = MBM^T$ is the relationship between the matrices of the quadratic forms in x, y and x', y', respectively. If we think of M as inducing a change in basis, we may consider A as the matrix of the quadratic form with respect to the given basis and B that with respect to the transformed basis. If the transformation T is an isometry, then the two quadratic forms will have the same coefficients. Thus we can say that T *takes the form into itself*. From this point of view it is sometimes called an *automorph*. However, we shall see that it is usually easier to deal with isometries without reference to the matrices.

First we refer back to Theorem 5.29 on normal matrices. We can show that an isometry has, with some modifications, the first property listed but not the second. We list three properties, prove them, and then show by an example that what corresponds to part 2 of Theorem 5.29 does not hold:

1. $TT^* = T^*T = I$.
2. If $T(\alpha) = a\alpha$, then $T^*(\alpha) = a^{-1}\alpha$.
3. If α is a characteristic vector of T, then $(\alpha,\beta) = 0$ implies $(\alpha,T(\beta)) = 0$.

To prove property 1 notice that T an isometry implies

$$(\alpha,\beta) = (T(\alpha),T(\beta)) = (\alpha,T^*T(\beta)) \qquad \text{for all } \alpha \text{ and } \beta \text{ in } V$$

Thus $(\alpha, (I - T^*T)\beta) = 0$ for all α and β. Since the vector product is nonsingular over V, $I = T^*T$. This implies that T and T^* are nonsingular and thus that $TT^* = I$ and all the characteristic roots of T are different from zero. To prove property 2, apply T^* to both sides of the equality $T(\alpha) = a\alpha$. For property 3, note that $(\alpha,\beta) = 0$ and $T^*(\alpha) = a^{-1}\alpha$ imply

$$0 = a^{-1}(\alpha,\beta) = (T^*(\alpha),\beta) = (\alpha,T(\beta))$$

To show that part 2 of Theorem 5.29 is not, in this setting, true, let F be the field of complex numbers, V the vector space of dimension 2, and define T by

$$T(1,i) = (1,i) \qquad T(1,-i) = -(1,-i)$$

In this case the characteristic roots of T are 1 and -1, and the accompanying characteristic vectors are of length 0, that is, isotropic. Thus there is no orthogonal basis of V consisting of characteristic vectors of T.

However, we have a completion theorem for isometries, called *Witt's theorem*, as follows:

Theorem 5.33 (***Witt's Theorem***) Let V_1 be a vector subspace of V over a field F in which $1 + 1 \neq 0$, and let a symmetric vector product function be nonsingular over V_1 as well as over V. Let T_1 be an isometry over V_1. Then there is an isometry T of V which, when restricted to V_1, is T_1.

Proof From Theorem 5.2 we know there is an orthogonal basis of V_1, α_1, α_2, . . . , α_k. There is no loss in generality if we consider the isometries with respect to this basis. Let $T_1(\alpha_i) = \beta_i$ for $1 \leq i \leq k$ and have

$$(\alpha_i,\alpha_j) = (\beta_i,\beta_j) \qquad \text{for } 1 \leq i \leq k \text{ and } 1 \leq j \leq k$$

We start with $k = 1$ and later proceed to an induction. Now

$$(\alpha_1 - \beta_1, \alpha_1 - \beta_1) + (\alpha_1 + \beta_1, \alpha_1 + \beta_1) = 4(\alpha_1,\alpha_1) \neq 0$$

Hence $\alpha_1 - \beta_1$ and $\alpha_1 + \beta_1$ are not both isotropic. Assume that $\alpha_1 + \beta_1$ is not isotropic. Then, by Sec. 5.2, it has an orthogonal complement, call it V_0.

Now $\alpha_1 - \beta_1$ is in V_0 since calculation shows that it is orthogonal to $\alpha_1 + \beta_1$. So we can form a basis of V_0 (not an orthogonal basis):

$$\alpha_1 - \beta_1, \omega_1, \omega_2, \ldots, \omega_{n-2}$$

Now define a transformation T on V as follows:

$$T(\alpha_1 - \beta_1) = -(\alpha_1 - \beta_1) \qquad T(\alpha_1 + \beta_1) = \alpha_1 + \beta_1$$
$$T(\omega_i) = -\omega_i \qquad \text{for } 1 \leq i \leq n - 2$$

This transformation T is an isometry on V since it is on V_0 and since it takes $\alpha_1 + \beta_1$ into itself and $\alpha_1 + \beta_1$ is orthogonal to all vectors of V_0. Furthermore,

$$T(2\alpha_1) = T(\alpha_1 - \beta_1) + T(\alpha_1 + \beta_1) = -\alpha_1 + \beta_1 + \alpha_1 + \beta_1 = 2\beta_1$$

and similarly, $T(\beta_1) = \alpha_1$. So T, restricted to $\text{sp}(\alpha_1)$, is T_1.

The case when $\alpha_1 - \beta_1$ is not isotropic can be similarly dealt with.

Now we assume the theorem to be true for k replaced by $k - 1$. We want to prove it for k. Thus we have an isometry T over V which, over $V_2 = \text{sp}\{\alpha_1, \alpha_2, \ldots, \alpha_{k-1}\}$ restricts to T_1, that is,

$$T(\alpha_i) = T_1(\alpha_i) \qquad \text{for } 1 \leq i \leq k - 1$$

where the $k - 1$ α's form an orthogonal basis for V_2. Then let K be the orthogonal complement of V_2, and, by the first part of this proof, determine an isometry T_0 over K such that $T_0(\alpha_k) = T_1(\alpha_k)$. Then defining T' to be T over V_2 and T_0 over K, we have an isometry with the desired properties. This completes the proof.

The characteristic polynomial of an isometry takes a special form which may be shown by a little calculation. Suppose T is an isometry. Then $T^*T = I$ implies

$$T^*(xI - T) = (xT^* - I) = x(T^* - x^{-1}I) \tag{5.70}$$

If we let M be the matrix of T with respect to a certain basis and M^* be the matrix of T^*, we have

$$|M^*|\,|xI - M| = x^n|M^* - x^{-1}I| \tag{5.71}$$

Now we know from Sec. 5.3 that the characteristic polynomials of M and M^* are the same. This implies that the determinants are the same, which, with $MM^* = I$, implies $|M|^2 = 1$. Furthermore,

$$|yI - M| = |yI - M^*|$$

Thus

$$f(x) = |xI - M| = \pm x^n|x^{-1}I - M^*| = \pm x^n|x^{-1}I - M| \tag{5.72}$$

So that if $f(x)$ is the characteristic polynomial of M, we have

$$f(x) = \pm x^n f(x^{-1}) \qquad \pm 1 = (-1)^n|M| \tag{5.73}$$

This means that $f(x)$ must take the following form:

$$f(x) = (x+1)^r(x-1)^s \prod_{i=1}^{k} (x-a_i)(x-a_i^{-1}) \qquad \text{where } a_i^2 \neq 1 \tag{5.74}$$

Compare this with Theorem 5.30.

Though we do not in general have a splitting of an isometry into a direct sum as we had for orthogonal and unitary transformations, an isometry can be expressed as a product of reflections (see Sec. 5.13). We shall state the theorem and give two references to proofs:

Theorem 5.34 An isometry over a vector space V of dimension n, with respect to a symmetric vector product function which is nonsingular over V, can be expressed as a product of n or fewer reflections, if in the field $1 + 1 \neq 0$.

See E. Artin, *Geometric Algebra*, pp. 129ff., listed in the bibliography. Also, P. Scherk, On the Decomposition of Orthogonalities into Symmetries, *Proc. Am. Math. Soc.*, vol. 1, pp. 481–491, 1950.

EXERCISES

1. Let $\alpha_1 = (1,0,0)$, $\alpha_2 = (0,1,0)$, $\alpha_3 = (0,0,1)$ be a basis of V and let R be a linear transformation taking α into $\frac{1}{3}(1,2,2) = \beta$. Extend R to an isometry of V; that is, find an isometry T of V which takes α_1 into β.

2. Using the notation of Exercise 1, let $\gamma = \frac{1}{3}(2,-2,1)$. Find an isometry T of V taking α_1 into β and α_2 into γ.

3. What connection do Exercises 1 and 2 have with theorems on orthogonal bases?

4. Let T be an isometry over V^3 having a symmetric nonsingular vector product function. Show that for some number c its characteristic polynomial is

$$x^3 + cx^2 \pm (cx + 1) \qquad \text{where } \pm 1 = -|T|$$

5. Show that the characteristic polynomial of an isometry over V^4, with a symmetric nonsingular vector product function, takes the form

$$x^4 + cx^3 + dx^2 \pm (cx + 1)$$

where $\pm 1 = |T|$ and $d = 0$ whenever $|T| = -1$.

6. In Exercise 5 for the case in which $|T| = 1$, show that if $T + T^{-1} = S$ and $g(y) = y^2 + cy + d - 2$, then $g(S) = 0$.

7. Let A be the (symmetric) matrix of a quadratic form and Q a skew-symmetric matrix (that is, $Q^T = -Q$) such that $A + Q$ is nonsingular. Prove that $(A + Q)^{-1}(A - Q)$ is an automorph of A. (This form is due to Cayley.)

8. Prove that if P is an automorph of a symmetric matrix A for which $P + I$ is nonsingular, then there exists a skew-symmetric matrix Q such that

$$(A + Q)P = A - Q$$

9. Is the result of Exercise 8 the converse of that of Exercise 7? If so, show why this is so; if not, show why it is not.

6

THE JORDAN FORM AND APPLICATIONS

6.1 INTRODUCTION AND RECAPITULATION

In this chapter we return to the consideration of a vector space V over an unrestricted field F and linear transformations of V into itself. Recall that in Sec. 3.12 and earlier we noted that if M and M' are matrices of a transformation relative to two different bases (in each case using the same basis for the domain and image spaces), then M and M' are similar matrices; that is, for some nonsingular matrix C, $M' = C^{-1}MC$. The chief object of this chapter is to find a canonical form for similar matrices, that is, to choose from all the matrices similar to a given one, that which is in some sense "simplest." We shall see that, just as a canonical form for symmetric matrices was of assistance in acquiring information about quadratic forms, so a canonical matrix of a transformation T yields useful information about T. In fact we shall demonstrate how such information can be used.

There is one case in which we already know a canonical form. In Theorem 3.18 it was stated that if the minimum polynomial of a transformation is a product of distinct linear factors, then there is a set of characteristic vectors which spans V. In this case it was shown that by choosing a set of characteristic vectors as a basis we obtain a diagonal matrix which represents the transformation. Before starting to deal with the more general case, we consider an application of the result just stated. Later in the chapter, after we have developed a general canonical form, we shall consider more general applications along the same lines.

6.2 AN APPLICATION

Let us first phrase the problem in somewhat frivolous terms. Suppose a President of the United States was undecided whether or not to run for a second term. When he made up his mind, he told it to a certain gossip in Washington. Now the gossip was known to report something correctly one-third of the time and incorrectly two-thirds of the time. This situation could be represented by the following matrix:

$$\begin{array}{c} \text{Gossip} \\ \begin{array}{cc} & \begin{array}{cc} Y & N \end{array} \\ \text{President} \begin{array}{c} Y \\ N \end{array} & \begin{bmatrix} \frac{1}{3} & \frac{2}{3} \\ \frac{2}{3} & \frac{1}{3} \end{bmatrix} \end{array} \end{array}$$

The first row means that if the President said "yes" the probability that the gossip reports "yes" is $\frac{1}{3}$ and for "no," $\frac{2}{3}$. The second row gives the corresponding probabilities if the President said "no." Now the gossip had a crony who was just as unreliable as he (or she) was. Let us see what the probabilities are for the report of the crony. If the President said "yes" there are two sequences which can lead to a "yes" from the crony. First, the gossip might say "yes" with a probability of $\frac{1}{3}$ and the crony "yes" with a probability of $\frac{1}{3}$, giving a probability of $(\frac{1}{3})^2 = \frac{1}{9}$. Second, the gossip might say "no" and the crony "yes" with a probability of $(\frac{2}{3})^2 = \frac{4}{9}$. Hence the probability that a "yes" from the President leads to a "yes" from the crony is $\frac{1}{9} + \frac{4}{9} = \frac{5}{9}$. Along these lines we can find that the matrix for the report of the crony is

$$\begin{array}{c} \text{Crony} \\ \begin{array}{cc} & \begin{array}{cc} Y & N \end{array} \\ \text{President} \begin{array}{c} Y \\ N \end{array} & \begin{bmatrix} \frac{5}{9} & \frac{4}{9} \\ \frac{4}{9} & \frac{5}{9} \end{bmatrix} \end{array} \end{array}$$

It is not hard to see that this matrix is the square of the first matrix. It is interesting to note that the probability of a "yes" from the President leading to a "yes" from the crony is more than one-half, which indicates that the crony is more to be trusted than the gossip. Now the final question is: Suppose the crony has a crony, and so on, what in the limit is the final outcome? You might guess and work it out on the basis of the numbers used.

Now we generalize this and show how a solution can be found using Theorem 3.28. So we have the matrix

$$P = \begin{bmatrix} a & b \\ c & d \end{bmatrix}$$

where a, b, c, d are nonnegative and $a + b = 1 = c + d$. The question is:

$$\text{What is the } \lim_{n\to\infty} P^n?$$

We should be precise about what we mean by $\lim_{n\to\infty} P^n = Q$, where P and Q are matrices. We mean that by choosing n large enough we can make every element of P^n arbitrarily close to the corresponding element of Q. We can express this notationally as follows:

Let $p_{ij}^{(n)}$ denote the element in the ith row and jth column of P^n and let $Q = (q_{ij})$. Then

$$\lim_{n\to\infty} P^n = Q$$

means $$\lim_{n\to\infty} p_{ij}^{(n)} = q_{ij} \qquad \text{for all } i \text{ and } j$$

Notice that if $\alpha = (1,1)$, then $P\alpha^T = \alpha^T$. This shows that 1 is a characteristic root of P. Since $a + d$ is the sum of the characteristic roots, we know that the second characteristic root is $a + d - 1$. Now, $a + b = 1 = c + d$ and a, b, c, d, nonnegative imply that if $e = a + d - 1$, the absolute value of e is not greater than 1. Also if $|e| = 1$, then $a + d = 0$ or $a + d = 2$; that is,

$$P = \begin{bmatrix} 0 & 1 \\ 1 & 0 \end{bmatrix} \quad \text{or} \quad P = \begin{bmatrix} 1 & 0 \\ 0 & 1 \end{bmatrix}$$

In the former case the limit does not exist, and in the latter case it is I.

Then excluding these two cases, we see that 1 and e are the distinct characteristic roots of P and hence, by Theorem 3.28, P is similar to the matrix

$$B = \begin{bmatrix} 1 & 0 \\ 0 & e \end{bmatrix} \qquad \text{for } |e| < 1$$

Now it is easy to see that the limit of B^n as n becomes infinite is

$$C = \begin{bmatrix} 1 & 0 \\ 0 & 0 \end{bmatrix}$$

Since P is similar to B, there is a nonsingular matrix R such that $P = RBR^{-1}$. Then

$$P^n = (RBR^{-1})^n = RB^nR^{-1}$$

shows that $\lim_{n\to\infty} P^n$ exists and is equal to $Q = RCR^{-1}$. Thus, to recover the matrix Q, we must determine what R is. We showed in Sec. 3.12 that if we choose the columns of R to be characteristic vectors of P, then $R^{-1}PR$ will be a diagonal matrix. Now a characteristic vector associated with the characteristic root is (1,1). Let (x_0,y_0) be some characteristic vector associated with e. Then, for $c \neq 0$, (cx_0,cy_0) is also a characteristic vector associated with e. So we can choose

$$R = \begin{bmatrix} 1 & cx_0 \\ 1 & cy_0 \end{bmatrix} \tag{6.1}$$

The inverse of R will be simplest if we choose c so that 1 is the determinant of R, that is, if $cy_0 - cx_0 = c(y_0 - x_0) = 1$. Such a choice is possible since if y_0 and x_0 were equal then (cx_0,cy_0) would be a characteristic vector associated with 1 instead of e. So let $x = cx_0$ and $y = cy_0$ for this choice of c and define

$$R = \begin{bmatrix} 1 & x \\ 1 & y \end{bmatrix} \qquad \text{with } y - x = 1$$

Thus, letting $Q = RCR^{-1}$, we see that

$$R^{-1} = \begin{bmatrix} y & -x \\ -1 & 1 \end{bmatrix}$$

implies

$$Q = \begin{bmatrix} y & -x \\ y & -x \end{bmatrix}$$

where (x,y) is a characteristic vector associated with e such that $y - x = 1$. The vector (x,y) may of course be calculated in terms of a, b, c, d of matrix P. This is left as an exercise. However, notice that since every element of P is nonnegative, the same must be true of all its powers and its limit Q. Hence y is nonnegative and x is nonpositive. In the numerical example given at the beginning of this section the diagonal elements are equal and the matrix is symmetric. In that case it follows that $y = -x = \frac{1}{2}$. Hence, the outcome in the limit is that it made no difference what the President said in the first place. In fact, the eventual outcome would be the same if $\frac{1}{3}$ were replaced by 0.99 and $\frac{2}{3}$ by 0.01; it would merely approach the limit more slowly.

The example can be phrased in less frivolous terms. Using the numbers given, suppose that one-third of the sons of Republicans were Republicans and two-thirds were Democrats, and vice versa. If these proportions continued for a number of generations, the support of the two parties would gradually approach equilibrium. One could also phrase the problem as a much simplified problem in genetics having to do, say, with the color of the eyes.

Matrices whose elements are nonnegative numbers such that the sum of the elements in each row is 1 are called *stochastic matrices*. In Sec. 6.7 we shall have more to say about general stochastic matrices as an application of the canonical form for similar matrices.

6.3 SOME SPECIAL CANONICAL FORMS

We have seen in the Corollary of Theorem 4.12 that if the roots of the characteristic equation of a transformation are distinct, then the minimum and characteristic polynomials are the same. But the minimum and characteristic polynomials can be the same without the roots being distinct. Since it is convenient to learn about such transformations, it is helpful to give them a name:

Definition A matrix or transformation is said to be *nonderogatory* if its minimum and characteristic polynomials are the same. Such a transformation is sometimes called *cyclic*. If a transformation is not nonderogatory, it is called *derogatory*.

We begin by finding more information about nonderogatory transformations. At this point you are asked to look back to Theorem 4.18 where a matrix of a nonderogatory transformation was exhibited. Remember that a basis of a certain kind will lead to such a matrix. So suppose V has a basis $\alpha_1, \alpha_2, \ldots, \alpha_n$ and that the transformation T is defined by

$$T(\alpha_i) = \alpha_{i+1} \qquad \text{for } 1 \leq i \leq n - 1 \tag{6.2}$$

$$T(\alpha_n) = -(a_{n-1}\alpha_n + a_{n-2}\alpha_{n-1} + \cdots + a_1\alpha_2 + a_0\alpha_1) \tag{6.3}$$

Since (6.2) is equivalent to

$$T^i(\alpha_1) = \alpha_{i+1} \qquad \text{for } 0 \leq i \leq n - 1 \tag{6.2a}$$

we can write (6.3) in the form

$$f(T)(\alpha_1) = (T^n + a_{n-1}T^{n-1} + a_{n-2}T^{n-2} + \cdots + a_1T + a_0I)(\alpha_1) = \theta \tag{6.3a}$$

Thus $\qquad T^if(T)(\alpha_1) = f(T)T^i(\alpha_1) = f(T)(\alpha_{i+1}) = \theta \qquad \text{for } 0 \leq i \leq n - 1$

This last shows that $f(T) = 0$. Since the α's are linearly independent, so are

$$\alpha_1, T(\alpha_1), \ldots, T^{n-1}(\alpha_1) \tag{6.4}$$

Hence no nontrivial linear combination of the set (6.4) can be the zero vector. Thus no polynomial of degree lower than n has T as a zero. This shows that $f(x)$ is the minimum polynomial of T and, since it is of degree n, it must be also the characteristic polynomial.

Thus we have shown the following

Theorem 6.1 If a transformation T is defined by (6.2) and (6.3), it is nonderogatory, its minimum polynomial is $f(x)$ as defined in (6.3a), and its matrix with respect to this basis has the form

$$
(6.5) \qquad \begin{bmatrix} 0 & 0 & 0 & \cdots & 0 & -a_0 \\ 1 & 0 & 0 & \cdots & 0 & -a_1 \\ 0 & 1 & 0 & \cdots & 0 & -a_2 \\ \cdots & \cdots & \cdots & \cdots & \cdots & \cdots \\ 0 & 0 & 0 & \cdots & 1 & -a_{n-1} \end{bmatrix}
$$

which is a repetition of the companion matrix of T described in Theorem 4.18.

Note We have *not* yet shown that if a transformation is nonderogatory with characteristic polynomial $f(x)$, then there is a basis with properties (6.2) and (6.3). The existence of such a basis depends on the existence of a vector which has the property that its images under powers of T span the space. We call such a vector a *primitive vector* (or *cyclic vector*) for T; that is, if the set (6.4) spans the space V of dimension n, we call the vector α_1 *primitive* or *cyclic* and say that the set (6.4) is a basis *generated* by the vector α_1. There is indeed such a vector for each nonderogatory transformation, but at this point we can prove it only for a special case, leaving the proof of the general result (Theorem 6.6) for Sec. 6.4.

Recall that in Sec. 3.7 we defined the minimum polynomial of a vector α with respect to T as the polynomial of least positive degree with leading coefficient 1, such that $m_\alpha(T)(\alpha) = \theta$. We also proved Theorem 3.12a which, for ease of reference, we restate as follows:

Theorem 6.2 If $f(x)$ is a polynomial such that $f(T)(\alpha) = \theta$, then $f(x)$ is divisible by $m_\alpha(x)$, the minimum polynomial of α. In particular $m_\alpha(x)$ is a factor of the minimum polynomial of T.

Remark Indeed, the minimum polynomial of a transformation is the least common multiple of the minimum polynomials of the vectors of the space.

We leave the proof of this remark as an exercise. These results do not depend on T being nonderogatory.

Recall that an irreducible polynomial over a field F is one which cannot be expressed as a product of two polynomials with coefficients in F and of positive degree (see Appendix C). We can prove the following:

Theorem 6.3 If $p(x)$ is an irreducible polynomial over F and if a transformation T of V into itself has $p^s(x)$ as its minimum polynomial, then there is a vector α which has $p^s(x)$ as its minimum polynomial. If T is nonderogatory, α is a primitive vector.

Proof Since $p^{s-1}(T) \neq 0$, there must be some vector α such that $p^{s-1}(T)(\alpha) \neq \theta$. Then $m_\alpha(x)$ is a factor of $p^s(x)$, and $p^{s-1}(x)$ is not divisible by $m_\alpha(x)$. Since $p(x)$ is irreducible, $m_\alpha(x) = p^s(x)$. If T is nonderogatory, the degree of $p^s(x)$ is n, the dimension of V. Thus $m_\alpha(x) = p^s(x)$ implies that

$$\alpha,\ T(\alpha),\ T^2(\alpha),\ \ldots\ ,\ T^{n-1}(\alpha)$$

are linearly independent. Hence α is a primitive vector of T.

Corollary If a transformation T is nonderogatory and has as its minimum polynomial a power of an irreducible polynomial over F, then for α_1 a primitive vector there is a basis $\alpha_1, \alpha_2, \ldots, \alpha_n$ defined by Equations (6.2a) such that the matrix of T with reference to this basis takes the form of Theorem 6.1.

This corollary follows since α_1 being primitive implies that the vectors α_i defined by (6.2a) are linearly independent.

In case the irreducible polynomial referred to above is linear, there is another matrix associated with the transformation which has some advantages over the companion matrix. This is obtained by starting with a primitive vector α_1 as before, but instead of defining the rest of the basis by (6.2a) choose the α_i as follows:

$$(T - aI)^i(\alpha_1) = \alpha_{i+1} \qquad \text{for } 0 \leq i \leq n-1 \tag{6.6}$$

where $(x - a)^n$ is the minimum polynomial of the nonderogatory transformation T. We have the following:

Theorem 6.4 If T is a nonderogatory transformation of a vector space V into itself whose minimum polynomial is $(x - a)^n$ and if α_1 is a primitive vector of T, then Equations (6.6) define a basis of V; with respect to this basis the matrix of T is

$$M(a) = \begin{bmatrix} a & 0 & 0 & 0 & \cdots & 0 & 0 \\ 1 & a & 0 & 0 & \cdots & 0 & 0 \\ 0 & 1 & a & 0 & \cdots & 0 & 0 \\ \cdot & \cdot & \cdot & \cdot & \cdots & \cdot & \cdot \\ 0 & 0 & 0 & 0 & \cdots & 1 & a \end{bmatrix} \tag{6.7}$$

Proof First we must show that the α_i defined by (6.6) are linearly independent. Thus, suppose

$$\theta = \sum_{i=1}^{n} c_i\alpha_i = \sum_{i=1}^{n} c_i(T - aI)^{i-1}(\alpha_1)$$

Then if we define $g(x)$ by

$$g(x) = \sum_{i=1}^{n} c_i(x - a)^{i-1}$$

we see that for $g(x)$, a polynomial of degree less than n, $g(T)(\alpha_1) = \theta$. But $(x - a)^n$ of degree n is the minimum polynomial of α_1. This is a contradiction unless all the c_i are zero. Thus we have shown linear independence.

To show that the matrix takes the given form, notice that Equations (6.6) can be written

$$(T - aI)(\alpha_i) = \alpha_{i+1} \qquad \text{for } 1 \leq i \leq n - 1 \tag{6.8}$$

that is,

$$\begin{aligned} T(\alpha_i) &= a\alpha_i + \alpha_{i+1} \qquad \text{for } 1 \leq i \leq n - 1 \\ (T - aI)(\alpha_n) &= (T - aI)(T - aI)^{n-1}(\alpha_1) = (T - aI)^n(\alpha_1) = \theta \\ T(\alpha_n) &= a\alpha_n \end{aligned}$$

This completes the proof.

We call the matrix (6.7) an *elementary Jordan matrix.* The term *elementary* is not used in the sense of pertaining to the beginning of matrix education, but in that of being a kind of element of a larger class. More precisely, we shall see that if the characteristic polynomial of a linear transformation can be expressed as the product of linear factors, then, by a proper choice of basis, the matrix of the transformation becomes a direct sum of elementary Jordan matrices. This sum is called the *Jordan form.*

It should be noted that the matrix (6.7) can be written as $aI + N$, where N is the matrix of order n whose elements are all zero except for 1s immediately below the principal diagonal. Then we leave as an exercise the calculation leading to the following result:

$$(aI + N - aI)^n = N^n = Z$$

the zero matrix, and no smaller exponent n gives this result. The matrix N is a special example of what is called a *nilpotent matrix,* namely, a matrix M such that some positive integral power of M is the zero matrix. A nilpotent matrix can also be characterized by the property that its minimum polynomial $m(x)$ is a power of x.

EXERCISES

1. Compute P, P^2, and P^3 for the following two matrices P:

$$\begin{bmatrix} 0.9 & 0.1 \\ 0.1 & 0.9 \end{bmatrix} \qquad \begin{bmatrix} 0.1 & 0.9 \\ 0.9 & 0.1 \end{bmatrix}$$

2. Show that $(x + 1)^3$ is the minimum polynomial of the matrix

$$M = \begin{bmatrix} -1 & -1 & 1 \\ 1 & 0 & 0 \\ 0 & 1 & -2 \end{bmatrix}$$

Then show that $(1,0,0)^T$ is a primitive vector. First find a basis which yields the companion matrix (6.5) and second, one which produces the matrix (6.7).

3. Let α be a characteristic vector of M in Exercise 2. Can α be a primitive vector? Using Theorem 6.3 or by other means, show how one can find all the primitive vectors of M.

4. Suppose $(x^2 + 1)^2$ is the minimum polynomial of a transformation T over V^4 and the field of real numbers. Show that if a vector α is not a characteristic vector of T^2, it is a primitive vector of T.

5. Under what conditions, if any, can a characteristic vector of T^2 in Exercise 4 be a primitive vector of T?

6. If in Exercise 4 the field is that of the complex numbers, is the stated conclusion valid?

7. Let $P = \begin{bmatrix} a & b \\ b & a \end{bmatrix}$ where a and b are positive and $a + b = 1$. Find P^2 and P^3. From these guess a formula for P^k and establish your result by mathematical induction or by other means.

8. Prove that if N takes the form of the paragraph at the end of this section, then N^n is the zero matrix and no smaller power of N is the zero matrix.

9. Prove that if the number a in the matrix M displayed in (6.7) is less than 1 in absolute value, then $\lim_{k \to \infty} M^k = Z$, the zero matrix.

10. Prove the Remark immediately following Theorem 6.2.

11. In the example of Sec. 6.2, find the numbers x and y in terms of the elements a, b, c, d of the given matrix, and show that if the matrix P is symmetric, then $y = -x = \frac{1}{2}$.

12. Prove that if, in the matrix P of Sec. 6.2, the element a is strictly between 0 and $\frac{1}{2}$, and if P is symmetric, then the number in the first row and column of P^k is less than $\frac{1}{2}$ if k is odd and greater than $\frac{1}{2}$ if k is even.

13. What is the conclusion for Exercise 12 if the element a is strictly between $\frac{1}{2}$ and 1?

14. Let T be a linear transformation whose matrix with respect to the canonical basis is

$$\begin{bmatrix} 1 & 2 & -2 \\ 2 & -2 & 1 \\ 2 & 1 & -2 \end{bmatrix}$$

Find the characteristic roots of T, show that it is nonderogatory, and find a primitive vector α.

15. If for the transformation whose matrix is as in Exercise 14 we take as a basis $\alpha_1 = \alpha$, $\alpha_2 = T(\alpha_1)$, $\alpha_3 = T^2(\alpha_1)$, what will be the matrix with respect to this basis?

16. Let a matrix of a linear transformation T with respect to a certain basis be

$$\begin{bmatrix} 1 & 1 & 3 \\ 0 & 1 & 3 \\ 1 & 0 & 1 \end{bmatrix}$$

Find a primitive vector of T.

17. Let V be of dimension 3, and for a basis $\alpha_1, \alpha_2, \alpha_3$, let $T(\alpha_1) = 3\alpha_1$, $T(\alpha_2) = 3\alpha_2$, $T(\alpha_3) = 2\alpha_3$. Prove that there is no primitive vector β.

18. Let $T(\alpha) = a\alpha$, $T(\beta) = b\beta$, where $a \neq b$, and suppose the dimension of V is 2. Then prove that $\alpha + \beta$ is a primitive vector of T.

19. Let $T(\alpha) = a\alpha$, $T(\beta) = b\beta$, $T(\gamma) = c\gamma$, where no two of a, b, c are equal, and suppose 3 is the dimension of V. Prove that $\alpha + \beta + \gamma$ is a primitive vector of T.

20. Generalize Exercises 18 and 19 to n dimensions.

21. Let T be a nonderogatory transformation whose characteristic polynomial is $(x - a)^n$, and denote by n_i the nullity of $(T - aI)^i$. Prove that $n_i = i$ for $1 \leq i \leq n$.

22. Let a transformation T of a vector space of dimension 3 into itself with respect to a basis have a matrix whose elements are rational numbers and suppose that no characteristic root is rational. Show that any nonzero triple of rational numbers can serve as a primitive vector of T.

23. Prove that if a transformation T^2 has a primitive vector, then T has one. Is the converse true?

6.4 THE JORDAN CANONICAL FORM

In the previous section we developed canonical forms for nonderogatory transformations. Here we consider the general case. We shall see first that a factorization of the characteristic polynomial into relatively prime factors results in a "splitting" of the space and the transformation in a sense which is made precise in the following:

Theorem 6.5 Let $f(x)$ be the characteristic polynomial of a linear transformation T of a vector space V into itself, and suppose $f(x) = f_1(x)f_2(x)$, where f_1 and f_2 are relatively prime (that is, have g.c.d. 1). Let K_i be the kernel of $f_i(T)$ and E_i the image space of $f_i(T)$. Then

1. $K_1 \cap K_2 = \theta$.
2. $E_1 \oplus E_2 = V$.
3. $K_1 = E_2$ and $K_2 = E_1$.
4. K_i (and E_i) are invariant subspaces of T; that is, $T(K_i) \subseteq K_i$.
5. If T_i is the transformation T restricted to K_i, then the characteristic polynomial of T_i is $f_i(x)$, for $i = 1, 2$, and, for a proper choice of basis, the matrix of T is the direct sum of the matrices of T_1 and T_2.

Proof At this point, you might look back at the statement and proof of Theorem 3.11. That theorem is a special case of this one, and our proof here follows the same pattern.

Since f_1 and f_2 have 1 as their g.c.d., from Appendix C we know that there are polynomials $g_i(x)$ such that

$$g_1(x)f_1(x) + g_2(x)f_2(x) = 1 \tag{6.9}$$

Then if, in (6.9) we replace x by T and apply to α, we have

$$g_1(T)f_1(T)(\alpha) + g_2(T)f_2(T)(\alpha) = \alpha \tag{6.10}$$

for all vectors α in V.

Now, to prove part 1, suppose α is an element of $K_1 \cap K_2$. Then α in K_1 implies $f_1(T)(\alpha) = \theta$, and α in K_2 implies $f_2(T)(\alpha) = \theta$. Thus if α is in the intersection of K_1 and K_2, the left-hand side of (6.10) is the null vector. This shows that α is the null vector and proves part 1.

Now, $f(T)(\alpha) = f_1(T)[f_2(T)(\alpha)] = \theta$ for all α in V. This implies the first part of the following, and the second part follows similarly.

$$E_2 \subseteq K_1 \qquad \text{and} \qquad E_1 \subseteq K_2 \tag{6.11}$$

On the other hand, for any α in V, $g_i(T)f_i(T)(\alpha) = f_i(T)[g_i(T)(\alpha)]$. Thus $g_i(T)f_i(T)(\alpha)$ is an element of $f_i(T)(V) = E_i$. Then, looking back at (6.10), we see that any α in V will be the sum of an element in E_1 and an element in E_2. This shows

$$V = E_1 + E_2 \tag{6.12}$$

But (6.11) and part 1 show that E_1 and E_2 are disjoint, and hence the sum in (6.12) is a direct sum and part 2 is established. Now since (6.12) is a direct sum

$$\dim E_1 = n - \dim E_2$$

From Theorem 3.7, $n - \dim E_2 = \dim K_2$. Thus

$$\dim E_1 = \dim K_2$$

This and the corresponding result for the subscript 2, together with (6.11), show that $E_1 = K_2$ and $E_2 = K_1$, which is part 3 of the theorem.

To see that K_1 is an invariant subspace of T, notice that α is in K_1 if and only if $f_1(T)(\alpha) = \theta$. But then

$$f_1(T)[T(\alpha)] = Tf_1(T)(\alpha) = \theta$$

This proves part 4 for $i = 1$; the proof for $i = 2$ goes similarly.

To prove part 5 notice that parts 2 and 3 imply that we may choose a basis of V so that $\alpha_1, \alpha_2, \ldots, \alpha_k$ is a basis of $K_1 = E_2$ and $\alpha_{k+1}, \ldots, \alpha_n$ is a basis of $K_2 = E_1$. Then part 4 shows that the matrix of T with respect to this basis is $M_1 \oplus M_2$ for some matrices M_i. Now remember that calling T_1 the transformation T "restricted to K_1" means that T_1 is a transformation applied to K_1, which has the same effect on the vectors of K_1 that T does. Hence T_1 has the matrix M_1 with respect to the given basis of K_1. The same will hold for T_2 and M_2. The characteristic polynomials of T_i are thus $|xI_i - M_i|$, where I_1 and I_2 are identity matrices of orders k and $n - k$, respectively. But

$$|xI_1 - M_1| \cdot |xI_2 - M_2| = |xI - (M_1 \oplus M_2)|$$

This shows that if $h_i(x)$ denotes the characteristic polynomial of T_i,

$$h_1(x)h_2(x) = f(x) = f_1(x)f_2(x) \tag{6.13}$$

Now if α is in K_1, it follows that $f_1(T)(\alpha) = \theta$, and hence that $f_1(T_1) = 0$. We know also that $h_1(T_1) = 0$.

Let $m_1(x)$ be the minimum polynomial of T_1. From Theorem 4.19 we know that, for some positive integer k, $h_1(x)$ is a factor of $m_1^k(x)$. Now $f_1(T_1) = 0$ implies that $m_1(x)$ is also a factor of $f_1(x)$. Since $f_1(x)$ and $f_2(x)$ are relatively prime,

$$1 = (m_1(x),f_2(x)) = (m_1^k(x),f_2(x)) = (h_1(x),f_2(x))$$

Thus (6.13) implies that $h_1(x)$ is a factor of $f_1(x)$ and, similarly, $h_2(x)$ is a factor of $f_2(x)$. Thus (6.13) implies that $h_i(x) = f_i(x)$ for $i = 1, 2$, and the proof is complete.

Corollary If a linear transformation T of V into itself has as its characteristic polynomial

$$f(x) = p_1^{a_1}(x)p_2^{a_2}(x) \cdots p_k^{a_k}(x)$$

where the $p_i(x)$ are distinct irreducible polynomials over the field F, then

$$V = E_1 \oplus E_2 \oplus \cdots \oplus E_k$$

where E_i is the kernel of $p_i^{a_i}(T)$, the E_i are invariant under T, and, if T_i is

T restricted to E_i, then the characteristic polynomial of T_i is $p_i^{a_i}(x)$. Also, for a proper choice of basis, the matrix of T is the direct sum of the matrices of T_i.

This corollary can be shown by first letting $f_1(x)$ of the theorem be $p_1^{a_1}(x)$ and $f_2(x) = f(x)/f_1(x)$. Then, in similar fashion (by an induction argument if you wish) the factor $p_2^{a_2}(x)$ can be "split off" from $f_2(x)$. This process can be continued to prove the corollary.

(Corollary)^2 The minimum polynomial of the transformation T of Theorem 6.5 is the product of the minimum polynomials of the T_i.

The proof is left as an exercise.

We call T_i the *component transformations* of T with respect to the splitting. We can show as follows that $p_i(T_j)$ is nonsingular if $i \neq j$. Suppose $p_i(T_j)(\alpha_j) = \theta$, where α_j is in E_j. But $p_j^{a_j}(T_j)(\alpha_j) = \theta$ since α_j is in the kernel E_j of $p_j^{a_j}(T_j)$. Since $p_i(x)$ and $p_j^{a_j}(x)$ are relatively prime, there are polynomials $h_1(x)$ and $h_2(x)$ such that

$$h_1(x)p_i(x) + h_2(x)p_j^{a_j}(x) = 1$$

Replace x by T_j and apply this to α_j to get

$$h_1(T_j)p_i(T_j)(\alpha_j) + h_2(T_j)p_j^{a_j}(T_j)(\alpha_j) = \alpha_j = \theta$$

Hence $p_i(T_j)(\alpha_j) = \theta$ implies $\alpha_j = \theta$, and $p_i(T_j)$ is nonsingular.

We shall show that there can be a further splitting. But before that we should show, as promised in Sec. 6.3, that any nonderogatory transformation has a primitive vector. We first prove the following:

Lemma If T is nonderogatory in Theorem 6.5, then so are T_1 and T_2; that is, $f_i(x)$ is the minimum polynomial of T_i, for $i = 1, 2$.

Proof Let $m_i(x)$ be the minimum polynomial of T_i. We know from the previous corollary that $m(x) = m_1(x)m_2(x) \cdots m_k(x)$ is the minimum polynomial of T, and since T is assumed to be nonderogatory,

$$m(x) = f(x) = f_1(x)f_2(x) \cdots f_k(x)$$

Now n, the dimension of V, is thus the sum of the degrees of the m_i, which is the same as the sum of the degrees of the f_i. Since $m_i(x)$ is a factor of $f_i(x)$ and any pair of f_i are relatively prime, the degrees of $m_i(x)$ and $f_i(x)$ must be the same for each i. Hence $f_i(x) = m(x)$ for each i, and T_i is nonderogatory. Notice that this is (Corollary)^2 for the case when T is nonderogatory.

Now we can prove

Theorem 6.6 If T is nonderogatory, then it has a primitive vector.

Proof If the minimum polynomial of T is a power of an irreducible polynomial, Theorem 6.3 establishes the present theorem.

Now we suppose the minimum polynomial is $f(x) = f_1(x)f_2(x)$, where f_1 and f_2 are relatively prime polynomials of positive degree. Since, by the lemma, T_i is nonderogatory over K_i (using the notation of Theorem 6.5) we may, using an induction argument on the degree of $f(x)$, assume that each T_i has a primitive vector α_i in K_i. Then define

$$\alpha = \alpha_1 + \alpha_2$$

We shall show that α is a primitive vector for T. Suppose $g(x)$ is a polynomial. Then

$$g(T)(\alpha) = g(T)(\alpha_1) + g(T)(\alpha_2) = g(T_1)(\alpha_1) + g(T_2)(\alpha_2)$$

since T_i is T restricted to K_i. Now suppose $g(T)(\alpha) = \theta$. Since V is the direct sum of K_1 and K_2, we see that $g(T)(\alpha) = \theta$ implies

$$g(T_1)(\alpha_1) = \theta = g(T_2)(\alpha_2)$$

Since $f_i(x)$ is the minimum polynomial of T_i, it follows that each $f_i(x)$ is a factor of $g(x)$ and, since f_1 and f_2 are relatively prime, their product $f(x)$ is a factor of $g(x)$. Thus if $g(T)(\alpha) = \theta$, it follows that $f(x)$ is a factor of $g(x)$. Since $f(T)(\alpha) = \theta$ it follows that $f(x)$ is the minimum polynomial of α. This completes the proof.

Now, to return to our search for a canonical form of a linear transformation, we see that Theorem 6.5 and its Corollary reduces the search to a consideration of transformations whose characteristic polynomials are powers of irreducible polynomials. If these transformations are nonderogatory, the canonical form will be the companion matrix of Sec. 6.3 or, if the irreducible polynomial is linear, the elementary Jordan matrix. So it remains to deal with derogatory (that is, not nonderogatory) transformations whose characteristic polynomials are $p^a(x)$ where $p(x)$ is irreducible. A natural way to begin would be to pick more or less at random a nonzero vector α and look at the set

$$\alpha,\ T(\alpha),\ T^2(\alpha),\ \ldots,\ T^r(\alpha)$$

where r is the maximum positive integer for which the set is linearly independent. We can call W the subspace spanned by this set of vectors. The space W will certainly be invariant under T. We can find a complementary space W' such that $V = W \oplus W'$. But the rub is that W' may not be invariant under T. For example, a transformation over a four-dimensional vector space might be defined as follows:

$$T(\alpha_1) = \alpha_2 \qquad T(\alpha_2) = \alpha_3 \qquad T(\alpha_3) = \alpha_4 \qquad T(\alpha_4) = \alpha_1 - \alpha_3 + \alpha_4$$

Since $T^2(\alpha_4) = T^3(\alpha_4)$ and $T^2(\alpha_1) = \alpha_3 = T^3(\alpha_1)$, it follows that $x^3 - x^2$ is the minimum polynomial. Then we can take W to be the invariant subspace spanned by α_1, α_2, α_3 and $V = W \oplus W'$, where $W' = \text{sp}(\alpha_4)$. But $T(\alpha_4) = \alpha_1 - \alpha_3 + \alpha_4$ is not in W'. The fundamental problem is to make an adjustment on W' so that it is invariant under T.

To accomplish this adjustment we need a new concept, that of a *quotient* (or difference) *vector space*. Suppose E is a subspace of V. We can think of the vectors of V divided into classes modulo E, that is, we say that α_1 and α_2 are in the same class modulo E if and only if $\alpha_1 - \alpha_2$ is in E, just as we can divide the set of integers into classes modulo m, for m some positive integer. We can find the sum of two classes by the following rule: Let α be a representative of (that is, a vector in) the class $\bar{\alpha}$ and β a representative of the class $\bar{\beta}$. Then we define the class $\bar{\alpha} + \bar{\beta}$ by

$$\bar{\alpha} + \bar{\beta} = \overline{(\alpha + \beta)}$$

This class is independent of the representatives chosen since if γ_1 and γ_2 are in E, then

$$\bar{\alpha} + \bar{\beta} = \overline{(\alpha + \gamma_1)} + \overline{(\beta + \gamma_2)} = \overline{(\alpha + \beta + \gamma_1 + \gamma_2)} = \overline{(\alpha + \beta)}$$

Similarly we can define the class $c\bar{\alpha}$ by $\overline{(c\alpha)}$, for α any representative of class $\bar{\alpha}$. We denote the set of classes by V/E and see from the above that V/E is a vector space of dimension $\dim(V) - \dim(E)$.

Now if T is any transformation of V into itself such that E is invariant under T, we can define a transformation $\bar{T}$ of V/E into itself "induced by T" in the following manner. To find $\bar{T}(\bar{\alpha})$, let α be a representative of $\bar{\alpha}$ and define $\bar{T}(\bar{\alpha})$ to be the class of $T(\alpha)$. This class is independent of the representative we choose since, if α is replaced by $\alpha + \gamma$ with γ in E, then

$$T(\alpha + \gamma) = T(\alpha) + T(\gamma) = T(\alpha) + \gamma_1$$

with γ_1 in E, since E is invariant under T. Furthermore, $\bar{T}$ is a linear transformation.

Now we can state and prove the following theorem, which gives a splitting of the vector space and leads to the general canonical form.

Theorem 6.7 Let $p^m(x)$, with $p(x)$ irreducible over F, be the characteristic polynomial of a transformation T of V into itself and let $p^s(x)$ be its minimum polynomial, where e is the degree of $p(x)$. Then

$$V = E_1 \oplus E_2 \oplus \cdots \oplus E_k$$

where E_i is of dimension $a_i e$ and the a_i form a nonincreasing sequence, with $s = a_1$. Each E_i is invariant under T, and if T_i denotes T restricted to E_i, $p^{a_i}(x)$ is the characteristic and minimum polynomial of T_i and

$$a_1 + a_2 + \cdots + a_k = m$$

Proof We prove this by induction on the dimension of V. If $\dim(V) = 1$, there is nothing to prove. By Theorem 6.3 there is a vector α of V whose minimum polynomial is $p^s(x)$. Then the set

$$\alpha,\ T(\alpha),\ T^2(\alpha),\ \ldots,\ T^{se-1}(\alpha) \tag{6.14}$$

is a linearly independent set since the degree of the minimum polynomial is se. Denote by E the vector space spanned by the vectors of (6.14), and consider the quotient space $\bar{V} = V/E$. As noted above, since E is an invariant subspace of T, there is a linear transformation $\bar{T}$ of $\bar{V}$ into itself induced by T. Since $p^s(T) = 0$, it follows that $p^s(\bar{T}) = \bar{0}$, and hence the minimum polynomial of $\bar{T}$ must be a factor of $p^s(x)$; call it $p^r(x)$ with $r \leq s$.

Since $\bar{V}$ has smaller dimension than V, our induction hypothesis and Theorem 6.3 show that there is a primitive vector $\bar{\beta}$ of $\bar{V}$ such that

$$\bar{\beta},\ \bar{T}(\bar{\beta}),\ \ldots,\ \bar{T}^{re-1}(\bar{\beta}) \tag{6.15}$$

is a linearly independent set of vectors of $\bar{V}$ which span a subspace of $\bar{V}$. Let β be any vector in V representing the class $\bar{\beta}$ of $\bar{V}$. Then the set

$$\beta,\ T(\beta),\ T^2(\beta),\ \ldots,\ T^{re-1}(\beta) \tag{6.16}$$

will be linearly independent since any linear combination of the vectors of (6.16) will correspond to the same linear combination of the vectors of (6.15). But the trouble is that the space spanned by (6.16) may not be invariant under T. If it happens that $r = s$, then $p^r(T)(\beta) = \theta$ and we have no difficulty. But otherwise we may have to make an adjustment in β. In fact, we shall replace β in (6.16) by $\beta - \gamma$, where γ is a suitably chosen element in E. Such a choice will not change the class of $\bar{\beta}$ in $\bar{V}$. So suppose $p^r(T)(\beta) = \delta \neq \theta$. Since $p^r(\bar{T})(\bar{\beta}) = \bar{0}$, we see that δ must be in E. If we can find a γ in E such that $p^r(T)(\gamma) = \delta$, then $p^r(T)(\beta - \gamma) = \theta$ and we will have what we want.

So now we must find a γ in E such that $p^r(T)(\gamma) = \delta$. Now since δ is in E spanned by (6.14), there is some polynomial $h(x)$ such that $h(T)(\alpha) = \delta$. But $\delta = p^r(T)(\beta)$ implies that $p^{s-r}(T)(\delta) = \theta$, that is, $p^{s-r}(T)h(T)(\alpha) = 0$. Since α is a primitive vector, $p^s(x)$ divides $p^{s-r}(x)h(x)$; that is, $p^r(x)u(x) = h(x)$ for some polynomial $u(x)$. So $p^r(T)u(T)(\alpha) = \delta$, and we can choose $\gamma = u(T)(\alpha)$ to get $p^r(T)(\gamma) = \delta$.

Then for the γ chosen in the previous paragraph, replace β in (6.16) by $\beta - \gamma$ and see that $p^r(T)(\beta - \gamma) = \theta$. Such a replacement does not change $\bar{\beta}$

since γ is in E, but now the space E_2 spanned by (6.16) with β replaced by $\beta - \gamma$ is invariant under T.

Next in the above proof, replace E by $E \oplus E_2$ and find an E_3 invariant under T. So proceeding we can complete the proof.

To clarify the proof, let us follow it through for a particular example. Let $\alpha_1, \alpha_2, \alpha_3, \alpha_4$ be a basis of V and let the transformation T be defined by

$$T(\alpha_1) = \alpha_2 \qquad T(\alpha_2) = -4\alpha_1 + 4\alpha_2$$
$$T(\alpha_3) = -2\alpha_1 + \alpha_2 + 2\alpha_3 \qquad T(\alpha_4) = 2\alpha_1 - \alpha_2 + 2\alpha_4$$

Then E is the subspace spanned by α_1 and α_2 and is invariant under T. Since calculation shows that

$$(T - 2I)^2(\alpha_i) = 0 \qquad \text{for } i = 1, 2$$

we see that $(x - 2)^2$ is the minimum polynomial of T_1, that is, of T restricted to E. Then $\bar{V} = V/E$ is the set of vectors spanned by $\bar{\alpha}_3, \bar{\alpha}_4$, classes modulo E. Now

$$\bar{T}(\bar{\alpha}_3) = 2\bar{\alpha}_3 \qquad \bar{T}(\bar{\alpha}_4) = 2\bar{\alpha}_4$$

which shows that the minimum polynomial of $\bar{T}$ is $x - 2$. If we choose α_3 to be a representative of $\bar{\alpha}_3$, that is, in the notation of the proof, $\beta = \alpha_3$, we see that $\text{sp}(\alpha_3)$ is not invariant under T. Thus, in accordance with the proof, we let $\delta = p(T)(\beta) = (T - 2I)(\beta)$; we wish to determine γ in E such that $p(T)(\gamma) = (T - 2I)(\gamma) = \delta$. Now calculation shows that $\delta = (T - 2I)(\alpha_1)$. Hence we may take $\gamma = \alpha_1$ and see that $(T - 2I)(\alpha_3 - \alpha_1) = \theta$. So we choose $E_2 = \text{sp}(\alpha_3 - \alpha_1)$. Similarly $E_3 = \text{sp}(\alpha_4 + \alpha_1)$ and

$$V = E_1 \oplus E_2 \oplus E_3 \qquad T = T_1 \oplus T_2 \oplus T_3$$

where $(x - 2)^2$ is the minimum polynomial of T_1 and $x - 2$ is the minimum polynomial of T_2 and T_3. Notice that $(x - 2)^2$ is the minimum polynomial of T also.

By far the most important case of Theorem 6.7 is when the irreducible polynomial is linear, first, because in the field of complex numbers, every polynomial can be expressed as a product of linear factors. Second (see Appendix C), given any polynomial over a field F, it is possible to extend (that is, enlarge) the field to a larger one F' in which the polynomial is a product of linear factors. For this case we can use the elementary Jordan matrix defined in Theorem 6.4 and prove the following:

Theorem 6.8 If $(x - a)^n$ is the characteristic polynomial of a transformation T of a vector space V into itself and if $(x - a)^s$ is the minimum polynomial of T, then

$$V = E_1 \oplus E_2 \oplus \cdots \oplus E_k$$

where each E_i is of dimension a_i and the a_i form a nonincreasing sequence with $s = a_1$. Each E_i is invariant under T, and if T_i denotes T restricted to E_i, $(x - a)^{a_i}$ is the characteristic and minimum polynomial of T_i and

$$a_1 + a_2 + \cdots + a_k = n$$

We can choose a basis of E_i so that T_i has as its matrix an elementary Jordan matrix of the form (6.7) in Theorem 6.4.

We call the $(x - a)^{a_i}$ of this theorem the *elementary divisors* of this transformation T.

Proof This theorem is, except for the last sentence, Theorem 6.7 for the special case in which $p(x) = x - a$. The last sentence follows from Theorem 6.4. This is the proof.

Then, bringing into play Theorem 6.5, we have:

Corollary Every linear transformation T of V into itself over a field such that the characteristic polynomial is a product of linear factors, has, for a proper choice of basis, a matrix which is a direct sum of elementary Jordan matrices.

The direct sum described in the corollary is called *the Jordan canonical form* of the matrix of the transformation. The word *the* indicates that such a form is unique for a given transformation. We shall prove this in the next section.

We close this section with an example.

Example Classify all the linear transformations over the field of real numbers whose characteristic polynomial is $(x - 1)(x - 3)^4$. In each case find the minimum polynomial of the transformation T.

From Theorem 6.5 we see that we can write $V = E_1 \oplus E_2$, where the dimension of E_1 is 1, T_1 has the characteristic polynomial $x - 1$, the dimension of E_2 is 4, and T_2 has the characteristic polynomial $(x - 3)^4$. Then we have to consider the various possibilities for T_2. To simplify notation, let $T_2 = S$ and $E_2 = U$. We list the possibilities according to the minimum polynomial of S.

1. If $(x - 3)^4$ is the minimum polynomial of S, there is no further splitting of the vector space, and $(x - 1)(x - 3)^4$ is the minimum polynomial of T.
2. If $(x - 3)^3$ is the minimum polynomial of S, then $U = U_1 \oplus U_2$, where U_1 and U_2 are of dimensions 3 and 1, respectively, S_1 (over U_1) has

$(x - 3)^3$ as its minimum polynomial, and S_2 has minimum polynomial $x - 3$. Here the minimum polynomial of T is $(x - 1)(x - 3)^3$.

3. If $(x - 3)^2$ is the minimum polynomial of S, then there are two possibilities. First, U may be the direct sum of two subspaces of dimension 2, with the corresponding S_1 and S_2 each having minimum polynomial $(x - 3)^2$ as minimum polynomials. Second, U may be the direct sum of one subspace of dimension 2 and two subspaces of dimension 1, with S_1 having $(x - 3)^2$ as its minimum polynomial and S_2 and S_3 having $x - 3$ as their minimum polynomials. In either case the minimum polynomial of T is $(x - 1)(x - 3)^2$.

4. If $x - 3$ is the minimum polynomial of S, then U is the direct sum of four subspaces of dimension 1 and each S_i has $x - 3$ as its minimum polynomial. Here $(x - 1)(x - 3)$ is the minimum polynomial of T.

Notice that only in case 3 are there two possibilities for the splitting of the vector space and the transformation. This however does show that though knowing the minimum polynomial of a transformation gives more information about it, there are cases in which the transformation cannot be completely described in terms of the minimum and characteristic polynomials.

EXERCISES

1. Classify all linear transformations for each of the following characteristic polynomials. In each case give the minimum polynomial of the transformation.

(*a*) $(x - 1)^3(x + 1)^2$.

(*b*) $(x - 2)(x + 3)(x - 4)^3$.

(*c*) $(x^2 + x + 1)^2(x + 4)^3$; first for the field of real numbers, second for the field of complex numbers.

(*d*) $(x^2 - 3)^2(x + 1)^3$; first for the field of rational numbers, second for the field of real numbers.

2. In Exercise 1(*a*), find the ranks of the following for each of the possibilities.

$$T - I \qquad (T - I)^2 \qquad (T - I)^3 \qquad T + I \qquad (T + I)^2$$

3. Prove (Corollary)^2 of Theorem 6.5.

4. Let the field of the vector space be that of the complex numbers, and suppose a transformation T of the space into itself has the property that all its characteristic roots are less than 1 in absolute value. Prove that $\lim_{s \to \infty} T^s = Z$, the zero transformation. Hint: Show this first for transformations T with characteristic polynomial $(x - c)^k$, where c is a complex characteristic root; then apply Theorem 6.5.

5. A linear transformation T with respect to the canonical basis is defined in each case below by the matrix given. Find the minimum polynomial of T. If T is nonderogatory, find a primitive vector. If T is derogatory, find a basis of a subspace E such that T restricted to E is nonderogatory and find a primitive vector for T restricted to E. In each case find the Jordan canonical matrix of T.

$$(a)\ \begin{bmatrix} 4 & 0 & -1 \\ -2 & 1 & 1 \\ 6 & 0 & -1 \end{bmatrix} \qquad (b)\ \begin{bmatrix} 1 & 2 & 2 \\ 2 & -2 & 1 \\ 2 & 1 & -2 \end{bmatrix}$$

$$(c)\ \begin{bmatrix} 3 & 0 & 0 & 0 \\ 0 & 1 & 2 & 2 \\ 0 & 2 & -2 & 1 \\ 0 & 2 & 1 & -2 \end{bmatrix} \qquad (d)\ \begin{bmatrix} 1 & 2 & 4 & -4 \\ -2 & -3 & -6 & 6 \\ 0 & 2 & -2 & 1 \\ 0 & 2 & 1 & -2 \end{bmatrix}$$

6. Let a linear transformation T of a vector space V into itself have the matrix

$$\begin{bmatrix} 3 & 0 & 0 \\ 1 & 3 & 0 \\ 0 & 0 & 2 \end{bmatrix}$$

Let W be the kernel of $T - 3I$. Prove that there is no subspace U such that $V = W \oplus U$ and $T(U) \subseteq U$.

7. Let M be the matrix

$$\begin{bmatrix} 1 & 3 & -3 \\ 3 & 1 & -3 \\ 3 & 3 & -5 \end{bmatrix}$$

Find a matrix P with real elements such that $P^{-1}MP$ is in Jordan normal form.

8. Prove that if M and M' are two matrices of order 3 a necessary and sufficient condition that they be similar is that they have the same characteristic polynomial and minimum polynomial. Show that this is false for matrices of order 4.

9. Prove that two nilpotent matrices of order 3 are similar if and only if they have the same minimum polynomial.

10. Suppose a transformation T of a vector space V into itself has the characteristic polynomial $(x - 4)^4$ and the minimum polynomial $(x - 4)^2$. Let α be some vector which is not a characteristic vector of T. Show that α and $T(\alpha)$ span a two-dimensional subspace E which is invariant under T. Find a basis of the quotient space V/E. What are the possibilities for the transformation $\bar{T}$ over V/E induced by T?

11. Let V be a vector space V of dimension greater than 1 and let a transformation of V into itself be of rank 1. Prove that the transformation either is nilpotent or, for a proper choice of basis, has a diagonal matrix, but not both.

12. Let M be a real matrix of order n such that $M^2 + I = Z$, the zero matrix. Prove that if $n = 2k$, where k is an integer, then there is a matrix C with real elements such that

$$C^{-1}MC = \begin{bmatrix} 0 & -I \\ I & 0 \end{bmatrix}$$

where I is the identity matrix of order k.

13. Let T be a nonderogatory transformation. Prove that if R is a transformation such that $RT = TR$, then R can be expressed as a polynomial in T. Hint: Let α be a primitive vector of T and express $R(\alpha)$ as a linear combination of the basis generated by α. The coefficients of this linear combination will be the coefficients of the polynomial required. (Compare Exercise 22 of Sec. 5.12.)

14. Let T be a linear transformation with the property that, for any transformation R, $RT = TR$ implies that R is a polynomial in T. Prove that T has a primitive vector. (Compare Exercise 13.)

15. Let V be a finite-dimensional vector space over a field F and T a linear transformation of V into itself. Prove that every nonzero vector in V is a primitive vector of T if and only if the characteristic polynomial of T is irreducible over F. (Compare Exercise 22 of Sec. 6.3.)

16. Prove that every square matrix is similar to a matrix $D + N$, where D is a diagonal matrix, N is nilpotent, and $DN = ND$.

17. What is wrong with the following proof?† If A is a matrix whose elements are complex numbers such that $A^T = -A$, then $A = Z$, the zero matrix. "Proof": Let J be the Jordan form of A. Since $A^T = -A$, then $J^T = -J$. But all elements of J above the diagonal are zero; hence every element of J is 0. This shows that $A = Z$.

18. Let N be a nilpotent matrix of order 3 over the field of complex numbers.‡ Prove that if $A = I + \frac{1}{2}N - \frac{1}{8}N^2$, then $A^2 = I + N$. Use the binomial series for $(c + t)^{\frac{1}{2}}$ to obtain a similar formula for a square root of $cI + N$, where N is any nilpotent matrix of order n over the field of complex numbers.

19. Using Exercise 18 or by other means, prove that every nonsingular matrix over the field of complex numbers has a square root.

20. Follow through the proof of Theorem 6.7 for the following transformation defined for a basis $\alpha_1, \alpha_2, \alpha_3, \alpha_4$ as follows:

$$T(\alpha_1) = \alpha_2 \qquad T(\alpha_2) = \alpha_3 \qquad T(\alpha_3) = \alpha_1 - 3\alpha_2 + 3\alpha_3 \qquad T(\alpha_4) = \alpha_1 + \alpha_2 - 2\alpha_3 + \alpha_4$$

† Hoffman and Kunze, *Linear Algebra*, p. 210, listed in the bibliography.

‡ Hoffman and Kunze, *Linear Algebra*, p. 211, listed in the bibliography.

6.5 THE UNIQUENESS OF THE JORDAN FORM

For many of the applications it is not necessary to know that the Jordan form is unique. But in the process of establishing the uniqueness, we find relationships which enable us to determine what the Jordan form is, with less computation than would otherwise be required. We have means of finding the characteristic and minimum polynomials of a transformation T, but we saw in the example at the close of the previous section that that was not enough to determine T. There should be some property of the transformation itself which is decisive in determining the splitting of the space and the accompanying transformations T_i of the theorems of Sec. 6.4.

Let us start with the case when the transformation T is nonderogatory and has the minimum polynomial $(x - a)^n$. If we look at the elementary Jordan matrix (6.7) of Theorem 6.4, we see immediately that the rank of $M - aI$ is $n - 1$, that is, the nullity is 1. We know from that theorem that, for the given choice of basis, the matrix of T *must* be of the form given. Hence if T is nonderogatory with the minimum polynomial $(x - a)^n$, then the nullity of $T - aI$ must be 1. It is also true, though it can be shown most easily in the context of the more general case, that if the nullity of $T - aI$ is 1 and if T has the characteristic polynomial $(x - a)^n$, then T is nonderogatory. As it happens, we need information about the nullities of all the powers of $T - aI$. This is contained in the following

Theorem 6.9 If a transformation T of an n-dimensional vector space V into itself has $(x - a)^n$ as its minimum polynomial, then the nullity of $(T - aI)^i$ is i, for $0 \leq i \leq n$, and n, for $i > n$.

Proof Since $(T - aI)^n = Z$, the zero transformation, we see that the theorem holds for $i \geq n$. Since $(T - aI)^0 = I$, the theorem holds for $i = 0$. We have shown just above that the theorem is true for $i = 1$.

To prove the theorem for other values of i, one may look at the matrix or the transformation. Since the former is a little more concrete and the latter easier notationally, we do both. The matrix associated with $T - aI$ in Theorem 6.4 is N, where N is a matrix of order n in which all the elements are zero except for 1s immediately below the principal diagonal. Now for any matrix B, NB is obtained from B by replacing the first row of B by zeros, the second row by the first row, the third row by the second, and so on, to replacing the last row by the next to the last row. One can think of the effect as making the first row zero and "moving down" every other row except the last one. Thus N^2 will have the first two rows zero and the identity matrix of order $n - 2$ in the lower left-hand corner; N^3 will have the first three rows zero and the identity matrix of order $n - 3$ in the lower left-hand corner; and so on, to N^{n-1} whose elements are all zero except for a 1 in the lower left-hand corner. Hence the rank of N^{n-k} is k for $0 \leq k \leq n$, and the nullity is $n - k$.

To prove the result from the point of view of the transformation, notice first that Equations (6.8) imply $N(\alpha_i) = \alpha_{i+1}$ for $1 \leq i \leq n - 1$ and that $N(\alpha_n) = \theta$. We can simplify notation by agreeing that α_i is to be the null vector for $i > n$. Then,

$$N(\alpha_i) = \alpha_{i+1} \qquad \text{for } 1 \leq i \tag{6.17}$$

Now (6.17) implies that $N^t(\alpha_i) = \alpha_{i+t}$. But, for a given t, $\alpha_{i+t} = \theta$ if $i + t \geq n + 1$. So for N^t, the number of nonzero images of the basis vectors is $n - t$. This shows that the rank of N^t is $n - t$ and thus its nullity is t for $1 \leq t \leq n$. This completes the proof.

Now we are ready for the general theorem.

Theorem 6.10 Let $(x - a)^n$ be the characteristic polynomial and $(x - a)^s$ the minimum polynomial of a transformation T of a vector space V into itself Let the elementary divisors of T as given in Theorem 6.8 be

$$(x - a)^{a_1}, (x - a)^{a_2}, \ . \ . \ . \ , (x - a)^{a_k}$$

where $s = a_1 \geq a_2 \geq \cdot \cdot \cdot \geq a_k \geq 1$. If n_j denotes the nullity of $(T - aI)^j$, then $n_1 = k$ and $n_{j+1} - n_j$, for $1 \leq j \leq s - 1$, is equal to the number of a_i which are greater than j.

Proof Since the matrix of T is the direct sum of the matrices of the component transformations T_i, we see that the nullity of $T - aI$ is the sum of the nullities of $T_i - aI$, for $1 \leq i \leq k$, that is, k.

The nullity of $(T - aI)^2$ is the sum of the nullities of $(T_i - aI)^2$. But the nullity of $(T_i - aI)^2$ is 1 more than that of $T_i - aI$ if $a_i \geq 2$, while if $a_i = 1$, the nullities of $(T_i - aI)^2$ and $T_i - aI$ are the same. Thus the total increase in nullity of $(T - aI)^2$ over that of $T - aI$ is the number of a_i which are greater than 1.

So we can continue until the last step where $a_1 = s$. Here the nullity of $(T - aI)^s$ is the nullity of $(T - aI)^{s-1}$ plus the number of a_i which are equal to s. Since $(x - a)^s$ is the minimum polynomial of T, higher powers of $T - aI$ have nullity n. This completes the proof.

What we are aiming for is to approach the problem from the other direction: given the nullities, to determine the exponents a_i. To do this a certain diagram is convenient. Mark k columns of x's corresponding to the a_i, so that column i has a_i x's. Then the number of x's in the various rows is, from the above, the differences of the nullities. So schematically we have the following array:

	$s = a_1$	a_2	a_3	$\cdots$	a_{k-1}	a_k
n_1	x	x	x	$\cdots$	x	x
$n_2 - n_1$	x	x	x	$\cdots$	x	
$\cdots\cdots$	$\cdots$	$\cdots$	$\cdots$	$\cdots$	$\cdots$	$\cdots$
$n_s - n_{s-1}$	x	x				

That is, the number of x's in the first row is k, the nullity of $T - aI$. The number of x's in the second row is the number of a_i greater than 1; this is equal to $n_2 - n_1$. And so we may continue until the number of x's in the last row is the number of a_i which are equal to $s = a_1$.

We can perhaps clarify the diagram by matching it with an array of vectors as follows:

	$s = a_1$	a_2	$\cdots$	$\cdots\cdots\cdots$ a_{k-1}	a_k
n_1	α_1	α_2	$\cdots$	$\cdots\cdots\cdots$ α_{k-1}	$\alpha_k = \alpha_{n_1}$
$n_2 - n_1$	α_{n_1+1}	α_{n_1+2}	$\cdots$	$\cdots\cdots$ α_{n_2}	
$\cdots\cdots$	$\cdots$	$\cdots$	$\cdots$	$\cdots\cdots$	
$n_s - n_{s-1}$	$\alpha_{n_{s-1}+1}$	$\alpha_{n_{s-1}+2}$	$\cdots$	$\alpha_{n_s} = \alpha_n$	

where the α's in the first row form a basis for the kernel of $T - aI$, those in the first two rows are a basis of the kernel of $(T - aI)^2$, and, in general, the α's in the first j rows form a basis of the kernel of $(T - aI)^j$.

So now, if, instead of knowing the a_i, we know the nullities we can construct the diagram by rows. In the first row there will be n_1 symbols x, in the second row $n_2 - n_1$ symbols x, in the third row $n_3 - n_2$ symbols x, and so on, until in the last row there are $n_s - n_{s-1}$ symbols x. Then by counting the number of x's in the various columns we can determine the numbers a_i.

Notice that the sum of the positive integers a_i is n. So the set of a_i is called a *partition* of n. The following is also a partition of n:

$$n_1, n_2 - n_1, n_3 - n_2, \ldots, n_s - n_{s-1}$$

since $n_s = n$, that is, $(T - aI)^n$ and $(T - aI)^s$ are both zero and hence have nullity n, and all the differences are positive. The two partitions are called *conjugate partitions*.

For future reference we list four auxiliary results which we have stated above or which follow directly:

1. The number n_i is the number of x's in the diagram which are above and in the ith row.
2. We have $n_{j+1} > n_j$ for $1 \leq j \leq s - 1$.
3. Also, $n_r = n_s = n$ if $r \geq s$.
4. Finally $n_1 = k \leq n$.

Now we illustrate these relationships by a specific example. Suppose $(x - 2)^6(x + 1)^{12}$ is found to be the characteristic polynomial of a transformation T. Suppose we also have found by computation the following:

The nullities of $T - 2I$ and $(T - 2I)^2$ are 4 and 5, respectively, and the nullity of $(T - 2I)^3$ is 6.

The nullities of $T + I$, $(T + I)^2$, $(T + I)^3$, and $(T + I)^4$ are 5, 9, 11, and 12, respectively.

Our problem is to compute the exponents of the elementary divisors for each of the factors $(x - 2)^6$ and $(x + 1)^{12}$. We can do these separately by Theorem 6.5.

The diagram for the powers of $T - 2I$ is

$$\begin{array}{cccc} x & x & x & x \\ x & & & \\ x & & & \end{array}$$

First, there are six symbols x, and since 6 is the exponent of $x - 2$ in the characteristic polynomial, higher powers of $T - 2I$ than the third will lead to no increase in the nullity. From the diagram, $k = 4$, $a_1 = s = 3$, and a_2, a_3, and a_4 are all 1.

The diagram for the powers of $T + I$ is

$$\begin{array}{ccccc} x & x & x & x & x \\ x & x & x & x & \\ x & x & & & \\ x & & & & \end{array}$$

From the diagram, $k = 5$, $a_1 = s = 4$, $a_2 = 3$, $a_3 = a_4 = 2$, and $a_5 = 1$.

We also know that the minimum polynomial of T is

$$(x - 2)^3(x + 1)^4$$

Finally, justifying the title of the section, we can prove without difficulty

Theorem 6.11 Given a linear transformation T of a vector space into itself over a field in which the characteristic polynomial is a product of linear factors, the Jordan canonical form of the matrix of T is unique except for the order in which the matrices of the direct sum occur.

Proof Essentially we wish to show that, with the exception noted, the Jordan canonical form is independent of any basis. First, the characteristic polynomial is independent of a basis. This leads to the splitting according to powers of irreducible polynomials as described in the Corollary to Theorem 6.5. Then, for a given linear power factor of the characteristic polynomial, we have shown that the exponents of the elementary divisors are determined by the nullities of the powers of the transformation $T - aI$, where a is a characteristic root. These nullities are again independent of the basis. The elementary divisors in their turn determine the Jordan elementary matrices. This completes the proof.

The following theorem is useful:

Theorem 6.12 Let $f(x) = (x - a)^r g(x)$, with $g(a) \neq 0$, be the characteristic polynomial of a linear transformation T of a vector space V into itself and let n_i denote the nullity of $(T - aI)^i$. Then

$$n_i \geq i \qquad \text{for } 1 \leq i \leq r \tag{6.18}$$

and, if $m(x) = (x - a)^s h(x)$, where $h(a) \neq 0$, is the minimum polynomial of T, then

$$n_i = r \qquad \text{for } i \geq s$$

Proof First notice the auxiliary result 2 after the proof of Theorem 6.10. Since a is a characteristic root of T, we see that $n_1 \geq 1$, and thus $n_{j+1} > n_j$ for $1 \leq j \leq s - 1$ implies (6.18) with r replaced by s. Now, referring back to the splitting of V as given in Theorem 6.5, we see that $V = E_1 \oplus E_2$, where T_1 (T restricted to E_1) has $(x - a)^r$ as its characteristic polynomial. Hence the nullity of $(T - aI)^r$ is r, the dimension of E_1. But $m(x) = (x - a)^s h(x)$, where $h(a) \neq 0$, implies that $(x - a)^s$ is the minimum polynomial of T_1. That is,

$$(T - aI)^s(\alpha) = \theta \qquad \text{for all } \alpha \text{ in } E_1$$

This shows that $n_i = r$ for $i \geq s$ and completes the proof.

A special case is worthy of mention:

Corollary If the characteristic polynomial of a transformation T is as given in the theorem, then the nullity of $(T - aI)^r$ is r.

This theorem gives a lower bound for the nullities. Theorem 4.16 gives an upper bound for n_1. This result can also be proved using this section and is left as an exercise below. One can also show, using the notation of Theorem 6.10, that

$$n_i \leq ik = in_1$$

EXERCISES

1. Suppose the elementary divisors of a transformation T are $(x + 4)^5$, $(x + 4)^2$, $x + 4$, and $x + 4$. Give the characteristic and minimum polynomials of T and the nullities of the various powers of $T + 4I$.

2. Suppose a linear transformation T has $x^5(x+3)^4(x-4)^7$ as its characteristic polynomial. Let the nullities for various powers be given by the following table:

	$j = 1$	2	3	4
x^j	2	4	5	5
$(x+3)^j$	3	4	4	4
$(x-4)^j$	3	5	6	7

Find the elementary divisors of T.

3. List all the possibilities for sets of elementary divisors for a transformation whose characteristic polynomial is $(x+1)^{10}$. Which of the partitions are conjugate to themselves?

4. Using the notation of Theorem 6.10, prove $n_i \leq ik$.

5. In Theorem 4.16 we proved that the nullity of $T - aI$, for T any linear transformation of a vector space into itself, is not greater than the highest power of $x - a$ in the characteristic polynomial of T. Using this section, give another proof of this fact.

6. Prove that if, for a transformation T, the nullities of $(T - aI)^t$ and $(T - aI)^{t+1}$ are the same, then all higher powers of $T - aI$ have the same nullity. There is a proof using the results of this section and one that is independent of this section.

7. Let A and B be two matrices of order n with elements in the field of complex numbers. Prove that A and B are similar if and only if the nullities of $(A - xI)^t$ and $(B - xI)^t$ are the same for all complex numbers x and all positive integers t.

8. Using Exercise 7 or by other means, prove that every square matrix is similar to its transpose.

9. Prove the following generalization of Theorem 4.16: In the notation of Theorem 6.12, $n_i \leq r$ for all i. (Compare Exercise 5 above.)

6.6 SOME APPLICATIONS

We have seen (Exercise 14 in Sec. 3.7) that if c is a characteristic root of a transformation T, then c^k is a characteristic root of T^k for any positive integer k. Using the Jordan form we can prove a much more general result. We have defined the spectrum of a transformation to be the set of its characteristic roots. For our purposes here it is convenient to define what we call the *complete spectrum* of a transformation in the following manner. Let

$$f(x) = (x - a_1)(x - a_2) \cdots (x - a_n)$$

be the characteristic polynomial of T. Then we call the set $a_1, a_2, \ldots, a_n$ the *complete spectrum.* The a's need not therefore be distinct. Each is repeated as many times as its associated linear factor occurs in the characteristic polynomial.

Theorem 6.13 Let T be a transformation of a vector space V into itself over a field F in which the characteristic polynomial of T is a product of linear factors. Let $g(x)$ be any polynomial with coefficients in F. Then the complete spectrum of $g(T)$ is $g(a_i)$, where the a_i range over the complete spectrum of T.

Proof We first prove this result for the case in which, by choice of basis, the matrix of T is a Jordan elementary matrix of order m. As we noted after Theorem 6.4, such a matrix can be written in the form $aI + N$, where N is the matrix of order m whose elements are all zero except for 1s just below the principal diagonal. Since N is commutative with the identity, we see that

$$g(aI + N) = g(a)I + h(N)$$

for some polynomial $h(x)$. Now any polynomial in N has the property that it is a matrix in which all elements on and above the principal diagonal are zero. Thus the characteristic polynomial of $g(aI + N)$ is $[x - g(a)]^m$. The complete spectrum of T has m numbers a and that of $g(T)$ has m numbers $g(a)$. This is the proof for this case.

To complete the proof we need merely note that we may consider the matrix of T to be a direct sum of elementary Jordan matrices. Hence the complete spectrum of T will be the numbers on the principal diagonal of the Jordan canonical form. That is, we can write the Jordan form as

$$D + M$$

where D is a diagonal matrix whose elements are the complete spectrum of T and M is a matrix whose elements on and above the principal diagonal are all zero. Thus $g(D + M) = g(D) + h(M)$, where $h(x)$ is some polynomial. Then $g(D)$ is a diagonal matrix whose elements are $g(a_i)$ as the a_i range over the complete spectrum of T, and $h(M)$ has zero elements on and above the principal diagonal. Hence we have the same conclusion in this case as in the special one of the first paragraph of the proof.

Note This theorem can be extended to rational functions of T as follows. Suppose $g(x) = h(x)/k(x)$, where h and k are polynomials such that $k(T)$ is nonsingular. Since the inverse of $k(T)$ can be expressed as a polynomial in $k(T)$, it follows that $g(T) = s(T)$ for some polynomial $s(x)$. Then apply the theorem to $s(x)$ in place of $g(x)$.

We can use the above result to obtain a method for calculating the characteristic polynomial of a transformation—one which seems to be somewhat practical for machine calculation. We know from Theorem 4.15 that if A is a matrix its trace is the sum of the numbers of its complete spectrum. It follows from the above that the trace of A^k is the sum of the kth powers of its complete spectrum. So by computing the kth power of the matrix A and getting its trace, we can find

$$s_k = a_1{}^k + a_2{}^k + \cdots + a_n{}^k \qquad \text{for } 1 \leq k \leq n \tag{6.19}$$

On the other hand, it is known from Theorem 4.15 that if a polynomial $f(x)$ is written in the form

$$x^n - c_1x^{n-1} + c_2x^{n-2} - \cdots + (-1)^n c_n \tag{6.20}$$

then c_1 is the sum of the zeros of $f(x)$, c_2 is the sum of the products of the zeros two at a time, c_3 is the sum of the products three at a time, and so on, to c_n which is the product of all the zeros. There are formulas, called *Newton's formulas*, which can be found in almost any theory of equations book, by which one can obtain the c's in (6.20) from the s_k in (6.19). Thus in this case one can compute the characteristic polynomial. We illustrate this by an example.

Example Suppose A is a matrix of order 3 and suppose s_1, s_2, and s_3 are found. Then $s_1 = c_1$. To compute c_2, notice that if a_1, a_2, a_3 are the characteristic roots, then

$$c_1{}^2 = (a_1 + a_2 + a_3)^2 = s_2 + 2(a_1a_2 + a_1a_3 + a_2a_3) = s_2 + 2c_2$$

Hence $c_2 = (-s_2 + c_1{}^2)/2$.

It remains to compute c_3. For this,

$$\begin{aligned} c_1{}^3 = (a_1 + a_2 + a_3)^3 &= s_3 + 3\Big(\sum_{i \neq j} a_i{}^2a_j\Big) + 6a_1a_2a_3 \\ &= s_3 + 3(a_1 + a_2 + a_3)s_2 - 3s_3 + 6c_3 \\ &= -2s_3 + 3c_1s_2 + 6c_3 \end{aligned}$$

Hence $c_3 = (c_1{}^3 + 2s_3 - 3c_1s_2)/6$.

A second application is to linear differential equations. Suppose y denotes a real function of t for which the first n derivatives exist, and let $D(y)$ denote the derivative of y with respect to t. Now D is a linear transformation of y into another function of t, as we have seen previously (Sec. 3.5). We adopt the convention that $D^2(y)$ means $D(D(y))$ (not the square of the first derivative), just as for any vector, $T^2(\alpha)$ means $T(T(\alpha))$, and we use the notation D^i for the ith derivative. Now consider the differential equation

$$h(D)y = D^ny + a_{n-1}D^{n-1}y + a_{n-2}D^{n-2}y + \cdots + a_1Dy + a_0y = 0 \tag{6.21}$$

In the language of linear algebra, the set of solutions of this differential equation is the kernel of the linear transformation $h(D)$ over the vector space of differentiable functions of t.

First consider the case when $h(x) = (x - a)^m$ with a real. We know from the theory of differential equations that the kernel K of $h(D)$ is of dimension m and hence that the transformation D is nonderogatory. Thus there is a primitive solution z such that

$$z, (D - a)z, (D - a)^2z, \ldots, (D - a)^{n-1}z \tag{6.22}$$

form a basis of the kernel; that is, every solution is a linear combination of these functions with real coefficients. Then, as in Theorem 6.4 we can let $z_1 = z$ and define z_{i+1} by

$$\begin{aligned} D(z_i) &= az_i + z_{i+1} \qquad 1 \leq i \leq n - 1 \\ D(z_n) &= az_n \end{aligned} \tag{6.23}$$

From the last equation of (6.23) we have $z_n = e^{at}$. To explore the possibilities for other solutions, let $z_{n-1} = g(t)z_n$ and try to find a function $g(t)$ which satisfies (6.23) for $i = n - 1$. Then

$$\begin{aligned} D(z_{n-1}) &= g'(t)z_n + g(t)D(z_n) = g'(t)z_n + g(t)az_n \\ &= az_{n-1} + g'(t)z_n \end{aligned}$$

This satisfies (6.23) for $i = n - 1$ if $g'(t) = 1$, that is, $g(t) = t$. Thus it can be shown that

$$(D - aI)^k t^{k-1} e^{at} = 0 \qquad \text{for } 1 \leq k \leq n$$

These are equivalent to (6.23) by Equations (6.6) and (6.8). Hence the general solution of $(D - aI)^m z = 0$ is

$$c_1 e^{at} + c_2 t e^{at} + \cdots + c_m t^{m-1} e^{at}$$

for arbitrary constants c_i.

Second, suppose $h(x)$ is a product of distinct real linear factors $x - a_i$. Then the kernel K is a direct sum of n subspaces K_i as in Theorem 6.5, where K_i is the kernel of $D - a_i$. Now a solution of $(D - a)y = 0$ is e^{at}. Hence the kernel K, that is, the set of solutions of (6.21), is in this case

$$c_1 e^{a_1 t} + c_2 e^{a_2 t} + \cdots + c_n e^{a_n t}$$

Modifications are made in the study of differential equations to handle the cases when zeros are imaginary or when factors are repeated.

6.7 POSITIVE STOCHASTIC MATRICES

Here we extend to n dimensions the results obtained for two dimensions in Sec. 6.2. First we need definitions of some convenient terms.

Definitions A matrix is called *nonnegative* if all its elements are real numbers and none are negative; it is called *positive* if all its elements are positive (that is, actually greater than zero). A matrix is called *stochastic* if it is nonnegative and if 1 is the sum of the elements of each row. If, in addition, the sum of the elements of each column is 1, we call the matrix *doubly stochastic.*

In this section we shall denote by ν a row vector whose components are all 1. An important use of this vector is contained in the following

Theorem 6.14 A matrix M is stochastic if and only if $M\nu^T = \nu^T$. The product of two stochastic matrices is a stochastic matrix.

Proof Multiplying any row vector on the right by ν^T serves to add the components of the vector. Hence if the product is 1, the sum of the elements of the row is 1, and conversely. This proves the first part of the theorem. To prove the second part, suppose A and B are two stochastic matrices. Then

$$AB\nu^T = A\nu^T = \nu^T$$

This completes the proof.

In the rest of this section we shall confine our attention to positive stochastic matrices since the proofs are easier and the results simpler. Our aim is first to show that if P is a positive stochastic square matrix, then $\lim_{k\to\infty} P^k$ exists; and then to find what the limit is. We defined precisely what this means in Sec. 6.2.

We can take our cue for accomplishing this from the two-dimensional case in Sec. 6.2. There it was important to get information first about the characteristic roots. If we can show that the characteristic polynomial of P is $(x - 1)g(x)$, where all the zeros of $g(x)$ are less than 1 in absolute value, then from Theorem 6.5 we know that P is similar to a matrix

$$1 \oplus L$$

where all the characteristic roots of L are less than 1 in absolute value. Then, from Exercise 4 of Sec. 6.4 we know that $\lim_{k\to\infty} L^k = Z$, the zero matrix. From this information we will know that the limit of P^k exists and will have a means of finding what the limit is.

So the first part of our task is to prove the following:

Lemma If P is a positive stochastic matrix, all its characteristic roots except 1 itself are less than 1 in absolute value. Furthermore, 1 is the nullity of $P - I$.

Proof Let c be a characteristic root of $P = (p_{ij})$. Then if $\xi = (x_1, x_2, \ldots, x_n)$ is an associated characteristic vector, we have

$$\sum_{j=1}^{n} p_{ij}x_j = cx_i \qquad \text{for } 1 \leq i \leq n \tag{6.24}$$

By a permutation of subscripts we may assume $|x_1| \geq |x_j|$ for $1 \leq j \leq n$. Since ξ a characteristic vector associated with the characteristic root 1 implies that $r\xi$ is also, for any nonzero number r, we can let $y_j = x_j/x_1$ and have, from (6.24),

$$\sum_{j=1}^{n} p_{1j}y_j = cy_1 \qquad \text{where } y_1 = 1 \geq |y_i| \text{ for } 1 \leq i \leq n \tag{6.25}$$

Then from (6.25) we have

$$|c| = \left| \sum_{j=1}^{n} p_{1j}y_j \right| \leq \sum_{j=1}^{n} p_{1j}|y_j| \leq \sum_{j=1}^{n} p_{1j} = 1 \tag{6.26}$$

This shows that every characteristic root of P has absolute value less than or equal to 1.

Now we shall show that $|c| = 1$ implies $c = 1$ and $y_j = 1$, for $1 \leq j \leq n$. First, the equality in (6.26) implies that $|y_j| = 1$. If we can show that $y_j = 1$ for all j, then it will follow that c in (6.25) is 1. Let us look at

$$1 = \left| \sum_{j=1}^{n} p_{1j}y_j \right| \qquad \text{where } \sum_{j=1}^{n} p_{1j} = 1 \text{ and } |y_j| = 1 \tag{6.27}$$

from a geometric point of view. Notice that each y_j is a complex number and thus represents a vector of length 1. Also

$$p_{11}y_1 + p_{12}y_2 + \cdots + p_{1n}y_n = \alpha \tag{6.28}$$

is a vector of length 1. It is a vector sum of vectors whose lengths are the positive numbers $p_{11}, p_{12}, \ldots, p_{1n}$, respectively, where the sum of the lengths of the vectors is 1. Thus by the triangle inequality [see inequality (5.27)] the points defined by $p_{1j}y_j$ are all collinear. Since $y_1 = 1$, the line on which all these points lie is the axis of real numbers. The sum on the left of Equation (6.28) will therefore be less than 1 in absolute value unless all the y_j are 1. Thus, the only characteristic vectors of P associated with the characteristic root 1 are scalar multiples of ν. This proves that the nullity of $P - I$ is 1 and completes the proof of the lemma.

We still have more to do to prove that $(x - 1)^2$ is not a factor of the characteristic polynomial of P, for though (see Theorem 4.16) it is true that if

$(x - 1)^2$ is not a factor of the characteristic polynomial, then $P - I$ is of nullity 1, the converse of this statement is not true.

We show that $(x - 1)^2$ is not a factor of the characteristic polynomial of P on the way to proving

Theorem 6.15 If P is a positive stochastic matrix, then

$$\lim_{k \to \infty} P^k = B$$

where B is a positive stochastic matrix of rank 1.

Proof First we show that $(x - 1)^2$ is not a factor of the characteristic polynomial of P. Suppose this were the case, since $P - I$ is of nullity 1, we can see from Theorems 6.5 and 6.8 that P is similar to a direct sum

$$P_1 \oplus P_2$$

where $P_1 = I + N$, N is the matrix whose elements are all zero except for 1s immediately below the principal diagonal, P_1 is of order at least 2, and all the characteristic roots of P_2 are less than 1 in absolute value. Now

$$P_1{}^k = I + kN + \binom{k}{2} N^2 + \cdots$$

Thus the elements immediately below the principal diagonal of $P_1{}^k$ are all k. Hence the elements of the powers of P are unbounded as k increases. This means that for some C the elements of CP^kC^{-1} are unbounded as k increases. This is not possible since every element of P and therefore of P^k is nonnegative and not greater than 1. Thus we have shown that $(x - 1)^2$ is not a factor of the characteristic polynomial of P.

It follows that for some nonsingular matrix C,

$$CPC^{-1} = 1 \oplus P_3$$

where all the characteristic roots of P_3 are less than 1 in absolute value. Thus, by Exercise 4 of Sec. 6.4,

$$\lim_{k \to \infty} P_3{}^k = Z$$

the zero matrix. Hence

$$\lim_{k \to \infty} P^k = C^{-1}(1 \oplus Z)C = B \tag{6.29}$$

and B is of rank 1 since $1 \oplus Z$ is. Since all powers of P are positive stochastic, B must be also.

It remains to find the precise form of B. One could do this by first finding D, but it is easier to do it another way. To this end let ρ denote the first row of B. It must be positive stochastic. Since B is of rank 1, every row of B is a scalar multiple of ρ. But $d\rho$ can be positive stochastic only if $d = 1$. Hence every row of B is the same vector ρ. But (6.29) implies that by choosing k large enough one can make the absolute value of every element of $P^k - B$ arbitrarily small and similarly for every element of $BP - P^k$. Hence, by proper choice of k one can make every element of $BP - B$ arbitrarily small. But the elements of BP and B are independent of k. Hence $BP - B = 0$ and $BP = B$. This is equivalent to $\rho P = \rho$. This determines B as is expressed formally in our last theorem.

Theorem 6.16 The limiting matrix B as defined in Theorem 6.15 has, as every row, the positive stochastic vector ρ determined by $\rho P = \rho$.

There is a theorem of Frobenius which gives information about matrices whose elements are nonnegative real numbers. For this, see F. R. Gantmacher's book listed in the bibliography.

EXERCISES

1. Find the limiting matrix Q for each of the positive stochastic matrices P:

(a) $P = \begin{bmatrix} \frac{1}{2} & \frac{1}{2} \\ \frac{2}{3} & \frac{1}{3} \end{bmatrix}$

(b) $P = \dfrac{1}{4}\begin{bmatrix} 1 & 1 & 2 \\ 2 & 1 & 1 \\ 1 & 2 & 1 \end{bmatrix}$

(c) $P = \dfrac{1}{9}\begin{bmatrix} 2 & 3 & 4 \\ 1 & 2 & 6 \\ 5 & 2 & 2 \end{bmatrix}$

2. Using the method in Sec. 6.6, find the characteristic polynomial of the matrices in Exercise 1.

3. Suppose that two-ninths of the sons of Republican fathers are Republicans, two-ninths are Democrats, and five-ninths are Independents; and similarly for the other two rows of the following stochastic matrix. The rows are according to the party of the father and the columns that of the son:

$$\begin{array}{cccc} & R & D & I \\ P = \dfrac{1}{9} & \left[\begin{array}{ccc} 2 & 2 & 5 \\ 4 & 2 & 3 \\ 4 & 3 & 2 \end{array}\right] & & \begin{array}{c} R \\ D \\ I \end{array} \end{array}$$

What is the limit of P^k as k becomes infinite? What is the meaning of this result in terms of political alignment?

4. In Exercise 3, change the matrix P to

$$P = \frac{1}{3}\begin{bmatrix} 1 & 2 & 0 \\ 1 & 1 & 1 \\ 0 & 2 & 1 \end{bmatrix}$$

Since this matrix has zero elements, it is not covered by Theorem 6.16. Show that nevertheless the results stated in the Lemma (following Theorem 6.14) and Theorem 6.16 hold. Find the limit of P^k as k becomes infinite.

5. Show that $f(x) = x^3 - 9x^2 + 18x + 8$ is the characteristic polynomial of the following matrix:

$$\begin{bmatrix} 1 & 2 & -1 \\ 2 & 3 & 0 \\ -1 & 0 & 5 \end{bmatrix}$$

Since $f(x)$ is the characteristic polynomial of a symmetric real matrix, all its zeros are real (why?). Let the zeros be d_1, d_2, and d_3. Assuming (or proving) that the zeros are distinct, find, in terms of the d_i, the solutions of the differential equation $f(D)y = 0$.

6. Solve the following differential equation:

$$f(D)y = 0 \qquad \text{where } f(x) = (x - 2)^3(x + 1)^2$$

7. Indicate how the proof of Theorem 6.15 can be shortened if we assume that P is positive and doubly stochastic. In this case, what will be the matrix Q?

8. Let P be a square matrix whose elements are all nonnegative and such that no row has the sum of its elements zero. Prove that there is a diagonal matrix D such that DP is stochastic.

9. Prove that all the characteristic roots of a stochastic matrix (without the requirement that it be positive) are less than or equal to 1 in absolute value.

APPENDIX *A*

GROUPS, RINGS, AND FIELDS

A.1 GROUPS

In this section, and indeed in this appendix, we shall for the most part be merely defining terms and illustrating them. Throughout the field of algebra and in other branches of mathematics as well runs the concept of a group, which can be defined as follows:

Definition A *group* is a set S of elements and an operation with the following properties, where $a, b, c, \ldots$ denote elements of S and $*$ denotes the operation:

1. The closure property: $a * b$ is in S for all a and b in S.
2. The associative property: $(a * b) * c = a * (b * c)$.
3. S contains an identity element e with the property that

$$a * e = e * a = a$$

for all elements a in S.

4. Every element b of S has an inverse element, call it $\bar{b}$, with the property that $b * \bar{b} = e = \bar{b} * b$.

If, in addition, $a * b = b * a$ for all a and b in S, the group is called *Abelian* or commutative.

There are other properties of a group which follow easily from the above. Some of the more important ones are:

5. The identity and inverse elements are unique.
6. For any two elements a and b of S, the following equations are true for unique elements x and y of S:

$$a * x = b \qquad \text{and} \qquad y * a = b$$

7. The inverse of $(a * b)$ is $(b * a)$.
8. The cancellation property: If $a * c = b * c$ or $c * a = c * b$, then $a = b$.

In dealing with groups, we usually use the terminology of multiplication. Thus we write ab instead of $(a * b)$ and b^{-1} instead of $\bar{b}$ and we denote the identity element by 1. The set of nonsingular matrices of given order form a multiplicative group, as well as the set of nonsingular transformations of a given space onto itself. The set of rational numbers without zero also form a multiplicative group. The set of orthogonal matrices of given order form a group. In fact, most of the material in this book is more or less involved with groups of transformations and their properties.

Additive groups are also important. Here $(a * b)$ becomes $a + b$, $\bar{b}$ is written $-b$, and 0 is the identity element. The set of matrices with r rows and s columns forms an additive group. The set of integers is an additive group, but not a multiplicative one.

In a group the operation need not be either addition or multiplication. The rotations about the origin in the plane form a group. In Appendix B we deal with another type of group: a permutation group.

In Felix Klein's so-called Erlanger Program, geometry is thought of as a study of properties left unchanged by a group of transformations. For different groups one gets different geometries.

A.2 RINGS

For a *ring* we have two operations. They do not necessarily need to be addition and multiplication, but we use these terms in describing them. Indeed for our purposes in this book they are addition and multiplication. So we have the following:

Definition A *ring* is a set S of one or more elements with two operations $+$ and with the following properties:

1. The elements of S form an Abelian group under addition.
2. Multiplication has the closure and associative properties.
3. The distributive properties hold; that is,

$$a \cdot (b + c) = (a \cdot b) + (a \cdot c) \qquad \text{and} \qquad (b + c) \cdot a = (b \cdot a) + (c \cdot a)$$

Using the usual conventions of algebra, we may write the distributive properties more briefly as

$$a(b + c) = ab + ac \qquad \text{and} \qquad (b + c)a = ba + ca$$

The notations for identity elements and inverses are as given in Sec. A.1.

If, in addition to the three properties above, S contains an identity element for multiplication, we call it a *ring with an identity element.* If S contains an identity element for multiplication and multiplication is commutative, we call S a *commutative ring.*

If S is a ring with an identity element, the following two properties can be proved:

1. $0 \cdot a = a \cdot 0 = 0$ for all a in S.
2. $(-a)b = a(-b) = -(ab)$ and $(-a)(-b) = ab$.

The set of square matrices of the same dimension form a ring with an identity element but it is not a commutative ring.

The set of integers is a commutative ring that has one important property which the set of matrices does not have, namely,

Property D *If $ab = 0$, then one of a and b is zero.* (We use the letter "D" because it defines an integral domain.) A commutative ring which has Property D is called an *integral domain.* In some texts this term includes the existence of a multiplicative identity and in others it does not. In any case, the set of matrices does not form an integral domain. Another property that can be shown without much difficulty to be equivalent to Property D (assuming the previous properties) is

Property D′ **(*Cancellation Property*)** If $ax = ay$ with $a \neq 0$, then $x = y$, and if $xb = yb$ with $b \neq 0$, then $x = y$.

An element b of a ring S is called a *divisor of zero* (or zero divisor) if it is not zero and yet there is a number c in S such that $bc = 0$. In these terms we could describe Property D by "S has no divisors of zero."

A.3 FIELDS

A field is a commutative ring which, except for the element zero, is closed under division as well as addition and multiplication; that is, in a field, multiplication has an inverse. But it is probably more satisfactory to define it without reference to a ring.

Definition A *field* is a set of elements S with two operations $+$ and $\cdot$, called addition and multiplication, with the following properties:

1. The elements of S form an Abelian group under addition.
2. The set S^*, namely, S with zero excluded, forms an Abelian group under multiplication.
3. The distributive property holds:

$$a(b + c) = ab + ac$$

Notice here that because of the commutativity of multiplication the distributive property can be more briefly stated. Property D holds for a field since if $ab = 0$, with $a \neq 0$, then

$$0 = a^{-1}(ab) = (a^{-1}a)b = 1 \cdot b = b$$

If all the properties of a field hold except for the commutativity property for multiplication, the system is called a *division ring* or a skew field. Some authors use the term *field* when the commutativity property is lacking.

The set of matrices of order n does not form a field; in fact, it does not even form an integral domain. The best-known fields are the set of rational numbers, the set of real numbers, and the set of complex numbers.

Consider some field F. We know it contains the element 1 (the identity for multiplication) and, from the property of closure for addition, it contains $1 + 1$, $1 + 1 + 1$, and, in fact, the sum of any number of 1s. In this regard, two things may happen:

(*a*) The sum of r 1s is equal to the sum of s 1s only if $r = s$.

(*b*) There are two positive integers r and s, with $r > s$ such that the sum of r 1s is equal to the sum of s 1s.

Let us consider the two cases and explore some of the consequences. If case (*a*) holds, every integer corresponds to a unique sum of 1s; that is, for each positive integer r there is a unique element of F, namely, the sum of r 1s. Then one can either consider that F has a subset isomorphic to the positive integers or one can call $1 + 1$ the integer 2, $1 + 1 + 1$ the integer 3, and so forth, and thus say that F contains the set of positive integers. If we take the latter point of view, the existence of an additive inverse shows that

F contains the negative integers and zero as well. Then the existence of a multiplicative inverse except for zero shows us that the set of rational numbers must be a subset—indeed a subfield—of F. So any field which contains the set of rational numbers as a subfield is said to be of *chararteristic zero.* (Sometimes the term is *characteristic infinity.*) We shall see the reasons for this term a little later in this section.

In case (*b*) it must be true that, by the cancellation property for addition, the sum of $r - s$ 1s is zero. Let m be the least positive integer such that the sum of m 1s is zero. Then 1 is equal to the sum of $m + 1$ 1s, $1 + 1$ equal to the sum of $m + 2$ 1s, and so on. In general the sum of r 1s will be equal to the sum of s 1s if $r - s$ is divisible by m. So we can consider a subset of elements of F to be the set of integers modulo m. Now if m is not a prime number, it can be written $m = tu$, where t and u are positive integers less than m. So we will have $tu = 0$ with neither t nor u zero. This is impossible for a field. Our conclusion is that m must be a prime number p. It can be shown that for a prime p the set of numbers modulo p form a field. This number p is called the *characteristic* of the field F.

Recapitulating, our conclusion is, with the terminology above: Any field F either contains as a subfield the set of rational numbers or the set of integers modulo p, for some prime number p. In the former case we say that F is of *characteristic zero*, and in the latter, of *characteristic p*. A reason for the term characteristic zero is that for characteristic p two numbers are the same if they differ by a multiple of p, and hence one might say that for characteristic zero two numbers are the same if they differ by a multiple of zero. From another point of view one could defend the term characteristic infinity since no finite sum of 1s is zero for such a field.

Many of the results in linear algebra are independent of what the field is, but the applications are mostly for the real and complex fields. The field of characteristic 2 can be troublesome since for such fields $1 = -1$. It has its advantages, too, since in a field of characteristic 2, $(x + y)^2 = x^2 + y^2$.

In Appendix C we shall indicate how one may construct fields from other fields so that certain desired properties come into being. For further development of the theory of groups, see books on modern algebra, including that of Birkhoff and MacLane listed in the bibliography.

APPENDIX **B**

PERMUTATION GROUPS

B.1 PERMUTATION GROUPS

The object of this appendix is, first, to give an example of a group in which the operation is neither multiplication nor addition of numbers and, second, to prove Theorems 4.1 and 4.2 needed in the definition of a determinant.

Given n symbols $x_1, x_2, \ldots, x_n$. Any ordering of these n symbols is called a *permutation*. There are $n!$ permutations of these n symbols since there are n different symbols which can occupy the first place, then $n - 1$ for the second place, $n - 2$ for the third place, and so on. Actually, the word *permutation* has two meanings: a static meaning in which it refers to an ordered set, and a dynamic one which refers to the process of changing the order. It is fortunate that usually the context makes it clear which of the meanings is intended. Of special importance is the interchange of two elements, which we call a *transposition*.

Instead of choosing the general symbols $x_1, x_2, \ldots, x_n$, it simplifies notation to use the symbols 1, 2, 3, . . . , n. Then, in the static sense,

$$p = (j_1, j_2, \ldots, j_n)$$

is a permutation of the numbers 1, 2, . . . , n if every j_i is one of these numbers and no two j_i are equal. In the dynamic sense we can write p as

$$p = \begin{pmatrix} 1 & 2 & 3 & \cdots & n \\ j_1 & j_2 & j_3 & \cdots & j_n \end{pmatrix}$$

which "takes 1 into j_1, 2 into j_2, 3 into j_3, . . . , n into j_n." Technically, it is a one-to-one mapping of the set 1, 2, . . . , n onto itself. If $j_i = i$ for $1 \leq i \leq n$, we have the identity mapping, that is, the *identity permutation.*

Notice that a given permutation can be written in a number of ways according to the order in the upper row. For instance, two representations of the same permutation are

$$\begin{pmatrix} 1 & 3 & 2 \\ 3 & 1 & 2 \end{pmatrix} \qquad \begin{pmatrix} 2 & 3 & 1 \\ 2 & 1 & 3 \end{pmatrix}$$

However, for a given permutation, once the order in the upper row is settled on, that in the lower row is determined.

On our way to showing that the set of all permutations on n symbols form a group, we now can define an operation which we call a *product,* using an old term in a new sense. Let us first illustrate the process. Let permutations p and q be defined as follows:

$$p = \begin{pmatrix} 1 & 2 & 3 \\ 3 & 2 & 1 \end{pmatrix} \qquad q = \begin{pmatrix} 1 & 2 & 3 \\ 2 & 1 & 3 \end{pmatrix}$$

The permutation pq is defined to be that obtained by applying first p and then q. (You will find that in some texts the other order is used.) Thus p takes 1 into 3 and q takes 3 into 3; hence pq takes 1 into 3. Also p takes 2 into 2 and q takes 2 into 1; hence pq takes 2 into 1. Finally, p takes 3 into 1 and q takes 1 into 2; hence pq takes 3 into 2. Thus we have

$$pq = \begin{pmatrix} 1 & 2 & 3 \\ 3 & 1 & 2 \end{pmatrix}$$

Similarly qp takes 1, 2, 3 into 2, 3, 1, respectively.

By this means we can define the product of any two permutations p and q on n symbols and the product will be a permutation on n symbols. Hence the set S_n of all permutations on n symbols is closed under the operation of *product.*

The inverse of a permutation is obtained by interchanging the two rows in the representation. Thus, all that is left to make S_n a group is the property of associativity. Here one might look at the product as a mapping. If p, q, and r are three permutations, then pqr will be the mapping first of p, then of q, and finally of r. So $(pq)r = p(qr)$, and we have proved

Theorem B.1 The set S_n of all permutations of n symbols under the operation of product is a group. It is called the *symmetric group* on n symbols.

It is not hard to show that any group with n elements is isomorphic to a subgroup of S_n, but we shall not prove it here.

B.2 EVEN AND ODD PERMUTATIONS

Here we shall be dealing with a subgroup of S_n which has special importance. The application with which we are concerned is to the definition of determinants.

Let p be the permutation, in the static sense, of the integers 1, 2, 3, . . . , n:

$$p = (j_1, j_2, j_3, \ldots, j_n) \tag{B.1}$$

Any pair satisfying the condition

$$j_k > j_i \qquad \text{and} \qquad k < i$$

is called an *inversion*, and the number of such pairs is called the *number of inversions in p*. We denote this by $N(p)$. We now give the first of two equivalent definitions of an *even permutation*. It is the second of these definitions which is stated in Sec. 4.3. We shall find that the two supplement each other.

Definition A permutation p given by (B.1) is called *even* if $N(p)$ is even and *odd* if $N(p)$ is odd; that is, an *even permutation* is one which has an even number of inversions, and an *odd permutation* is one which has an odd number of inversions.

An advantage of this definition is that one can find whether a given permutation is even or odd by counting. For instance, the permutation (1,3,4,2,5) is even because there are two inversions (3,2) and (4,2); but (5,4,1,2,3) is odd since there are seven inversions: (5,4), (5,1), (5,2), (5,3), (4,1), (4,2), and (4,3).

Now, leading to the other definition, we think of the permutation in dynamic form and prove the following:

Theorem B.2 An even permutation can be represented as a product of an even number of transpositions, and an odd permutation can be represented as a product of an odd number of transpositions.

Proof We wish to show that a permutation p can be represented as a product of $N(p) + 2k$ transpositions for some integer k. We do this by induction on n. Suppose in (B.1) the number n is in the rth place. Then there are $n - r$ inversions (n,j_i), where $i > r$. By a succession of $n - r$ interchanges of n with the number on its right, we take p into a permutation p' whose last number is n. In p' the only inversions involve pairs of numbers from 1 to $n - 1$ inclusive, and the relative order of the numbers 1 through $n - 1$ will be the same in p and p'. Then

$$N(p) = (n - r) + N(p')$$

where $$p' = (j_1, j_2, \ldots, j_{r-1}, j_{r+1}, \ldots, j_n, n)$$

Then if we let p'' be p' with the last number n omitted, we see that p'' is a permutation on the integers from 1 to $n - 1$ inclusive. Thus, by the induction hypothesis, p'' can be represented as a product of $N(p'') + 2k$ transpositions. Hence p' can be represented as a product of $N(p') + 2k$ transpositions. This shows that p can be represented as a product of

$$(n - r) + N(p') + 2k = N(p) + 2k$$

transpositions, and our proof is complete.

Notice that we have shown in the process that every permutation can be represented as a product of transpositions. Also, while it is true that a permutation can be represented as many different products of transpositions, yet for any permutation the number of transpositions in all such products is even for an even permutation and odd for an odd one. We see now the advantage of starting with the idea of an inversion which depends only on the permutation in the static sense. An alternative definition for an even permutation (that of Sec. 4.3) is the following:

Definition A permutation is even if it can be represented as a product of an even number of transpositions and odd if it can be represented as a product of an odd number of transpositions.

This definition is logically equivalent to the previous one. One could look on the first one as a static definition and the second as a dynamic one. The first definition is easier from a computational point of view and the second from a conceptual viewpoint. We can use it to prove the following:

Theorem B.3 The set A_n of even permutations of the symmetric group S_n on n symbols is a subgroup with $(n!)/2$ elements.

Proof We look at the four properties of a group in succession. If p and q are two even permutations, each can be represented as a product of an even number of transpositions and hence so can their product. This proves closure.

The permutation (1,2) followed by itself is the identity, which shows that the identity is an even permutation.

To show that the inverse of an even permutation is even, let the permutation p be written

$$p = t_1 t_2 t_3 \cdots t_{2k}$$

where each t_i is a transposition. Then, from property 7 of Sec. A.1,

$$p^{-1} = t_{2k} t_{2k-1} \cdots t_2 t_1$$

and, since a transposition its own inverse, p^{-1} is also the product of an even number of transpositions.

The associative property holds since A_n is a subset of S_n. It remains to establish the number of elements of A_n. Let

$$p_1, p_2, \ldots, p_k \tag{B.2}$$

be the permutations of A_n. Then, if t is the transposition (1,2), we see that the permutations

$$tp_1, tp_2, \ldots, tp_k \tag{B.3}$$

are all odd permutations of S_n. Furthermore, if q is any odd permutation of S_n, then tq is an even permutation and hence one of the set (B.2); that is, $tq = p_i$ for some i between 1 and k inclusive. Thus $t(tq) = tp_i$ and $q = tp_i$. Hence the set (B.3) is the complete set of odd permutations. Thus there are just as many odd permutations as even ones and $n! = 2k$. This completes the proof.

Definition The subset of all even permutations of n symbols is called the *alternating group on n symbols* and is denoted by A_n.

APPENDIX *C*

POLYNOMIALS AND EXTENSION FIELDS

C.1 INTRODUCTION

As we all know, a *polynomial* is an expression of the following form:

$$f(x) = a_n x^n + a_{n-1}x^{n-1} + \cdots + a_1 x + a_0$$

We call a_n the *leading coefficient* and, if $a_n \neq 0$, n is the degree of the polynomial. If $f(x)$ is a number different from zero, we call its degree 0; while if $f(x)$ is the number zero, we say that it has no degree.

We have familiar rules for adding and multiplying two polynomials. The symbol x is often called an *indeterminate* or dummy variable. Its function is to help us to keep the manipulations straight. It is commutative with the coefficients, the a_i. One can get by up to a point if the a's are in a ring, but it is a little more convenient if the ring is commutative and even more so if it is an integral domain. We shall here go all the way and assume the a's are in a field. So, with this restriction, we shall call $f(x)$ a *polynomial over a field F*, that is, the coefficients are to be in F. It is not hard to see that the set of all polynomials over a field F forms a ring which is commutative, has an identity, and no divisors of zero. That is, $F[x]$, the set of all polynomials over F, is an integral domain. But it has other properties beside. In fact it behaves very much like the set of integers with respect to factorization. To emphasize this we shall in the next section state parallel theorems in integers and polynomials without proving the former. Indeed the proofs for integers and polynomials are much alike. The set of all integers is designated by **Z**.

C.2 FACTORIZATION

First we list without proof some parallel definitions and properties. (Formal proofs may be found in the references mentioned at the end of this appendix.)

1. An integer b is a *factor* or *divisor* of an integer c if there is an integer d such that $bd = c$. We express this relationship by $b|c$. (Note that the vertical line indicates a relationship, whereas if we use a slanting line as in b/c we have a fraction.) A polynomial $f(x)$ is called a *factor* or *divisor* of $g(x)$ if there is a polynomial $h(x)$ such that $f(x)h(x) = g(x)$. We write $f(x)|g(x)$. We call two polynomials equal if they are identical.
2. An integer which is a factor of 1 is called a *unit*. The units of **Z** are 1 and -1. A polynomial which is a factor of 1 is called a unit. The units of $F[x]$ are the nonzero numbers of F.
3. If $a|b$ and $a|c$, then $a|(b + c)$. If $f(x)|g(x)$ and $f(x)|h(x)$, then $f(x)|[g(x) + h(x)]$.
4. If $a|b$ and $b|c$, then $a|c$. If $f(x)|g(x)$ and $g(x)|h(x)$, then $f(x)|h(x)$.
5. If $b|c$ and $c|b$, then $b = uc$ for some unit u. If $f(x)|g(x)$ and $g(x)|f(x)$, then $f(x) = ug(x)$ for some unit u of $F[x]$, that is, for some nonzero number of F.
6. An integer p is called a *prime number* if it is positive and if its only factors less in absolute value than p are units. A polynomial $f(x)$ is called an *irreducible* or *prime polynomial* in F if its only factors in $F[x]$ of lesser degree are units.

The first five of the properties of polynomials above are paraphrases of the corresponding ones for integers. But there is a difference in the sixth which deserves special notice. Whether or not a polynomial is prime depends on the field. Thus $x^2 - 2$ is irreducible over the rational field but not over the real field, while $x^2 + 3$ is irreducible over the field of reals but not over the field of complex numbers.

Now we state two parallel theorems with only an indication of the proof, using the letter "n" for that about integers and "p" for polynomials.

Theorem C.1n If a and b are two integers with $b \neq 0$, there are unique integers q and r such that

$$a = qb + r \qquad 0 \leq r < |b|$$

Theorem C.1p If $f(x)$ and $g(x)$ are two polynomials with $g(x) \neq 0$ over a field F, then there are polynomials $q(x)$ and $r(x)$ over F such that

$$f(x) = q(x)g(x) + r(x)$$

where $r(x) = 0$ or the degree of $r(x)$ is less than that of $g(x)$.

Theorem C.1p is a formal statement of the well-known property that one can divide any polynomial by another and have a quotient and a remainder of lesser degree than the divisor. The theorem can be proved by indicating the process of division.

Now we come to the idea of greatest common divisor. Here, for our convenience, we shall give a definition which may be a little more sophisticated than the one you are used to but which can easily be shown equivalent to it.

Definition n If b and c are two integers, the number g is called a *greatest common divisor* (g.c.d.) of b and c if it has the two following properties:

1. The integer g is a factor of both b and c.
2. If d is a factor of both b and c, then d is a factor of g.

For example, under this definition, 6 and 9 have two greatest common divisors: 3 and -3.

Definition p If $f(x)$ and $h(x)$ are two polynomials in $F[x]$, the polynomial $g(x)$ is called a greatest common divisor (g.c.d.) of f and h if it has the two following properties:

1. The polynomial $g(x)$ is a factor of both $f(x)$ and $h(x)$.
2. If the polynomial $d(x)$ is a factor of both $f(x)$ and $h(x)$, then $d(x)|g(x)$.

In neither case above is the g.c.d. unique but it is "almost" unique in a sense which we shall discuss below.

Now we state a pair of fundamental theorems and give the proof for the polynomial case.

Theorem C.2n If b and c are two integers not both zero and if g is a greatest common divisor of b and c, then there are integers s and t such that

$$bs + ct = g$$

The integers s and t are not unique, but g is, except for a unit factor; that is, the quotient of any two greatest common divisors is a unit.

Theorem C.2p If $f(x)$ and $h(x)$ are two polynomials over F, not both the zero polynomial, and if $g(x)$ is a g.c.d. of f and h, then there are polynomials

$s(x)$ and $t(x)$ such that

$$f(x)s(x) + h(x)t(x) = g(x)$$

The polynomials $s(x)$ and $t(x)$ are not unique, but $g(x)$ is unique except for a unit factor.

Proof Consider the set S of all polynomials

$$f(x)s(x) + h(x)t(x)$$

as $s(x)$ and $t(x)$ range over all the polynomials in $F[x]$. Let $g(x)$ be a polynomial in S of least degree and

$$f(x)s_0(x) + h(x)t_0(x) = g(x) \tag{C.1}$$

We shall prove that $g(x)$ is a g.c.d. of $f(x)$ and $h(x)$. To do this we refer back to the definition of the g.c.d. and see that from the properties listed at the beginning of this section any polynomial $d(x)$ which is a factor of both $f(x)$ and $h(x)$ must, by (C.1), be a factor of $g(x)$. Thus property 2 of the g.c.d. holds. Now, to show property 1, use Theorem C.1p and write

$$f(x) = q(x)g(x) + r(x)$$

with $r(x)$ the zero polynomial or of lower degree than $g(x)$. We proceed to exclude the second possibility.

From (C.1) we have

$$-q(x)[f(x)s_0(x) + h(x)t_0(x)] = -q(x)g(x) = -f(x) + r(x) \tag{C.2}$$

and

$$f(x)[1 - q(x)s_0(x)] + h(x)[-q(x)t_0(x)] = r(x) \tag{C.3}$$

Thus Equation (C.3) shows that $r(x)$ is in S. But $r(x)$ is either zero or of lesser degree than $g(x)$. Since $g(x)$ was assumed to be a polynomial of least degree in S, we see that $r(x)$ must be zero, and hence $g(x)$ is a factor of $f(x)$. Similarly we could show that $g(x)$ is a factor of $h(x)$. This completes the proof except for the last remark about $g(x)$.

Suppose we have two greatest common divisors g and g'. By property 2 of the g.c.d., each must divide the other. Then property 5 at the beginning of this section shows that g and g' are the same except for a unit factor. This completes the proof.

It should be remarked that in the case of integers one may make the g.c.d. unique by requiring that it be positive. For polynomials we can get uniqueness by specifying that the leading coefficient be 1.

Two integers whose g.c.d. is a unit are called *relatively prime*. Two polynomials whose g.c.d. is a unit are also called *relatively prime*. Two polynomials or two integers can, of course, be relatively prime without either being a prime.

An immediate consequence of this theorem is the following pair.

Theorem C.3n If two integers b and c are relatively prime, then $b|cd$ implies $b|d$.

Theorem C.3p If two polynomials $f(x)$ and $h(x)$ are relatively prime, then $f(x)|h(x)k(x)$ implies $f(x)|k(x)$.

Proof Since $f(x)$ and $h(x)$ are relatively prime they have a unit as a g.c.d., and hence the number 1 is also a g.c.d. So there are polynomials $s(x)$ and $t(x)$ such that

$$f(x)s(x) + h(x)t(x) = 1$$

Multiplication by $k(x)$ gives

$$k(x)f(x)s(x) + k(x)h(x)t(x) = k(x)$$

Since $f(x)$ divides both terms on the left side, it must divide the right side, and the proof is complete.

Corollary If $f(x)$ and $h(x)$ are two polynomials over F and if $f(x)$ is irreducible over F, then either $f(x)$ is a factor of $h(x)$ or $f(x)$ and $h(x)$ are relatively prime.

This follows since any common factor of $f(x)$ and $h(x)$ is either a unit, when $f(x)$ and $h(x)$ are relatively prime, or is a polynomial of positive degree. But any polynomial factor of $f(x)$ of positive degree must be a unit multiple of $f(x)$; in that case $f(x)$ is a factor of $h(x)$.

We state without proof the fundamental theorem of arithmetic and prove its analog for polynomials.

Theorem C.4 (Fundamental Theorem of Arithmetic) Any positive integer can be expressed as a product of powers of distinct prime numbers uniquely except for the order of the factors.

Theorem C.5 (Theorem of Unique Decomposition of Polynomials) If $f(x)$ is a polynomial in $F[x]$ with leading coefficient 1, it can be expressed as a product of powers of distinct irreducible polynomials over F with leading coefficients 1. This factorization is unique except for the order of the factors; it depends, however, on the field F.

Proof First, to show that $f(x)$ can be expressed as a product of powers of distinct irreducible polynomials over F, we proceed by induction on the degree n of f. If $f(x)$ is irreducible, we have nothing to prove. If $f(x)$ is reducible it can be written as a product $g(x)h(x)$ of polynomials of positive degree.

Since each of $g(x)$ and $h(x)$ is of degree less than n, the induction hypothesis shows that they are each expressible as products of irreducible polynomials. This shows that $f(x)$ is also.

Second, to prove that the product is unique except for the order of factors, we use an induction argument also, this time on the number of irreducible factors. Suppose

$$f(x) = p_1(x)p_2(x) \cdots p_r(x) = q_1(x)q_2(x) \cdots q_s(x)$$

where the p's and q's are irreducible polynomials. Since $p_1(x)$ is a factor of $f(x)$, it is a factor of the product of q's. If $p_1(x)$ is a factor of $q_1(x)$, since both are irreducible, they must be the same except for a unit factor, that is, a factor which is in F. In this case

$$\frac{f(x)}{p_1(x)} = p_2(x) \cdots p_r(x) = uq_2(x) \cdots q_s(x)$$

for some unit u. Then the induction hypothesis shows that the p's and q's are equal in some order.

If $p_1(x)$ does not divide $q_1(x)$, it must be relatively prime to $q_1(x)$ and hence, by Theorem C.3p, must divide the product of the remaining q's. Repeating this argument shows that $p_1(x)$ must divide one of the q's. Then, permuting the subscripts puts us back in the previous case. This completes the proof.

It should be noted that though the irreducibility of a polynomial depends on the field, the g.c.d. of two polynomials over a field F is not changed by the enlargement of the field. To see this, consider the identity (C.1) where all the polynomials have coefficients in a field F. Let F' be a field which contains F. Let $g'(x)$ be the g.c.d. of $f(x)$ and $h(x)$ in F'. Since all the polynomials in (C.1) have coefficients in F', we see that $g'(x)$ a factor of $f(x)$ and $h(x)$ implies that it is a factor of $g(x)$. On the other hand, since $g(x)$ is a common factor of $f(x)$ and $h(x)$, it must divide $g'(x)$. This shows that $g(x)$ and $g'(x)$ are the same except for a unit factor.

C.3 NEW FIELDS FROM OLD

Given any polynomial $f(x)$ in $F[x]$. One can find a field F' containing F such that in F', $f(x)$ is *completely decomposable*, that is, can be expressed as a product of linear factors. We show how this can be done for a special case and indicate how it can be managed in general.

Let F be the field of real numbers. Then $z^2 + 1$ is an irreducible polynomial in F. The usual method of enlarging F so that all polynomials are completely decomposable is to introduce the square root of -1. Here we do it in a different way, which, in the end, amounts to the same thing. We consider all the polynomials in z modulo $z^2 + 1$. That is, if $g(z)$ is any polynomial in z with real coefficients, we replace it by its remainder after dividing by $z^2 + 1$. For instance,

$$z^4 + 3z^2 + z + 1 = (z^2 + 2)(z^2 + 1) + z - 1$$

Hence, we identify the polynomial on the left with $z - 1$. Since the remainders are always linear in z or are numbers of F, we see that taking the polynomials modulo $z^2 + 1$ is equivalent to considering the polynomials

$$az + b$$

where a and b are real numbers. We define $(az + b) + (cz + d)$ as $(a + c)z + (b + d)$ and find the product as follows:

$$\begin{aligned}(az + b)(cz + d) &= acz^2 + (bc + ad)z + bd \\ &= ac(z^2 + 1) + (bc + ad)z + bd - ac\end{aligned}$$

Hence the product of $az + b$ and $cz + d$ is defined to be

$$(bc + ad)z + (bd - ac)$$

Under these definitions of addition and product, it is easy to show that the elements $az + b$ satisfy all the properties of a field except, perhaps, that of division. To show that division is possible except by zero, consider $az + b$ and $z^2 + 1$ in $F[z]$, the set of polynomials with real coefficients. Since $z^2 + 1$ is irreducible, it cannot divide $az + b$ unless a and b are both zero. Hence the two are relatively prime. We know by Theorem C.2p that there are polynomials $s(z)$ and $t(z)$ such that

$$(az + b)s(z) + (z^2 + 1)t(z) = 1$$

This implies that, modulo $z^2 + 1$,

$$(az + b)s(z) = 1$$

and hence $s(z)$ is the multiplicative inverse of $az + b$. So now we have shown

The polynomials in z with real coefficients modulo $(z^2 + 1)$ form a field. Call it F'.

Now we assume a well-known property of real polynomials: The only irreducible polynomials over the field of real numbers are linear and quadratic. We prove

Theorem C.6 Every irreducible quadratic polynomial with real coefficients is the product of two linear factors in the field F' defined above.

Proof We know from the quadratic formula that the zeros of $ax^2 + bx + c$, with $a \neq 0$, are

$$\frac{-b \pm \sqrt{D}}{2a} \qquad \text{where } D = b^2 - 4ac$$

If D is nonnegative, the zeros are real and the quadratic is a product of two real linear factors. If $D < 0$, we have

$$ax^2 + bx + c = a\left(x + \frac{b + z\sqrt{-D}}{2a}\right)\left(x + \frac{b - z\sqrt{-D}}{2a}\right)$$

This completes the proof.

Corollary Every real polynomial in x can be expressed as a product of linear factors in F'.

Actually, while we are about it, we can show the following stronger theorem:

Theorem C.7 Every polynomial with coefficients in F' can be expressed as a product of linear polynomials in F'.

Proof Here $f(x)$ is a polynomial each of whose coefficients is of the form $rz + s$, where r and s are real numbers and one or both of r and s may be zero. Denote by $\bar{f}(x)$ the polynomial obtained from $f(x)$ by replacing each z in a coefficient by $-z$. Then we can write

$$f(x) = zg(x) + h(x) \qquad \text{and} \qquad \bar{f}(x) = -zg(x) + h(x)$$

where $g(x)$ and $h(x)$ are polynomials with real coefficients. Now

$$F(x) = f(x)\bar{f}(x) = h^2(x) - z^2g^2(x) = h^2(x) + g^2(x)$$

which shows that $F(x)$ is a polynomial with real coefficients. We know by Theorem C.6 that $F(x)$ can be expressed as a product of linear factors in F', and hence, by the theorem of unique decomposition of polynomials (Theorem C.5), each of $f(x)$ and $\bar{f}(x)$ is the product of linear factors.

Now we can sketch the proof of the general theorem.

Theorem C.8 Let $f(x)$ be a polynomial with coefficients in a field F; there is a field K containing F such that in K, $f(x)$ is the product of linear factors.

Sketch of the Proof Let $p(x)$ be an irreducible factor over F of $f(x)$. Then define F' to be the set of polynomials over $F[z]$ modulo $p(z)$. This set forms a field as we showed above for $z^2 + 1$. Now in F', $p(x)$ has a zero z, and hence it has a linear factor $x - z$ and

$$p(x) = (x - z)q(x) \qquad \text{in } F'$$

Now $q(x)$ may be reducible or irreducible in F'. If the latter, we can enlarge the field still further to get a linear factor of $q(x)$. This process can be continued until $p(x)$ is a product of linear factors in some field containing F. Then we can work on the other irreducible factors of $f(x)$ to arrive at the desired field K.

Notice the difference between what happens in the general case and in that covered by the two previous theorems. There the single field F' served to decompose $f(x)$ completely into a product of linear factors and then in F' every polynomial was completely decomposable. This will not be the case for polynomials over general fields.

Note that there is no reason why the field of the polynomial should not be a finite field. For instance, let F be the field with two elements 0 and 1 and consider the polynomial $x^2 + x + 1$. This is irreducible since if it were reducible either 1 or 0 would be a zero. But

$$0^2 + 0 + 1 = 1 \neq 0 \qquad \text{and} \qquad 1^2 + 1 + 1 = 1 \neq 0$$

So we can consider the field F' of polynomials over F modulo $(x^2 + x + 1)$. These form a field with four elements. It can be shown that if a field has a finite number of elements, the number of elements of the field is p^k for some prime number p and positive integer k. We shall not, however, prove this here.

For the development of polynomials and fields, see books on modern algebra, including that by Birkhoff and MacLane listed in the bibliography. For corresponding properties of integers, see books on the theory of numbers, including that by Niven and Zuckerman listed in the bibliography.

APPENDIX *D*

A PROPERTY OF BILINEAR FUNCTIONS

We here prove a statement made near the end of Sec. 5.1. It was to the following effect: Given a function of $V \times V$ into a field F, which is bilinear but not necessarily symmetric [that is, we do not assume that $(\xi,\eta) = (\eta,\xi)$], then, if for all vectors ξ, η of V,

(D.1) $$(\xi,\eta) = 0 \text{ implies } (\eta,\xi) = 0$$

it follows that

(D.2) $$\begin{aligned} &\text{Either } (\xi,\xi) = 0 && \text{for all } \xi \text{ in } V \\ &\text{Or } (\xi,\eta) = (\eta,\xi) && \text{for all } \xi \text{ and } \eta \text{ in } V \end{aligned}$$

The proof which we give is that of E. Artin in *Geometric Algebra*, pp. 110, 111 (see bibliography). We have

$$(\alpha,(\alpha,\gamma)\beta - (\alpha,\beta)\gamma) = (\alpha,\gamma)(\alpha,\beta) - (\alpha,\beta)(\alpha,\gamma) = 0$$

Hence by condition (D.1)

(D.3) $$0 = ((\alpha,\gamma)\beta - (\alpha,\beta)\gamma,\alpha) = (\alpha,\gamma)(\beta,\alpha) - (\alpha,\beta)(\gamma,\alpha)$$

First in (D.3) take $\xi = \alpha = \gamma$ and $\beta = \eta$ to get

$$(\xi,\xi)(\eta,\xi) = (\xi,\eta)(\xi,\xi) \tag{D.4}$$

that is,

$$(\xi,\xi) = 0 \qquad \text{or} \qquad (\xi,\eta) = (\eta,\xi) \tag{D.5}$$

Now we seek a contradiction by supposing that for some α, β, γ,

$$(\alpha,\beta) \neq (\beta,\alpha) \qquad \text{and} \qquad (\gamma,\gamma) \neq 0 \tag{D.6}$$

If we can show that this leads to a contradiction, we will have proved property (D.2). To this end in (D.5), take $\xi = \gamma$ and $\eta = \alpha$. Then (D.5) shows that $(\gamma,\gamma) \neq 0$ implies $(\gamma,\alpha) = (\alpha,\gamma)$. Then (D.3) shows that $(\alpha,\beta) \neq (\beta,\alpha)$ with $(\gamma,\alpha) = (\alpha,\gamma)$ imply $(\alpha,\gamma) = 0$. Similarly, if we interchange α and β, we have $(\beta,\gamma) = 0$. Thus

$$(\alpha,\gamma) = (\gamma,\alpha) = 0 = (\beta,\gamma) = (\gamma,\beta) \tag{D.7}$$

Then $$(\alpha + \gamma, \beta) = (\alpha,\beta) \neq (\beta,\alpha) = (\beta, \alpha + \gamma)$$

Using (D.5) with $\xi = \alpha + \gamma$ and $\eta = \beta$, we see that $0 = (\alpha + \gamma, \alpha + \gamma)$. But

$$\begin{aligned}(\alpha + \gamma, \alpha + \gamma) &= (\alpha,\alpha) + (\gamma,\alpha) + (\alpha,\gamma) + (\gamma,\gamma) \\ &= (\alpha,\alpha) + (\gamma,\gamma)\end{aligned}$$

from (D.7). Taking $\xi = \alpha$ and $\eta = \beta$ in (D.5), we see that $(\alpha,\beta) \neq (\beta,\alpha)$ implies $(\alpha,\alpha) = 0$. This, in turn, implies

$$(\alpha + \gamma, \alpha + \gamma) = (\gamma,\gamma) \neq 0$$

which contradicts $0 = (\alpha + \gamma, \alpha + \gamma)$ as shown above. This completes the proof of property (D.2).

Finally we show that if the field is not of characteristic 2, that is, if $1 + 1 \neq 0$, then the following two statements are equivalent:

$$\begin{aligned}(\gamma,\gamma) &= 0 && \text{for all } \gamma \text{ in } V \\ (\alpha,\beta) &= -(\beta,\alpha) && \text{for all } \alpha \text{ and } \beta \text{ in } V\end{aligned} \tag{D.8}$$

First, if $(\gamma,\gamma) = 0$ for all γ in V, we have

$$\begin{aligned}0 &= (\alpha + \beta, \alpha + \beta) = (\alpha,\alpha) + (\alpha,\beta) + (\beta,\alpha) + (\beta,\beta) \\ &= (\alpha,\beta) + (\beta,\alpha)\end{aligned}$$

Conversely, if $(\alpha,\beta) = -(\beta,\alpha)$ for all α and β in V, it holds in particular for $\alpha = \beta$, which shows that $(\alpha,\alpha) = 0$ for all α in V.

The results of this appendix could be described in the following terms: If for a bilinear function, symmetry (commutativity) is not assumed but if, in its place, orthogonality is assumed symmetric, then the bilinear form for the function is either symmetric or skew symmetric.

A SHORT BIBLIOGRAPHY

Artin, Emil: *Geometric Algebra*, Interscience Publishers, Inc., New York, 1957.
Birkhoff, Garrett, and Saunders MacLane: *A Survey of Modern Algebra*, 3d ed., The Macmillan Company, New York, 1965.
Eves, Howard: *Elementary Matrix Theory*, Allyn and Bacon, Inc., Boston, 1966.
Gantmacher, F. R.: *Applications of the Theory of Matrices*, trans. by J. L. Brenner, Interscience Publishers, Inc., New York, 1959.
Greub, K. H.: *Linear Algebra*, 2d ed., Academic Press, Inc., New York, 1963.
Halmos, P. R.: *Finite-Dimensional Vector Spaces*, 2d ed., D. Van Nostrand Company, Inc., Princeton, N.J., 1958.
Hoffman, Kenneth, and Ray Kunze: *Linear Algebra*, Prentice-Hall, Inc., Englewood Cliffs, N.J., 1961.
Jones, B. W.: *The Arithmetic Theory of Quadratic Forms*, John Wiley & Sons, Inc., New York, 1950.
MacDuffee, C. C.: *The Theory of Matrices*, corrected reprint, Chelsea Publishing Company, New York, 1946.
______: *Vectors and Matrices*, Mathematical Association of America, Buffalo, 1943.
Nering, E. D.: *Linear Algebra and Matrix Theory*, 2d ed., John Wiley & Sons, Inc., New York, 1970.

Niven, Ivan, and H. S. Zuckerman: *An Introduction to the Theory of Numbers*, John Wiley & Sons, Inc., New York, 1960.

Perlis, S.: *Theory of Matrices*, Addison-Wesley Publishing Company, Inc., Reading, Mass., 1952.

Schreier, O., and E. Sperner: *Modern Algebra and Matrix Theory*, trans. by M. Davis and M. Hausner, Chelsea Publishing Company, New York, 1955.

Stoll, R. R., and E. T. Wong: *Linear Algebra*, Academic Press, Inc., New York, 1968.

INDEX